ADVANCED SEPARATIONS BY SPECIALIZED SORBENTS

CHROMATOGRAPHIC SCIENCE SERIES

A Series of Textbooks and Reference Books

Editor:
Nelu Grinberg

Founding Editor:
Jack Cazes

1. Dynamics of Chromatography: Principles and Theory, J. Calvin Giddings
2. Gas Chromatographic Analysis of Drugs and Pesticides,
    Benjamin J. Gudzinowicz
3. Principles of Adsorption Chromatography: The Separation of Nonionic Organic
    Compounds, Lloyd R. Snyder
4. Multicomponent Chromatography: Theory of Interference,
    Friedrich Helfferich and Gerhard Klein
5. Quantitative Analysis by Gas Chromatography, Josef Novák
6. High-Speed Liquid Chromatography, Peter M. Rajcsanyi and
    Elisabeth Rajcsanyi
7. Fundamentals of Integrated GC-MS (in three parts),
    Benjamin J. Gudzinowicz, Michael J. Gudzinowicz, and Horace F. Martin
8. Liquid Chromatography of Polymers and Related Materials, Jack Cazes
9. GLC and HPLC Determination of Therapeutic Agents (in three parts),
    Part 1 edited by Kiyoshi Tsuji and Walter Morozowich, Parts 2 and 3
    edited by Kiyoshi Tsuji
10. Biological/Biomedical Applications of Liquid Chromatography,
    edited by Gerald L. Hawk
11. Chromatography in Petroleum Analysis, edited by Klaus H. Altgelt
    and T. H. Gouw
12. Biological/Biomedical Applications of Liquid Chromatography II,
    edited by Gerald L. Hawk
13. Liquid Chromatography of Polymers and Related Materials II,
    edited by Jack Cazes and Xavier Delamare
14. Introduction to Analytical Gas Chromatography: History, Principles,
    and Practice, John A. Perry
15. Applications of Glass Capillary Gas Chromatography,
    edited by Walter G. Jennings
16. Steroid Analysis by HPLC: Recent Applications, edited by Marie P. Kautsky
17. Thin-Layer Chromatography: Techniques and Applications,
    Bernard Fried and Joseph Sherma
18. Biological/Biomedical Applications of Liquid Chromatography III,
    edited by Gerald L. Hawk
19. Liquid Chromatography of Polymers and Related Materials III,
    edited by Jack Cazes

ADVANCED SEPARATIONS BY SPECIALIZED SORBENTS

EDITED BY

ECATERINA STELA DRAGAN

CRC Press
Taylor & Francis Group
Boca Raton London New York

CRC Press is an imprint of the
Taylor & Francis Group, an **informa** business

CRC Press
Taylor & Francis Group
6000 Broken Sound Parkway NW, Suite 300
Boca Raton, FL 33487-2742

First issued in paperback 2021

ISBN 13: 978-1-4822-2055-1 (hbk)
ISBN 13: 978-1-03-224057-2 (pbk)

Library of Congress Cataloging-in-Publication Data

Advanced separations by specialized sorbents / edited by Ecaterina Stela Dragan.
    pages cm. -- (Chromatographic science series)
    "A CRC title."
    Includes bibliographical references and index.
    ISBN 978-1-4822-2055-1 (hardcover : alk. paper) 1. Separation (Technology) 2. Sorbents. I. Dragan, Ecaterina Stela, editor.

TP156.S45A365 2015
660'.2842--dc23
                                    2014020396

To my daughters, Irina and Ioana

Ecaterina Stela Dragan

Contents

Preface

This book provides a comprehensive overview of the advanced techniques employed to create specialized sorbents with a wide range of functions, which can be used to enhance the separation and/or purification of useful bioactive species like proteins and cells, heavy metal ions, dyes, etc. It illustrates some of the most efficient materials promoted in recent decades for the separation processes. The main purpose of this book is to update the scientific information in a field of research that is growing dynamically. Thus, the latest information in the field of separation processes by specialized sorbents like monolith cryogels, composite hydrogels, magnetic composite adsorbents, metal-impregnated ion exchangers, molecularly imprinted polymers, and solid phase extraction by mixed mode sorbents are presented and compared with the authors' results. Biobased polymer composites occupy a unique place in the dynamic world of new sorbents, and this book provides novel information on them. Readers will get updated information and an in-depth perspective on the design strategies, characterization, and application of novel sorbents. The material will also help researchers in the design of their projects on specialized sorbents for the separation and/or purification of ionic species. The chapters in this book have been contributed by a team of renowned scientists from around the world whose expertise will enlarge the visibility of some of the most effective sorbents and will provide readers an overall view on the efficiency of different separation techniques.

Chapter 1 presents composite hydrogel materials consisting of cross-linked homo- and copolymers of acrylamide and N-isopropylacrylamide with embedded clay minerals, metal nanoparticles, drugs, and proteins. Special attention has been paid to the metal complexes of linear polyampholytes, cross-linked polybetaines, and macroporous amphoteric gels. Molecularly and ion-imprinted polymers focusing on selective recovery of transition and rare earth metal ions are presented. The potential applications of composite hydrogel materials in the oil industry for cleaning the internal surface of main pipes, in catalysis as metal nanoparticles immobilized within hydrogel matrices, and in medicine and biotechnology as controlled release of drugs and proteins are also outlined.

The progress during recent decades in the field of affinity chromatography is presented in Chapter 2. Affinity chromatography is a very efficient method of protein purification. Recently, dye-ligand affinity chromatography and immobilized metal affinity separation have gained considerable attention in the purification of proteins, both in laboratory and large-scale applications, assuring higher specificity, purity, and recovery in a single chromatographic step, as well as cost efficiency and safety. Lately, cryogel materials have been considered as a novel generation of stationary phases in separation science. They have proven to be highly efficient in protein purification with many advantages, including large pores, short diffusion path, low pressure drop, and very short residence time for both adsorption and elution. These unique features make them attractive matrices for the chromatography of biomolecules, viruses, plasmids, and even whole cells.

Monoliths are uniform matrices without interparticular voids, having significant importance as a stationary phase in different modes of chromatography. The pores in monoliths form interconnected channels across the matrix, which provides high permeability for the convective flow of the mobile phase and a large surface area for the binding of analytes. The advantages of macroporous monoliths are discussed in Chapter 3. Macroporous monoliths can be composed of silica, polymer, metal oxides, and carbon-based materials. Unlike conventional columns, they can easily be chemically modified, and a single monolith can have different functionalities in the separation of many analytes. Macroporous monolithic matrices provide fast, efficient, and easy separation of large biomolecules such as proteins, nucleic acids, bacteria, mammalian cells, or particulate matter with low mass transfer resistance. This chapter describes the different types of monoliths and their working principles and applications in particulate/cell separations.

Over the last decade, a special area of focus has been the removal of heavy metals and dyes from the environment because of their nonbiodegradability and long-term toxicity, which make them very dangerous for human health. Biosorbents derived from polysaccharides like chitosan and alginate attracted a strong interest as a cost-effective alternative to the existing sorbents like activated carbon and synthetic ion exchangers. Due to high adsorption capacity, chitosan and alginate have been extensively used as biosorbents in wastewater remediation. The advantages and perspectives of using specialized polysaccharide-based composites in the removal of heavy metals and dyes are presented in Chapters 4 through 6 and in Chapters 9 and 11.

Traditional hydrogels from synthetic and/or natural polymers often have some limitations, such as low mechanical stability and poor biodegradability, which restrict their practical applications. Recently, polysaccharide-based composite hydrogels, a new group of materials at the interface of hydrogels, polymer/clay nanocomposites, and polysaccharides, have attracted much attention due to their unique properties. The latest developments on this type of hydrogels are reviewed in Chapter 4. The applications of novel composite hydrogels in the removal of pollutants, including heavy metals, dyes, and ammonium nitrogen in water, are reviewed. Due to the synergistic effect among polysaccharides, vinyl monomers, and clay minerals, many of the physicochemical properties, such as swelling ratio and rate, thermostability, and gel strength of composite hydrogels, are superior to their counterparts.

Chapter 5 is focused on the sorption of heavy metals by magnetic adsorbent particles, the so-called magnetic beads. The facile separation of magnetic sorbents from the aqueous phase is the main advantage, which differentiates them from the traditional adsorbents. Their efficient removal in a magnetic field followed by regeneration and reuse decreases the overall cost of water treatment. Due to their high applicative potential, composite materials containing iron oxide incorporated in functional polymeric supports are intensely studied. This chapter presents recent developments in the very important field of the magnetic separation of heavy metals by composite biosorbents.

Synthesis and characterization of some biosorbents based on chitosan, alginate, and cellulose, as biopolymer matrix, embedded with synthetic or natural zeolites and their applications for the removal of heavy metal ions and the separation of aqueous–organic mixtures are summarized in Chapter 6. Removal of dyes by chitosan–zeolite composites are also discussed. The sorption capacities and the pervaporation separation performances of biopolymer–zeolite composites are compared with those of raw zeolite, pristine biopolymer, or other biopolymer-based composites.

In recent decades, the wastewater treatment industry has identified the discharge of nutrients, including phosphates and nitrates, into waterways as a risk to natural environments due to the serious effects of eutrophication of the water bodies. An abundance of algal blooming in eutrophic water bodies can deplete dissolved oxygen in water, causing fish deaths. Accordingly, it is necessary and urgent to explore effective techniques for phosphate removal from wastewater. The development and performance of new phosphate-selective sorbents, referred to as hybrid anion exchangers (HAIX), are presented in Chapter 7. HAIX combines the durability and mechanical strength of polymeric anion exchange resins with the high sorption affinity of hydrated ferric oxide toward phosphate.

Different chemicals like medicines, pesticides, plastics components, or industry pollutants, all toxic to the endocrine system, are found in natural waters. These substances are poorly removed from solutions by conventional methods. The use of molecularly imprinted polymers offers the possibility of removing them as they have a high affinity and selectivity toward templates. Chapter 8 presents the methods of synthesis of such sorbents with a focus on their use in hybrid systems, which seems to be a promising alternative for the removal of endocrine-disrupting compounds.

Chapter 9 describes the sorption mechanisms and performances of biopolymers (chitosan and alginate) as a function of the type of functional groups, the pH, the composition of the solution, as well as the size and morphology of particles. Sorption may proceed through chelation/complexation, ion exchange/electrostatic attraction, or the formation of a ternary complex. The choice of the biopolymer depends on the target metal and the metal speciation. The versatility of these materials is of great interest for developing novel sorbents with improved diffusion properties, enhanced hydrodynamic behavior, and innovative application modes. In addition, these biopolymers can be used for encapsulating reactive compounds (ionic liquids, extractants, ion exchangers) in order to improve the reactivity, selectivity, or sorption efficiency of these materials, profiting from the possibility to condition these composite sorbents under different forms (beads, membranes, foams, etc.). Hybrid materials (e.g., metal-loaded biopolymers) can also be used to design new materials and new applications. Some examples are discussed that show how biopolymers can be given fresh life after metal binding.

Mixed-mode polymeric sorbents that enhance selectivity and capacity of extraction in a single material are described in Chapter 10. Different aspects of these materials are described, including their synthesis, morphological and chemical properties, as well as their application in solid-phase extractions (SPE). SPE protocols for each type of mixed-mode sorbents (strong/weak and cation/anion-exchange materials) are also discussed, since the protocols are crucial for the success of this kind of material. Applications of sorbents in different types of matrices are presented compared with commercial sorbents.

Single-network hydrogels have poor mechanical properties and slow responses at swelling. Various strategies, including the preparation of interpenetrating polymer network (IPN) hydrogels, have therefore been developed to remediate these weak points. The most significant classes of IPN composite hydrogels and their applications, mainly in the separation processes of dyes, heavy metal ions, and liquids, are presented in Chapter 11. Synthesis parameters such as cross-linker ratio, monomer concentration, and synthesis temperature are the key factors that determine the properties of the semi-IPN and IPN hydrogels, such as interior morphology, swelling kinetics, mechanical strength, etc. Sorption kinetics and reusability of IPN composite hydrogels are further enhanced by the synthesis of IPN hydrogels under the freezing temperature of the solvent (cryogels).

A rational approach for building molecular channels in hybrid organic–inorganic materials via the inorganic (sol–gel) transcription of dynamic self-assembled superstructures is presented in Chapter 12. The basic and specific molecular information encoded in the molecular precursors results in the generation of tubular superstructures in solution and in a solid state, which can be frozen in a polymeric hybrid matrix by the sol–gel process. These systems have been successfully employed to design solid dense membranes that function as ion channels and to illustrate how a self-organized hybrid material performs interesting and potentially useful transporting functions.

Furthermore, the book contains numerous illustrations and tables that will guide readers in advanced separation procedures. In conclusion, this book focuses on a variety of advanced techniques available for separation and/or purification of target ionic species and addresses the needs and challenges for future research in this growing field.

Editor

Dr. Ecaterina Stela Dragan is the head of the Functional Polymers Department, "Petru Poni" Institute of Macromolecular Chemistry, Iasi, Romania.

Her research interests include the synthesis and characterization of ionic polymers, smart composite hydrogels, porous ionic nano- and microstructured materials, surface modification by polyelectrolytes, and separation by ion exchangers and biosorbents.

She is the author or coauthor of more than 140 publications in international journals, 12 book chapters, and 38 national patents. She has been involved in the publication of seven books, both as author and editor. She is also a reviewer for numerous international journals. Her individual impact factor is >77 and Hirsch index, h = 20.

Contributors

Mihail Barboiu
Adaptive Nanosystems Group
Institut Européen des Membranes
Montpellier, France

Nilay Bereli
Biochemistry Division
Department of Chemistry
Hacettepe University
Ankara, Turkey

F. Borrull
Department of Analytical Chemistry
    and Organic Chemistry
University of Rovira and Virgili
Tarragona, Spain

Marek Bryjak
Faculty of Chemistry
Division of Polymer and Carbon
    Materials
Wroclaw University of Technology
Wroclaw, Poland

Adil Denizli
Biochemistry Division
Department of Chemistry
Hacettepe University
Ankara, Turkey

Maria Valentina Dinu
Functional Polymers Department
"Petru Poni" Institute of
    Macromolecular Chemistry
Iasi, Romania

Gianina Dodi
Faculty of Chemical Engineering and
    Environmental Protection
"Gheorghe Asachi" Technical
    University of Iasi
Iasi, Romania

and

SCIENT
Research Center for Instrumental
    Analysis
Bucharest, Romania

Ecaterina Stela Dragan
Functional Polymers Department
"Petru Poni" Institute of
    Macromolecular Chemistry
Iasi, Romania

N. Fontanals
Department of Analytical Chemistry
    and Organic Chemistry
University of Rovira and Virgili
Tarragona, Spain

Igor Yuri Galaev
DSM Nutritional Products, DBC
Heerlen, The Netherlands

Eric Guibal
Center for Materials Research
Alès School of Mines
Alès, France

Doina Hritcu
Faculty of Chemical Engineering and
    Environmental Protection
"Gheorghe Asachi" Technical
    University of Iasi
Iasi, Romania

Zh. Ibrayeva
Laboratory of Engineering Profile
K.I. Satpaev Kazakh National Technical
    University
and
Institute of Polymer Materials and
    Technology
Almaty, Kazakhstan

S. Kabdrakhmanova
S. Amanzholov East-Kazakhstan State
    University
Ust-Kamenogorsk, Kazakhstan

S. Kudaibergenov
Laboratory of Engineering Profile
K.I. Satpaev Kazakh National Technical
    University
and
Institute of Polymer Materials and
    Technology
Almaty, Kazakhstan

Ashok Kumar
Department of Biological Sciences and
    Bioengineering
Indian Institute of Technology Kanpur
Kanpur, India

Rosa Maria Marcé
Department of Analytical Chemistry
    and Organic Chemistry
University of Rovira and Virgili
Tarragona, Spain

Ioana Moleavin
Centre of Advanced Research in
    Nanobioconjugates and Biopolymers
"Petru Poni" Institute of
    Macromolecular Chemistry
Iasi, Romania

Ricardo Navarro
Department of Chemistry
University of Guanajuato
Guanajuato, Mexico

Marcel Ionel Popa
Faculty of Chemical Engineering and
    Environmental Protection
"Gheorghe Asachi" Technical
    University of Iasi
Iasi, Romania

Arup K. SenGupta
Department of Civil and Environmental
    Engineering
Lehigh University
Bethlehem, Pennsylvania

Sukalyan Sengupta
Department of Civil and Environmental
    Engineering
University of Massachusetts, Dartmouth
North Dartmouth, Massachusetts

Akhilesh Kumar Shakya
Department of Biological Sciences and
    Bioengineering
Indian Institute of Technology Kanpur
Kanpur, India

Akshay Srivastava
Department of Biological Sciences and
    Bioengineering
Indian Institute of Technology Kanpur
Kanpur, India

G. Tatykhanova
Laboratory of Engineering Profile
K.I. Satpaev Kazakh National Technical
    University
and
Institute of Polymer Materials and
    Technology
Almaty, Kazakhstan

Deniz Türkmen
Biochemistry Division
Department of Chemistry
Hacettepe University
Ankara, Turkey

Thierry Vincent
Center for Materials Research
Alès School of Mines
Alès, France

Aiqin Wang
Center of Eco-material and Green
    Chemistry
Lanzhou Institute of Chemical Physics
Chinese Academy of Sciences
Lanzhou, Gansu, People's Republic of
    China

Joanna Wolska
Faculty of Chemistry
Division of Polymer and Carbon
    Materials
Wroclaw University of Technology
Wroclaw, Poland

M. Yashkarova
Semipalatinsk State Shakarim
    University
Semey, Kazakhstan

Handan Yavuz
Biochemistry Division
Department of Chemistry
Hacettepe University
Ankara, Turkey

Junping Zhang
Center of Eco-material and Green
    Chemistry
Lanzhou Institute of Chemical Physics
Chinese Academy of Sciences
Lanzhou, Gansu, People's Republic of
    China

1 Composite Hydrogel Materials

S. Kudaibergenov, Zh. Ibrayeva, M. Yashkarova,
S. Kabdrakhmanova, and G. Tatykhanova

CONTENTS

1

1.1 INTRODUCTION

At present, composite hydrogel materials have attracted considerable interest in research and industrial spheres (Kudaibergenov et al. 2007, Pavlyuchenko and Ivanchev 2009). Composite polymer hydrogels consist of at least two components that exhibit a synergistic effect. According to the canons of thermodynamic compatibility, there are many possible structures of composite hydrogels starting from complete phase separation and ending to formation of structures consisting of polymer matrix and nano-, micro-, and macrosized inclusions. The nature of interaction between the components can have covalent, ionic, and donor–acceptor character and can be stabilized by hydrogen bonds, hydrophobic interactions, and entanglement of macromolecular chains producing interpenetrating and semi-interpenetrating polymer networks (IPNs) (Wu et al. 2006, Zhang et al. 2005). Due to their composite structure and unique properties such as improved mechanical, thermal, electrical, and optical characteristics, they have been found to have a wide application in medicine, membrane technology, optical engineering, and catalysis (Frimpong et al. 2006, Lao and Ramanujan 2004, Lu et al. 2003, Sershen et al. 2000, 2005). This chapter is devoted to composite hydrogel materials based on cross-linked homo- and copolymers of acrylamide (AAm) and N-isopropylacrylamide (NIPA) within which inorganic nano- and microparticles, polymer-protected metal nanoparticles, proteins, drugs, and low-molecular-weight ligands are immobilized. Physicochemical, physicomechanical, and catalytic properties and volume-phase transition (VPT) of composite hydrogel materials have been studied. Application aspects of composite hydrogel materials in oil industry and catalysis, for wastewater purification, and as drug delivery systems are also outlined.

1.2 IMMOBILIZATION OF NANO- AND MICROSIZED CLAY MINERALS INTO THE HYDROGEL MATRIX

1.2.1 PREPARATION AND CHARACTERIZATION OF ORGANIC–INORGANIC COMPOSITE MATERIALS BASED ON POLY(ACRYLAMIDE) HYDROGELS AND CLAY MINERALS

The properties of hydrogels can be modified by embedding inorganic materials, such as montmorillonite (MMT), bentonite, mica, silica, titanium and aluminum oxides, and sericite, within the gel matrix (Avvaru et al. 1998, Cheng et al. 2007, Kabiri and Zohuriaan-Mehr 2003, Kurokawa and Sasaki 1982, Lee and Yang 2004, Lin et al. 2001, Ray and Okamoto 2003, Starodoubtsev et al. 2000). The pioneering works to strengthen the mechanical properties of gel specimen by adding inorganic components were done by Haraguchi and colleagues (Haraguchi et al. 2003, 2013, Haraguchi and Li 2006, Haraguchi and Takehisa 2002). Gel sample made from MMT and NIPA is elastically stretched to about 10 times its original length (Haraguchi et al. 2002). Osada and colleagues (Gong et al. 2003, Nakayama et al. 2004, Tanaka et al. 2005) designed a series of double-network hydrogels with extremely high mechanical strength. This kind of nanocomposite hydrogel exhibited high transparency, high deswelling rate, and extraordinary mechanical properties with elongation at break in excess of $10^3\%$. In an organic/inorganic network structure, the clay sheets will act as effective multifunctional cross-linkers through ionic or polar interactions. The layered structure of clay minerals and their ability to swell in water allow monomers and polymer chains to diffuse into clay layers and act as additional cross-linker. The overall stability of composite materials directly depends on whether exfoliation or intercalation process takes place and on the choice of monomer or initiator that can be adsorbed to the clay surface (Abdurrahmanoglu et al. 2008, Essawy 2008, Jia et al. 2008, Xiang et al. 2006). Preparation of lightweight porous materials by templating hydrogels with a range of hydrophilic and hydrophobic scaffolding materials was explored (Rutkevičius et al. 2012). Submillimeter hydrogel slurries of polyacrylamide (PAAm) and gellan gum were templated with aqueous slurries of cement, gypsum, and clay–cement mixtures or alternatively dispersed in curable polydimethylsiloxane. After the solidification of the scaffolding material, the evaporation of a structured hydrogel produced porous composite material whose pores mimic the hydrogel mesostructure. This versatile hydrogel templating method can be applied to yield lightweight porous materials with a great potential for use in the building industry in heat and sound insulation panels, lightweight building blocks, porous rubber substitutes, and foam shock absorbers and as an alternative to aerated concretes. The poly(acrylamide-*co*-acrylate)/rice husk ash hydrogel composites and a series of poly(acrylic acid-*co*-acrylamide)/kaolin composites are applied as soil conditioner and superabsorbent and serve as release carrier of urea fertilizer in agricultural industry (Cândido et al. 2013, Lianga and Liu 2007, Lianga et al. 2007).

The effect of silica nanoparticles on the linear viscoelastic response of model polyacrylamide hydrogel (PAAH) systems was examined (Kalfus et al. 2012). The removal of methylene blue (MB) cationic dye from its aqueous solution was performed with the help of chitosan-g-poly(acrylic acid) (CTS-g-PAAc)/MMT nanocomposites as adsorbent (Wang et al. 2008). The influence of pH value, MMT content (wt.%), weight ratio (w.r.) of acrylic acid (AAc) to CTS, and adsorption temperature on the adsorption capacity of the nanocomposite was investigated. The results showed that the w.r. of AAc to CTS of the nanocomposites has great influence on adsorption capacities and introducing a small amount of MMT could improve the adsorption ability of the CTS-g-PAAc. The adsorption behaviors of the nanocomposite showed that the maximum adsorption capacity is 1859 mg/g for CTS-g-PAAc/MMT with 30 wt.% and w.r. of 7.2:1. The desorption studies revealed that the nanocomposite provided the potential for regeneration and reuse after MB dye adsorption. The synthesis of poly(acrylic acid)–bentonite–FeCo (PAAc-B-FeCo) hydrogel nanocomposite via ultrasound-assisted *in situ* emulsion polymerization was carried out (Shirsath et al. 2011). Addition of exfoliated bentonite clay platelets and FeCo increased the strength and stability of the hydrogel and assisted the adsorption of an organic pollutant. The response of the nanocomposite hydrogel was evaluated using a cationic dye, crystal violet under a different temperature, pH, and cavitation environment. The optimum temperature was found to be 35°C, and basic pH at 11 was responsible for the higher adsorption of dye due to dissociation of COO^- ions at higher pH.

Amphoteric semi-IPN nanocomposite hydrogels were prepared by graft polymerization of AAc onto starch in cationic polyacrylamide (CPAM)/bentonite nanocomposite aqueous dispersion (Xu et al. 2008). CPAM was used as both an intercalating agent to enlarge interlayer space and a linear polymer chain to fabricate the semi-IPN structure. X-ray diffraction (XRD) and TEM confirmed a successful intercalation of CPAM into bentonite. The results showed that the hydrogel was of a high swelling and compressive strength even under water content of more than 99%.

Highly swollen AAm/2-acrylamido-2-methyl-1-propanesulfonic acid (AMPS) hydrogels and AAm/AMPS/bentonite composite hydrogels were prepared by free radical solution polymerization in aqueous solutions of AAm with AMPS and a clay such as bentonite and a multifunctional cross-linker such as ethylene glycol dimethacrylate (Kundakci et al. 2008). Highly swollen AAm/AMPS and AAm/AMPS/bentonite hydrogels were used in experiments on the sorption of water-soluble monovalent cationic dye such as Lauth's violet (LV) (thionine). Swelling of AAm/AMPS hydrogels was increased up to 2,282%–12,603% in water and 921%–3,575% in LV solutions, while AAm hydrogels swelled 927% in water, and swelling of AAm/AMPS/bentonite hydrogels was increased up to 3,225%–15,421% in water and 1,360%–4,189% in LV solutions, while AAm/bentonite hydrogels swelled 828% in water.

Both clay minerals embedded within neutral or charged hydrogel networks and linear charged macromolecules that stabilize clay minerals exhibit excellent absorbance capacity with respect to metal ions (Saber-Samandari and Gazi 2013) and dye molecules (Nakamura and Ogawa 2013, Shirsath et al. 2013, Yang and Ni 2012) and as a controlled-release drug carrier (Kevadiya et al. 2011). The nanocomposite hydrogels have much greater equilibrium swelling ratio, much faster response rate to pH, excellent thermal responsibility, and significantly improved tensile mechanical properties and high storage modulus (Xiang et al. 2006, Zhang et al. 2009).

The composite hydrogel materials based on clay minerals, TiO_2, SiO_2, and PAAH were obtained by one-step *in situ* polymerization (Svetlichnyy et al. 2009a). As a result, the flexible, elastic, and mechanically stable composite materials were designed. Swelling–deswelling behavior, VPT, and physicochemical, physicomechanical, and thermal properties of composite hydrogels have been studied (Ibrayeva 2010, Zhumaly et al. 2013). The mechanism of formation of the composite structures can be represented as diffusion of AAm monomers into the layered clay structure. After monomer intercalation into the space of minerals and polymerization with simultaneous cross-linking, composite hydrogel materials are formed where nano- and microsized clay particles play the role of additional physical cross-linking centers. It leads to a significant increase in mechanical properties of composite materials. The swelling degree of samples increases in the following order: PAAH/bentonite > PAAH/TiO_2 > PAAH/SiO_2 > PAAH/kaolin ≈ PAAH/MMT. For the PAAH/bentonite, PAAH/kaolin, PAAH/TiO_2, and PAAH/SiO_2 composites, the values of n that are between 0.6 and 0.94 correspond to an anomalous swelling mechanism, for example, non-Fickian diffusion. The effect of water–organic solvent mixture, pH, temperature, and ionic strength on the behavior of the composite materials was studied. Composite materials shrank in water–acetone and water–ethanol mixtures, as well as at high ionic strength of the solution, while changing of pH and temperature has no substantial influence. For the PAAH/kaolin and PAAH/bentonite composite hydrogels, the swelling degree decreased with increasing both the content of methylenebisacrylamide (MBAA) and bentonite, respectively. In the former case, it was connected with increasing of the density of chemical cross-links and, in the latter case, physical cross-links. Scanning electron microscopy (SEM) images revealed that the morphology of composite materials is represented as flat surface, cracks, and micropores with an average diameter of 5–10 μm. The XRD patterns are characterized by amorphous halo from PAAH followed by smaller peaks from clay minerals that are embedded within hydrogel matrix. The Fourier transform infrared spectroscopy (FTIR) and Raman spectroscopy results revealed that composite materials have not been simply a mechanical mixture of two components; in contrast, they were stabilized by hydrogen bonds between NH_2 groups of PAAH and oxygen groups of TiO_2, SiO_2, and aluminosilicates. The positive values of the enthalpy of mixing ΔH_m indicated that the swelling of PAAH/kaolin and PAAH/TiO_2 in water had endothermic character. It was shown that the thermal decomposition of composite hydrogel materials was shifted to a higher-temperature region in comparison with PAAH. The increase of kaolin quantity in PAAH volume led to reinforcing of mechanical properties of composite materials.

1.2.2 POTENTIAL APPLICATION OF COMPOSITE HYDROGEL MATERIALS AS "PIGS" FOR CLEANING OF THE INTERNAL SURFACE OF MAIN PIPES

Pipelines are used to transport the powders and fluids from one point to another. Pigging is an operation to remove debris or unwanted deposit buildup in a pipeline (Al-Yaari 2011, Jaggard and Allen 1977, Uzu et al. 2000). Debris, sand, and asphaltene–resin–paraffin depositions (ARPDs) in a pipeline will result in a pressure buildup, and if no pigging exists, their buildup could continue to rise and will create greater back pressure on the line, causing higher maintenance on pumps, and the line could eventually become blocked. It is forecasted that the composite hydrogel materials may bear more external load than that of pure hydrogel. In contrast to ordinary hydrogels, the composite materials consisting of hydrogels and clay minerals exhibit an improved physicomechanical property (Ibrayeva 2010, Svetlichnyy et al. 2009b). The mechanical stability of the PAAH/kaolin sample in comparison with pure PAAH is shown in Figure 1.1.

The laboratory device for study of the model oil pipeline is as follows: A slightly swollen hydrogel plunger (not miscible with oil) is immersed into the pipeline to

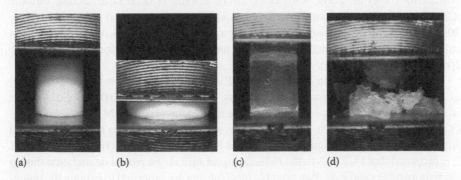

(a) (b) (c) (d)

FIGURE 1.1 Mechanical stability of PAAH/kaolin composite (a, b) and pristine PAAH (c, d) gels.

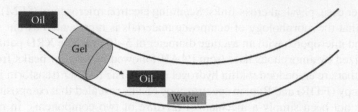

FIGURE 1.2   Schematic representation of cleaning of inner part of pipeline from APRD and water by hydrogel "pigs."

separate the oil flow. As the hydrogel "pig" moves along the pipe, it absorbs the water–saline solution and swells. The hydrogel swelling allows tight hydraulic sealing to the pipe wall. This, in turn, leads to efficient removal of gas accumulations, ARPD, mechanical impurities, and mineralized water from the pipeline inner cavity (Figure 1.2).

In cleaning a model pipeline from ARPD, the PAAH/kaolin composite hydrogel that showed the best elongation at break, tensile strength, and Young's modulus at 15 wt.% of kaolin was used (Zheksembayeva et al. 2012). The effectiveness of cleaning of deposited paraffins from Kumkol and Usen oil fields by composite hydrogel "pigs" ranges between 94% and 96% (Kudaibergenov et al. 2012a).

1.3   PHYSICOCHEMICAL AND CATALYTIC PROPERTIES OF POLYMER-PROTECTED AND HYDROGEL-IMMOBILIZED GOLD, SILVER, AND PALLADIUM NANOPARTICLES

1.3.1   STABILIZATION OF GOLD AND SILVER NANOPARTICLES BY HYDROPHILIC POLYMERS

Gold (AuNPs) and silver (AgNPs) nanoparticles have attracted significant attention of researchers due to their unique optical, electrical, biomedical, and catalytic properties (Balasubramanian et al. 2010, Motoyuki and Hidehiro 2009, Shan and Tenhu 2007, Zhou et al. 2009). A lot of polymers possessing nonionic (Chung et al. 2012, Dai et al. 2007, Morrow et al. 2009, Ram et al. 2011), anionic (Dorris et al. 2008), cationic (Chen et al. 2012a), and amphoteric (Li et al. 2010, Mahltig et al. 2010, Note et al. 2007) nature are widely used as protecting agents of AuNPs and AgNPs in aqueous solution or organic solvents for preventing nanoparticle aggregation (Bekturov et al. 2010, Ibrayeva et al. 2013).

The size of poly(N-vinylpyrrolidone) (PVP)-protected AuNPs ranging from 10 to 110 nm was easily controlled by varying the concentration (0.01–10 g/dL) (Ram et al. 2011) or the average-number molecular weight of PVP (M_n = 10–350 kDa) (Yesmurzayeva et al. 2013). The shape, size, and optical properties of the AuNPs and AgNPs are tuned by changing the employed PVP/metal salt ratio (Hoppe et al. 2006). It is proposed that PVP acts as the reducing agent suffering a partial degradation during the nanoparticle synthesis. Two possible mechanisms are proposed to explain the reduction step: direct hydrogen abstraction induced by the metal ion and/or reducing action of macroradicals formed during degradation of the polymer. The initial formation of the macroradicals might be associated with the metal-accelerated decomposition of low amounts of peroxides present in the commercial polymer. Gold catalysts have recently attracted rapidly growing interests due to their potential applicabilities to many reactions of both industrial and environmental importance (Haruta 1997). Typical examples are the low-temperature catalytic combustion, partial oxidation of hydrocarbons, hydrogenation of carbon oxides and unsaturated hydrocarbons, and reduction of nitrogen oxides (Haruta and Daté 2001). A recent review (Shiju and Guliants 2009) describes the size-, shape-, structure-, and composition-dependent behavior of AuNPs employed in alkylation, dehydrogenation, hydrogenation, and selective oxidation reactions for the conversion of hydrocarbons (with main emphasis on fossil resources) to chemicals. The perspectives of substituting platinum group metals for automobile emission control with gold were outlined by authors (Zhang et al. 2011).

1.3.2 IMMOBILIZATION OF POLYMER-PROTECTED AuNPs AND AgNPs WITHIN HYDROGEL MATRIX

Hydrogels are chemically stable and interlocked polymeric networks that retain vast amounts of water without dissolving; therefore, they are feasible for the preparation of metal nanoparticles *in situ* and readily applicable in the catalysis of various aquatic and nonaquatic reactions. The functional groups in the hydrogel network can act as both chelating and capping agents for metal nanoparticle preparation from metal ions and for their stabilization; thus, the metal particles are protected from the atmosphere hindering the oxidation/deactivation and aggregation, allowing an increase in their stability and longevity. Various synthesis methods have been reported to produce AuNPs–hydrogel composites (Dolya et al. 2013): (1) preparation of the nanoparticles and hydrogels, separate or in combination (Pardo-Yissar et al. 2001, Sheeney-Hai-Ichia et al. 2002); (2) mixing and polymerization of the preformed nanoparticles with monomer precursor(s) (Holtz and Asher 1997, Lee and Braun 2003, Sershen et al. 2000, 2001, Weissman et al. 1996); and (3) embedding of metal salts into a hydrogel matrix followed by a reduction process in the presence of reducing agents (Wang et al. 2004). The role of hydrophilic polymers in this system is to stabilize the metal nanoparticles and to prevent their aggregation, while the role of hydrogel matrix is restriction of diffusion of nanoparticles both inside of and outside from the gel matrix (Kudaibergenov 2008). A typical example of embedding of PVP-protected AuNPs, AgNPs, and palladium nanoparticles (PdNPs) within the hydrogel matrix is shown in Figure 1.3.

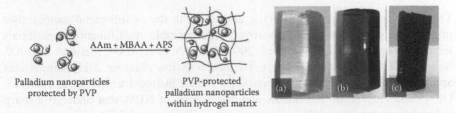

Palladium nanoparticles protected by PVP

AAm + MBAA + APS

PVP-protected palladium nanoparticles within hydrogel matrix

(a) (b) (c)

FIGURE 1.3 Immobilization protocol of polymer-protected nanoparticles within hydrogel matrix and PAAH samples with immobilized AgNPs (a), AuNPs (b), and PdNPs (c).

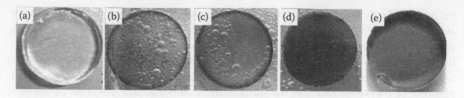

FIGURE 1.4 Swollen in water PAAH/PEI-HAuCl$_4$ (a) in the course of reduction by NaBH$_4$ (C = 0.1 mol/L) during 5 min (b), 15 min (c), 60 min (d), and 1 day (e).

The average size of AgNPs, AuNPs, and PdNPs in the volume of PAAH was equal to 20–30, 10–50, and 10–60 nm, respectively (Kudaibergenov et al. 2008). Metal ions with different oxidation states to be loaded into the hydrogel matrices can be reduced/precipitated to their metallic particle forms inside hydrogels of different dimensions using green chemicals or nontoxic chemical reducing agents such as NaBH$_4$, H$_2$, citrate, and ethylene glycol, depending upon the nature of the metal ions. Reduction of polyethyleneimine (PEI) protected and immobilized within PAAH AuNPs by NaBH$_4$ is shown in Figure 1.4 (Dolya 2009).

Reduction of PEI–Au^{3+} complexes to Au^0 within hydrogels is accompanied by the formation of a thin, colored layer on the gel surface that gradually moves into the gel volume. The driving force of this process is the constant diffusion of the reducing agent NaBH$_4$ deeply into the gel volume. Narrow-dispersed gold nanospheres and single crystals were prepared, respectively, by reducing HAuCl$_4$ within the hydrogel matrix (Kim and Lee 2007, Zhang et al. 2007). The authors (Kim and Lee 2007) described a unique strategy to prepare discrete composite nanoparticles consisting of a large gold core (60–150 nm) surrounded by a thermoresponsive hydrogel derived from the polymerization of NIPA or copolymerization with AAc. The growth of AuNPs in the presence of preformed spherical hydrogel particles allows a precise control of the size of composite nanoparticles between 200 and 550 nm. Most of the hydrogel-immobilized PdNPs exhibited good catalytic activity in both Heck and Suzuki reactions (Hagiwara et al. 2001, Kohler et al. 2001) and Suzuki–Miyaura cross-coupling reaction (Leadbeater and Marco 2002, Lu et al. 2004, Phan et al. 2004, Sivudu et al. 2008, Wu et al. 2011).

1.3.3 Catalytic Properties of Polymer-Protected PdNPs and AuNPs Immobilized within Hydrogels

The combination of natural catalytic abilities with the *in situ* metal nanocatalyst preparation capability makes hydrogels indispensable multifunctional materials for unique applications (Jiang et al. 2004, Kidambi et al. 2004, Metin et al. 2009, Sahiner 2004, Wunder et al. 2011). The recent review (Sahiner 2013) summarizes application aspects of metal nanoparticles within hydrogel templates in catalysis. Of special interest are the homo- and copolymers of NIPA that undergo a sharp volume transition around the body temperature (Peppas et al. 2006). Many researchers have examined the potential application of NIPA-based polymers for the immobilization of AuNPs (Echeverria and Mijangos 2010, Wang et al. 2004a,b).

Examples of catalytic system acting by "on-off" mechanism are NIPA-based hydrogels that reversibly swell or shrink in water–ethanol mixture (Wang et al. 2000) or reversibly turn "off" first and then "on" as the temperature is first raised and then lowered (Bergbreiter et al. 1998). The "smart" behavior of the PNIPA/PVP-Pd(0) system was demonstrated in the course of allyl alcohol hydrogenation (Dolya 2009, Dolya et al. 2008a,b, 2009). Swelling–deswelling of PNIPA at temperature interval 25°C–40°C causes the release or inflow of PVP-Pd(0) outside or inside of the hydrogel matrix. This in turn leads to periodic increase or decrease of the hydrogenation rate of allyl alcohol (Figure 1.5).

The catalytic activity of polymer-protected and PAAH-immobilized Pd(0) catalysts increased in the following order: PAAH/PVA-Pd(0) > PAAH/PVP-Pd(0) > PAAH/PEI-Pd(0) > PAAH/PAA-Pd(0). The catalytic activity of PAAH/PEI-Pd(0), PAAH/PVP-Pd(0), and PAAH/PVA-Pd(0) catalysts preserved up to hydrogenation of 12 sequential portions of allyl alcohol (Zharmagambetova et al. 2010). Turnover numbers (TONs) for PAAH/PEI-Pd(0) and PAAH/PVP-Pd(0) were equal to 4×10^3

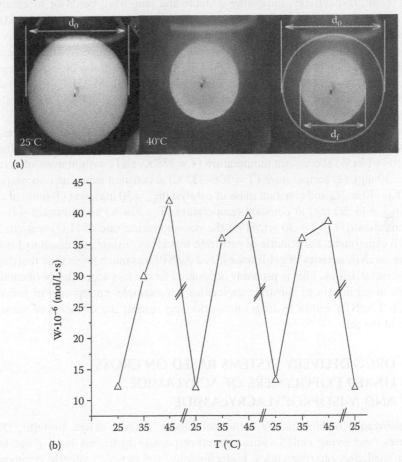

FIGURE 1.5 Reversible changing of size (a) and catalytic activity of PNIPA/PVP-Pd(0) (b) at 25 h 40°C.

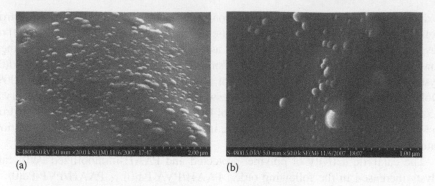

FIGURE 1.6   SEM pictures of PVP-protected PdNPs within the gel matrix of PAAH after hydrogenation of the 1st (a) and 12th (b) successive portions of allyl alcohol.

and 7×10^3, respectively, indicating a stable and long-lived behavior of catalysts. After hydrogenation of sequential portions of allyl alcohol, the amount of Pd(0) on the surface of gel matrix is considerably reduced (Figure 1.6). This is probably due to leaching out of Pd nanoparticles in the course of hydrogenation reaction. The average size of Pd nanoparticles was less than 100 nm, although the bigger aggregated particles were observed, while SEM micrographs of pristine PAAH/PVP-Pd(0) show spheres with an average diameter of about 60 nm that are related to PVP-stabilized spherical PdNPs or particle aggregates.

The catalytic activity of gel-immobilized AuNPs was evaluated with respect to hydrogen peroxide decomposition. The influence of (1) substrate concentration (C = 10–40 wt.%) at constant temperature (T = 328 K) and constant mass of catalyst (m_{cat} = 30 mg), (2) temperature (T = 308–323 K) at constant substrate concentration ($[H_2O_2]$ = 30 wt.%) and constant mass of catalyst (m_{cat} = 30 mg), and (3) mass of catalyst (m_{cat} = 15–50 mg) at constant temperature (T = 328 K) and constant substrate concentration ($[H_2O_2]$ = 30 wt.%) on the decomposition rate of H_2O_2 was studied. In each experiment, the volume of substrate was kept constant and equal to 1 mL.

The catalytic activity of gel-immobilized AuNPs was much lower than that deposited on metal oxides. This is probably accounted for the less accessibility of catalytic centers in gel matrix to substrate molecules, for example, entrapment of polymer-protected AuNPs within hydrogel networks may restrict the diffusion of substrate inside of the gel.

1.4   DRUG DELIVERY SYSTEMS BASED ON CROSS-LINKED COPOLYMERS OF ACRYLAMIDE AND N-ISOPROPYLACRYLAMIDE

Immobilization of biologically active substances, such as drugs, proteins, DNA, enzymes, and living cells, within stimuli-responsive hydrogels is of great interest for medicine, pharmaceutics, biotechnology, and bio- and genetic engineering (Hoffman and Stayton 2004, Lee and Yuk 2007, Liu et al. 2007, Peppas et al. 2006, Rzaev et al. 2007, Stein 2009). One of the serious problems of modern medicine is transportation of biologically and physiologically active substances to target places of organisms in a strictly definite dose. Presently, about 25% of drugs of leading pharmaceutical companies prepared for selling, production, and application are provided by transportation system. Hydrogel materials, due to excellent swellability in water, softness, elasticity, and biological compatibility, are widely applied for the design of drug delivery systems that are able to transport drugs to a target part of an organism by realization of positive feedback with environment providing afterward more reliable and controlled treatment of diseases (Anish and Abdul 2012, Eros et al. 2003, Galaev and Mattiasson 1999, Kumar et al. 2007, Manpreet et al. 2013). Among the well-known hydrogel systems of synthetic origin, the homo- and copolymers of AAc and NIPA are able to change morphology, size, and shape under the action of external stimuli (Bajpai et al. 2008, Feng et al. 2010, Hoare and Kohane 2008, Jagur-Grodzinski 2010, Qiu and Park 2012). pH medium and body temperature changes are the most widely used triggering signals for both site-specific therapy and pulsatile drug release (Anil 2007, Bajpai et al. 2008, Coughlan et al. 2004, Liusheng et al. 2011, Yoshida et al. 2013). In this connection, the development of thermo- and pH-responsive hydrogel materials that might realize "on-off" mechanism of drug delivery, that is, opening and closing the "thermo- or pH valve" to deliver the dosed amount of drug to the diseased part of the body, presents great interest (Chen et al. 2012b). The most significant weakness of external stimuli-sensitive hydrogels is that their response time is too slow. Therefore, the fast-acting hydrogels are necessary, and the easiest way of achieving that goal is to make thinner and smaller hydrogels. A method for making thermally responsive hydrogel scaffolds with a remarkably rapid response to temperature changes was developed by Cho et al. (2008). The recent remarkable review of Klinger and Landfester (2012) presents some of the important fundamental examinations on the influence of (tunable) network characteristics on loading and release profiles and basic synthetic concepts to realize these concepts and highlights several examples of different approaches to stimuli-responsive microgels for loading and release applications.

1.4.1 Hydrogel-Immobilized Local Anesthetic Drugs

Immobilization of local anesthetic drugs, such as lidocaine, novocaine, and bupivacaine, into stimuli-responsive hydrogel matrix is very important to solve the problems of "medicine of catastrophe" when first aid is needed after an earthquake and fire. Hydrogel-immobilized local anesthetic drugs can serve as wound dressing materials due to their versatility and unique properties, such as high water content and soft and rubbery consistency, that make them similar to natural tissues. Literature survey shows that lidocaine was loaded within IPNs based on PNIPA, PVP, and AMPS (Akdemir and Kayaman-Apohan 2007) and NIPA–itaconic acid (IA) copolymeric hydrogels (Taşdelen et al. 2004) by sorption immobilization. Lidocaine uptake of the IPNs was found to increase from 24 to 166 (mg lidocaine/g dry gel) with increasing amount of AMPS contents in the IPN structure, while lidocaine adsorption capacity of the NIPA-IA hydrogels was found to increase from 3.6 to 862.1 (mg lidocaine/g dry gel) with increasing amount of IA in the gel structure. In both cases, the electrostatic interactions between anionic groups of hydrogels and cationic groups of lidocaine are responsible for retarding drug release profile. The release characteristics of lidocaine from an anionic hydrogel composed of carbopol and a cationic hydrogel composed of chitosan were examined for optimizing hydrogel formulation as a sponge filler to stop the bleeding and as a carrier for delivering lidocaine to relief pain after a tooth extraction (Liu et al. 2007). The elasticity of the gel matrix and the ionic complexing effect between the anionic acid groups of hydrogels and cationic groups of lidocaine are two main factors influencing regulation of the diffusion coefficient for controlling drug release. Spherical nanoparticulate drug carriers made of poly(D,L-lactic acid) (Gorner et al. 1999, Polakovič et al. 1999) and of poly(D,L-lactic-*co*-glycolic acid) 50:50 mol/mol (Holgado et al. 2008, Zhang et al. 2008) with controlled size were designed for encapsulation of lidocaine and bupivacaine. Particles with sizes in the range of 250–820 nm and low polydispersity were prepared with good reproducibility; the large particles with a high loading (~30%) showed under *in vitro* conditions a slow release over 24–30 h, the medium-sized carriers (loading of ~13%) released the drug over about 15 h, and the small particles with small loading (~7%) exhibited a rapid release over a couple of hours. Two simple models, diffusion and dissolution, were applied for the description of the experimental data of lidocaine release and for the identification of the release mechanisms for the nanoparticles of different drug loading. The modeling results showed that in the case of high drug loadings (about 30% w/w), where the whole drug or a large part of it was in the crystallized form, the crystal dissolution could be the step determining the release rate. On the other hand, the drug release was diffusion controlled at low loadings (<10% w/w) where the solid drug was randomly dispersed in the matrix. The estimated values of the diffusion coefficient of lidocaine in these particles were in the range of $(5-7) \times 10^{-20}$ m^2/s. The efficacy and toxicity of bupivacaine loaded in biodegradable polymer poly(sebacic-*co*-ricinoleic acid) for producing motor and sensory block when injected near the sciatic nerve were evaluated (Shikanov et al. 2007). *In vitro* and *in vivo* bupivacaine release after injection in mice showed that 70% of the drug has been released during 1 week. Single injection of 10% bupivacaine in the polymer caused motor and sensory block that lasted 30 h. It was concluded that the poly(sebacic-*co*-ricinoleic acid) is a safe carrier for prolonged activity of bupivacaine. Richlocaine and richlocaine hemisuccinate are new local anesthetic drugs, invented by Kazakhstan chemists, that have been registered and approved for use in CIS countries (Sharifkanov et al. 2011) (Figure 1.7). In medicine, richlocaine is applied only as an isotonic injection solution. The anesthetic and antibacterial effectiveness of richlocaine is much higher than that of bupivacaine, novocaine, and lidocaine.

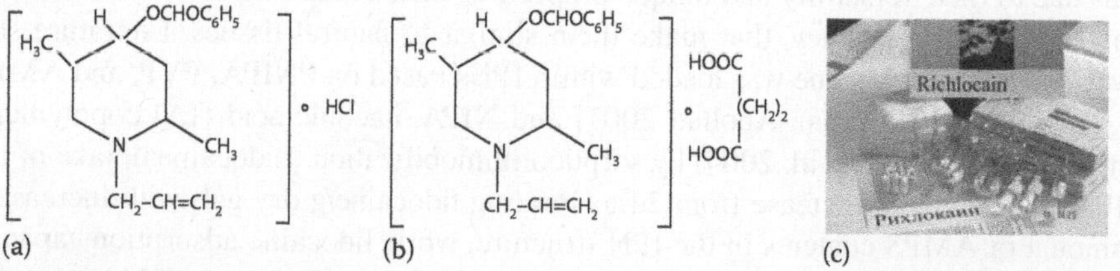

FIGURE 1.7 Structural formulas of (a) richlocaine, (b) richlocaine hemisuccinate, and (c) richlocaine as an injection solution.

Development of a prolonged drug dosage form would be beneficial. Richlocaine was immobilized into linear and weakly cross-linked PVP (Makysh et al. 2003), poly(sodium acrylate) (PSA), and betaine-type polyampholyte gels (Makysh et al. 2002). The properties of polymer–drug complexes were studied with respect to external factors, such as pH, temperature, and thermodynamic quality of water–ethanol mixture. The kinetics of richlocaine release from the PVP gel matrix into water was studied. At pH = 7.0, ~20% of richlocaine was released within 96 h. This quantity remained constant up to 384 h, indicating poor desorption of richlocaine. Comparatively, complexes of richlocaine with PSA and betaine-type polyampholyte gels displayed better desorption; the degree of release of richlocaine reached ~95% within 144 h and ~80% within 260 h, respectively. The quantity of released richlocaine increased up to 50% at pH = 8.0, obviously indicating the destruction of the PVP gel–richlocaine complex at this pH. The activation energies of drug release from the PVP gel matrix, PSA gel, and betaine-type polyampholyte gel were equal to 6.86, 5.26, and 17.14 kJ/mol, respectively. The effect of richlocaine on the swelling/deswelling kinetics and pulsatile drug release from the thermoresponsive hydrogels such as weakly cross-linked copolymers of AAm-AAc, hydrogels of PNIPA, and 3D networks of NIPA-AAc and NIPA-AMPSA was examined (Tatykhanova 2009). The richlocaine release profile exhibits a similar trend with the swelling–deswelling behavior of hydrogels (Figure 1.8). The initial release of the drug is due to the presence of surface-encapsulated components that are squeezed out during the first temperature pulse. The release of richlocaine at T < volume phase transition temperature (VPTT) is governed by diffusion. At T > VPTT, the surface of the hydrogel shrunk immediately and formed an impermeable "skin" layer restricting the release of immobilized bioactive molecules. The second and third temperature pulses lead to the decrease of the release rate due to the decrease in the concentration of richlocaine in the hydrogel volume.

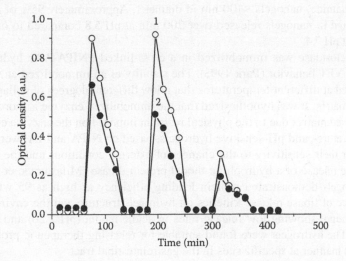

FIGURE 1.8 Time-dependent pulsatile release of richlocaine from PNIPAM hydrogel into phosphate buffer (1) and water (2) at 25°C and 40°C.

1.4.2 CONTROLLED RELEASE OF PROTEINS FROM STIMULI-RESPONSIVE HYDROGELS

The use of stimuli-sensitive hydrogels for the encapsulation and controlled release of proteins has received significant attention. The release of bovine serum albumin (BSA), a model drug, from a series of thermosensitive silk sericin (SS)/PNIPA and pH-responsive SS/poly(methacrylic acid) IPN hydrogels has been studied (Wen et al. 2013). The pulsatile releasing behavior of IPN hydrogels revealed that they can be made into microcapsules or thermo valves, which act as an on-off release control.

An efficient strategy to conjugate methacrylamide moieties to the lysine units of lysozyme for copolymerization and subsequent triggered release from hydrogels has been developed (Verheyen et al. 2011). Methacrylated dextran (dex-MA) was polymerized in the presence of native or modified lysozyme to yield hydrogels. The release of native and modified lysozyme from dex-MA hydrogels was studied in acetate buffer (pH 5, in the absence of any trigger), and only a minor fraction (~15%) of the modified lysozyme was released, whereas ~74% of the native lysozyme was released.

Horseradish peroxidase and alkaline phosphatase were immobilized into cellulose hydrogel prepared from an aqueous alkali–urea solvent (Isobe et al. 2011). Proteins were covalently introduced to cellulose gel by a Schiff base formation between the aldehyde and the amino groups of proteins and stabilized by a reduction of imines. The number of oxidized glucose per 100 glucose residues ranged between 3.3 and 18.6. The activity of the immobilized enzymes increased with aldehyde content, but the effect leveled off at a low degree of oxidation, at approximately 8.1 of oxidized glucose/100 glucose unit. The amount of immobilized peroxidase calculated from the activity was 8.0 ng/g for an aldehyde content of 0.18 mmol/g and 14.6 ng/g for both 0.46 and 1.04 mmol/g. Due to the high mechanical and chemical stability of cellulose, this technique and resulting materials are potentially useful in biochemical processing and sensing technologies.

Shi et al. (2008) studied the pH-sensitive release of lysozyme from the poly(N-vinyl formamide) nanogels ~100 nm in diameter. Approximately 95% of lysozyme encapsulated in nanogels released over 200 min at pH 5.8 compared to only ~15% released at pH 7.4.

β-Galactosidase was immobilized in a cross-linked PNIPA-AAc hydrogel that exhibits a VPT behavior (Park 1993). The stability of an immobilized enzyme was investigated at different temperatures that allow different degrees of collapse in the hydrogel matrix. It was hypothesized that the immobilized enzyme is more stable in the collapsed matrix due to the physical restraint imposed on the enzyme entrapped.

Temperature- and pH-sensitive hydrogels, based on NIPA and IA, were characterized for their sensitivity to the changes of external conditions and the ability to control the release of a hydrophilic model protein, lipase (Milasinovic et al. 2010). The hydrogels demonstrated protein loading efficiency as high as 95 wt.%. High dependence of lipase release kinetics on hydrogel structure and the environmental pH was found, showing low release rates in acidic media (pH 2.20) and higher at pH 6.80. The hydrogels were found suitable for releasing therapeutic proteins in a controlled manner at specific sites in the gastrointestinal tract.

Catalase was entrapped in PAAm, PSA, and poly(acrylamide-co-sodium acrylate) (PAAm-SA) gels (Jiang and Zhang 1993) and in thermally reversible poly(NIPA-*co*-hydroxyethylmethacrylate) (NIPA-HEMA) copolymer hydrogels (Arica et al. 1999) and on a cross-linked macromolecular carrier of a polysaccharide structure (gellan) (Popa et al. 2006). The percentage of entrapment was found to be about 85%. The enzyme immobilized in PAAm has very low activity, while the enzyme in PAAm-SA exhibits the highest activity. The kinetic behavior of the entrapped enzyme was investigated in a batch reactor. The apparent kinetic constant of the entrapped enzyme was determined by the application of the Michaelis–Menten model and indicated that the overall reaction rate was controlled by the substrate diffusion rate through the hydrogel matrix. Due to the thermoresponsive character of the NIPA-HEMA, the maximum activity was achieved at 25°C with the immobilized enzyme. The K_m value for immobilized catalase (28.6 mM) was higher than that of free enzyme (16.5 mM). Optimum pH was the same for both free and immobilized enzyme. Operational, thermal, and storage stabilities of the enzyme were found to increase with immobilization.

BSA and lysozyme were embedded into the hydrogel volume of AAm-AAc, PNIPA, and NIPA-AAc by *in situ* and sorption methods from aqueous and phosphate buffer solutions (pH = 7.4, μ = 0.15 M NaCl) (Kudaibergenov et al. 2011). Oscillating the "on-off" release mechanism of proteins from the volume of PNIPA and NIPA-AAc hydrogels was observed in the course of cyclic shrinking and swelling of hydrogels in water and phosphate buffer at 25°C and 40°C (Figure 1.9).

Sorption of catalase by AAm-AAc, NIPA-AAc, and PNIPA hydrogels proceeds via diffusion. Equilibrium swelling degree of dry samples in the course of catalase sorption and the activity of immobilized enzyme are changed in the following order: AAm-AAc > NIPA-AAc > PNIPA (Tatykhanova 2009). It is explained by the fact that binding of catalase by hydrogel matrix proceeds via electrostatic interaction with participation of carboxylic groups of the network and amine groups of enzyme.

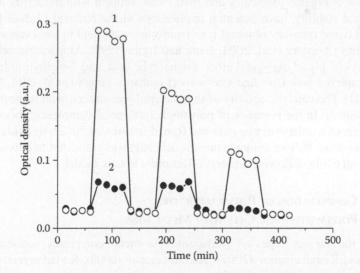

FIGURE 1.9 Time-dependent pulsatile release of BSA (1) and lysozyme (2) from PNIPAM hydrogel into phosphate buffer at 25°C and 40°C.

Maximal swelling and binding degree of catalase by hydrogels corresponds to neutral region. The relative activity of catalase encapsulated into AAm-AAc and NIPA-AAc networks after 74 days decreases two times, while the activity of catalase in solution decreases 46 times. The activity of immobilized and pristine catalase at temperature interval from 25°C to 70°C decreased 3 and 10 times, respectively. These results reveal that hydrogel-immobilized catalase preserves the catalytic activity for a long time and high temperature.

1.5 COMPLEXES OF LINEAR POLYAMPHOLYTES AND AMPHOTERIC GELS WITH TRANSITION METAL IONS

Renewed interest to polyampholyte–metal complexes is dictated by the fact that such complexes can model the protein–metal complexes and are relevant to catalysis (Bekturov and Kudaibergenov 1996, Casoloro et al. 2001, Khvan et al. 1985). For example, the kinetics and mechanism of complexation of AAc and vinylimidazole copolymers with Cu^{2+}, Co^{2+}, and Ni^{2+} ions are similar to the interaction of the carboxyl and imidazole groups of gelatin with the same metal ions (Annenkov et al. 2000, 2003). Polyampholyte–metal complexes are proved to exhibit catalase-like activity in decomposition of hydrogen peroxide (Bekturov et al. 1986, Lázaro Martínez et al. 2011, Sigitov et al. 1987) and to serve as hydrogenation or oxidation catalysts for organic substrates (Lázaro Martínez et al. 2008a,b, Xi et al. 2003). The ability of water-soluble or water-swelling polyampholytes to form stable chelate structure can be used for water treatment (Anderson et al. 1993) and recovery of metal ions from the wastewater (Ali et al. 2013, Chan and Wu 2001, Martınez et al. 2008, Rivas et al. 2006, Terlemezian et al. 1990, Xu et al. 2003) and polluted soils (Rychkov 2003). Amphoteric hydrogels, due to their high sorption and easy desorption of organic molecules and metal ions, coupled with durability and good mechanical stability, have potential applications in the removal of dyes (Dalaran et al. 2011) and recovery of metal ions from wastewater and in ion-exchange chromatography (Arasawa et al. 2004, Jiang and Irgum 1999). Amphoteric gel derived from ethylene glycol diglycidyl ether, methacrylic acid, and 2-methylimidazole has been complexed with Cu^{2+} and Co^{2+} ions (Lombardo Lupano et al. 2013, Martínez et al. 2011). The catalytic activity of this material was studied with respect to H_2O_2 decomposition. In the presence of polyampholyte–metal complexes, about 70% of methyl orange (model dye) was removed from distilled water in 2 h by oxidation with H_2O_2, and about 80% of epinephrine (model drug) was converted to adrenochrome in less than 6 min, following a pseudo-first-order kinetic model.

1.5.1 COMPLEXATION OF POLYBETAINIC OR POLYZWITTERIONIC GELS WITH METAL IONS

Among the various types of polyampholyte–metal complexes summarized in Ciferri and Kudaibergenov (2007), Kudaibergenov (2002), Kudaibergenov (2008), Kudaibergenov and Ciferri (2007), less attention has been paid to metal complexes of cross-linked polybetaines or polyzwitterions (Kudaibergenov et al. 2006). The polybetaines (or "polyzwitterions") are dipolar species, in which the cationic and

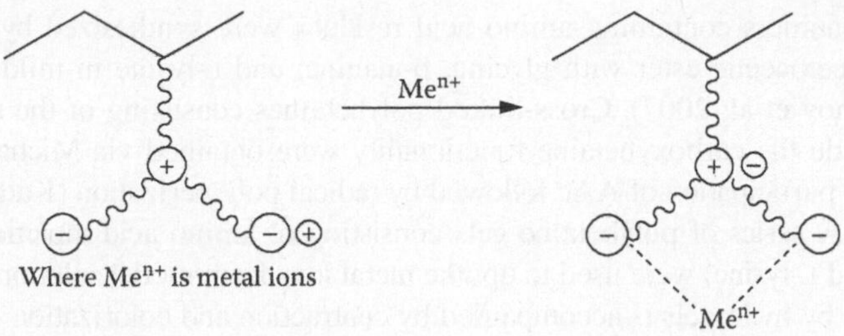

FIGURE 1.10 Simultaneous complexation of two units in CPZA is the driving force to capture Sr^{2+} ions.

anionic groups are separately bound to the same monomer unit and can be completely dissociated in a medium of sufficient dielectric permittivity. The most widespread chemical classes of polybetaines are carbo-, sulfo-, and phosphobetaines, that is, polymers with repeat units bearing simultaneously a quaternized ammonium group and a carboxylate, a sulfonate, or a phosphate group, respectively. As distinct from classical polybetaines, the research group of Ali (Ali and Haladu 2013, Ali and Hamouz 2012, Charles et al. 2012) developed novel polymers containing zwitterionic (±) and anionic (−) or cationic (+) groups such as poly(electrolyte–zwitterions) that have two negative and one positive charges (or two positive and one negative charges) in each monomer unit. The cross-linked polymer having zwitterionic/anionic group was synthesized via copolymerization of N,N-diallyl-N-sulfopropylammonioethanoic acid and sulfur dioxide in the presence of cross-linker 1,1,4,4-tetraallylpiperazinium dichloride followed by hydrolysis with NaOH to convert poly(zwitterions) into cross-linked polyzwitterion/anion (CPZA) (Ali and Haladu 2013). Simultaneous complexation of two units in CPZA is the driving force to capture Sr^{2+} ions (Figure 1.10).

The removal of 87% and 92% of Sr^{2+} ions at the initial concentrations of 200 ppb and 1 ppm was, respectively, observed. Excellent adsorption and desorption capacity of CPZA would enable its use in the treatment of radioactive nuclear waste containing Sr^{2+} ions.

New amphoteric gels based on NIPA and amino acid (L-ornithine) were prepared by free radical polymerization in aqueous solutions (Marcin et al. 2010). The presence of NIPA and amino acid moieties imparts their multiresponsive character to temperature, pH, and metal ion complexation. The gels were found to be most sensitive to concentrations of copper ions in the range 10^{-6} to 10^{-5} M. As the amount of amino acid in the polymer network increases, the gels gradually lose their temperature sensitivity and become more sensitive to copper ion concentration. The VPTT decreases significantly after the addition of copper ions. Analysis of the UV-Vis spectra and the swelling behavior indicates that both 1:1 and 1:2 complexes are present in the swollen state of the gels, whereas the latter complex is more dominant in the shrunken state. It is concluded that the metal ion sorption ability, the temperature, and the pH sensitivity of amphoteric hydrogels make them interesting materials in terms of the temperature- and pH-triggered swinging of the binding strength of heavy metal absorbers.

Novel monomers containing amino acid residues were synthesized by condensation of the acetoacetic ester with glycine, β-alanine, and L-lysine in mild conditions (Kudaibergenov et al. 2007). Cross-linked polybetaines consisting of the amino acid moieties beside the carboxybetaine functionality were obtained via Michael addition reaction with participation of AAc followed by radical polymerization (Kudaibergenov et al. 2007). A series of polybetaine gels consisting of amino acid moieties (glycine, β-alanine, and L-lysine) were used to uptake metal ions from model solutions. Sorption of metal ions by hydrogels is accompanied by contraction and colorization of samples. At first, the thin colored layer on the gel surface is formed and it gradually moves into the gel volume. The driving force of this process is "ion-hopping transportation" of metal ions through intra- and intermolecular chelate formation, for example, constant migration of metal ions deeply into the gel volume by exchanging of free ligand vacancies.

1.5.2 Metal Complexes of Amphoteric Cryogels

Cryogels are gel matrices that are formed in moderately frozen solutions of monomeric and polymeric precursors (Dinu et al. 2013, Mattiasson et al. 2010, Stein 2009). A system of large interconnected pores is a main characteristic feature of cryogels. The pore system in such spongelike gels ensures unhindered convectional transport of solutes within the cryogels, contrary to diffusion of solutes in traditional homophase gels. Semi-IPN cryogels based on cross-linked PAAm and anionic (Dragan and Apopei Loghin 2013) or cationic (Dragan and Dinu 2013) polyelectrolytes can serve as effective sorbents for the removal of dye molecules and metal ions. Amphoteric cryogels due to their response to temperature, pH, ionic strength, water–organic solvent composition, electric field, etc., belong to "smart" materials (Kudaibergenov et al. 2012b). A series of amphoteric cryogels with molar ratio of AAm, allylamine (AA), and methacrylic acid (MAA) (AAm:AA:MAA = 80:10:10, 60:20:20, 40:30:30, 20:40:40, and 0:50:50 mol.%/mol.%/mol.%) were synthesized (Kudaibergenov et al. 2012b, Tatykhanova et al. 2012). The structure and morphology of amphoteric cryogels and their complexation ability with respect to transition metal ions were evaluated. Cross and longitudinal sections of dry cryogels show spongelike porous structure with pore size ranging from 50 to 200 μm and the interconnected channels (Figure 1.11).

Complexation of amphoteric cryogels with transition metal ions is accompanied by colorization and slight shrinking of samples (Figure 1.12a). This is due to the formation of coordination and ionic bonds between metal ions and amine and/or

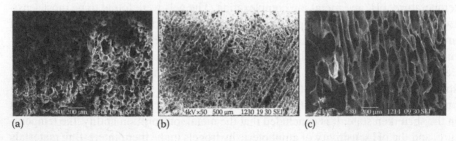

(a)   (b)   (c)

FIGURE 1.11 SEM images of cross- and longitudinal sections of cryogels with pore size (a) 50, (b) 100, and (c) 200 μm.

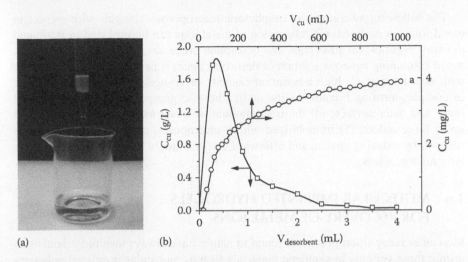

(a)   (b)

FIGURE 1.12 Sorption (a) and desorption (b) of copper ions by amphoteric cryogel ACG-334.

carboxylic groups of cryogels when aqueous solutions of metal salts pass through the gel specimen. The dynamic sorption capacity of amphoteric cryogels with respect to copper, nickel, and cobalt ions was evaluated. The amount of adsorbed metal ions varied from 99.17% to 99.55%. Dynamic exchange capacity of cryogels was in the range of 350–400 mg/L. Desorption of metal ions from cryogel volume was provided by disodium salt of ethylenediaminetetraacetic acid. The extracted amount of metal ions was equal to 75%–80%. Figure 1.12b demonstrates the adsorption and desorption curves of copper ions by amphoteric cryogel.

Preferentially, the adsorption of Cu^{2+} ions (79%) in comparison with Ni^{2+} (38%) and Co^{2+} ions (32%) from their mixture was also observed from aqueous solution containing 10^{-5} mol/L of metal ions indicating the specific binding of copper ions. High adsorption capacity of amphoteric macroporous gels with respect to metal ions may be perspective for purification of the wastewaters and analytical purposes. The reduction of cryogel–metal complexes by $NaBH_4$ leads to the formation of nano- and micron-sized particles of metals and/or metal oxides immobilized on the inner and surface parts of amphoteric cryogels (Figure 1.13). The chemical composition of the Ni-containing sample by energy-dispersive x-ray attached to SEM revealed that up to 34 wt.% of Ni particles is formed.

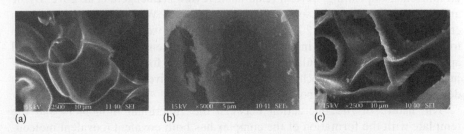

(a)   (b)   (c)

FIGURE 1.13 SEM pictures of pristine (a) ACG-334/copper(II) complexes, (b) ACG-334/nickel(II), and (c) ACG-334/cobalt(II) complexes reduced by $NaBH_4$.

The following advantages of amphoteric macroporous cryogels with respect to metal ions are outlined: (1) Adsorption of metal ions can be provided in static and dynamic regimes; (2) adsorption and desorption processes are simple, for example, metal containing aqueous solution or desorbing agent is passed through the sample with definite rate; (3) high adsorption capacity of cryogels is due to the presence of complex-forming ligands (amine and carboxylic groups) and highly developed inner and outer surface; (4) the trace amount of metal ions may be concentrated up to three orders; (5) immobilized within macropores, metal ions can easily be reduced by reducing agents, and afterward cryogels might be used as flowing catalytic microreactors.

1.6 MOLECULAR IMPRINTED HYDROGELS FOR RECOVERY OF METAL IONS

Molecular recognition processes found in nature have always inspired scientists to mimic these systems in synthetic materials such as molecular imprinted polymers (MIPs) (Bergmann and Nicholas 2008, Byrne et al. 2002). MIPs and molecular imprinted hydrogels (MIHs) are commonly accepted in literature as synthetic approaches to design a precise macromolecular architecture for the recognition of target molecules from an ensemble of closely related molecules, while molecular imprinted technology (MIT) or molecular recognition technology (MRT) can be defined as engineering applications of such materials. Molecular imprinting involves forming a prepolymerization complex between the template molecule and functional monomers or functional oligomers (or polymers) (Wizeman and Kofinas 2001) with specific chemical structures designed to interact with the template by either covalent (Wulff 1995) or noncovalent chemistry (self-assembly) (Mosbach and Ramstrom 1996, Sellergren 1997), or both (Kirsch et al. 2000, Whitcombe et al. 1995). In the last decade, there has been an exponential increase in the number of papers describing molecular imprinting technique that creates memory for template molecules within a flexible macromolecular structure (Byrne and Salian 2008). Cameron et al. (2006) comprehensively surveyed over 1450 original papers, reviews, and monographs, starting from the pioneering work of Polyakov (1931) to show the fundamental and engineering aspects of molecular imprinting science and technology for the years up to and including 2003. According to the Web of Knowledge database searched up to 2012, ca. 13,000 papers have been published on molecular imprinting. Several remarkable reviews (Buengera et al. 2012, Byrne and Salian 2008, Hendrickson et al. 2006, Mayes and Whitcombe 2005, Romana et al. 2012, Tokonami et al. 2009, Vasapollo et al. 2011) were published with the aim to outline the molecularly imprinted process and present a summary of principal application fields of molecularly imprinted polymers, focusing on chemical sensing, separation science, biochemical analysis, drug delivery, catalysis, microfluidic devices, and analytical purposes.

The nature of the interaction between the functional monomers and the template with the formation of the complex has both covalent (covalent molecular imprinting) and noncovalent (noncovalent molecular imprinting) characters. Covalent molecular imprinting refers to imprinting of preorganized systems where the monomer–template complex is formed by the covalent interactions. Pioneering works of Nishide and Tsuchida (Nishide et al. 1976) and Kabanov (Kabanov et al. 1977, 1979) served as the fundamental basis for the imprinting of metal ions to MIPs. Such kind of polymeric sorbents made from natural and synthetic materials is widely used for the recovery of metal ions from the wastewater (Ahmadi et al. 2010, Bessbousse et al. 2012, Birlik et al. 2007, Chauhan et al. 2005, 2009, Ge et al. 2012, Godlewska-Zyłkiewicz et al. 2012, Kowalczyk et al. 2013, Li et al. 2010, Orozco-Guareño et al. 2010, Panic et al. 2013, Wawrzkiewicz 2013). Noncovalent imprinting belongs to imprinting of self-organizing systems in which the prepolymerization complex is formed by hydrogen, ionic bonding, hydrophobic and $\pi-\pi$ interactions, as well as the van der Waals forces (Andersson and Mosbach 1990, Dunkin et al. 1993, Nicholls et al. 1995, Sellergren et al. 1985). The noncovalent imprinting approach seems to hold more potential for the future of molecular imprinting due to the vast number of compounds, including biological compounds, which are capable of noncovalent interactions with polymerizable monomers. These noncovalent interactions are easily reversed, usually by wash in aqueous solution of an acid, a base, or organic solvents, thus facilitating the removal of the template molecule from the network after polymerization.

The commonly accepted procedure for immobilization and leaching of imprinted metal ions is (a) *mixing* solutions of the functional monomer with a print molecule to afford the corresponding complex as the template, (b) *copolymerization* of the monomer–metal complex with the cross-linking agent in the presence of the initiator, (c) *washing* the crude copolymer to remove unreacted functional monomer, and (d) *leaching* the print molecule from the template to afford the MIP. Novel ion-imprinted polymers (IIPs) were used for selective solid-phase extraction of Cd(II) (Fan et al. 2012, Li et al. 2011, Singh and Mishra 2009), Pb(II) (Behbahani et al. 2013), Cu(II) (Chen and Wang 2009, Shamsipur et al. 2010), and Ni(II) (Saraji and Yousefi 2009) ions from aqueous solutions. The imprinted metal ions were completely removed by leaching with 1 M HNO_3 or 0.01 M EDTA in 0.5 M HNO_3. Compared with nonimprinted polymer particles, the IIP had higher selectivity for metal ions. New IIPs for selective sorption and separation of Cr(III) (Birlik et al. 2007), Fe(III) (Xie et al. 2012), Ru(III) (Godlewska-Zyłkiewicz et al. 2012), Nd(III) (Jiajia et al. 2009), and Au(III) (Ahamed et al. 2013) were synthesized. The IIPs for separation and preconcentration of UO_2^{2+} ions were obtained (Ahmadi et al. 2010, James et al. 2009). The applicability of IIP materials for the removal of emerging toxic pollutant uranium from uranium mining industry feed simulant solution is successfully demonstrated. An Al(III)-ionic imprinted polyamine functionalized silica gel sorbent was prepared by a surface imprinting technique for selectively adsorbing Al(III) from rare-earth solution (An et al. 2013). The adsorption of Th(IV) was studied using novel dibenzoylmethane MIPs, which was prepared using acryloyl-β-cyclodextrin as a monomer on surface-modified functional silica gel (Ji et al. 2013).

The Ni(II)-dimethylglyoxime (DMG)-IIP was encapsulated in polysulfone and electrospun into nanofibers with diameters ranging from 406 to 854 nm

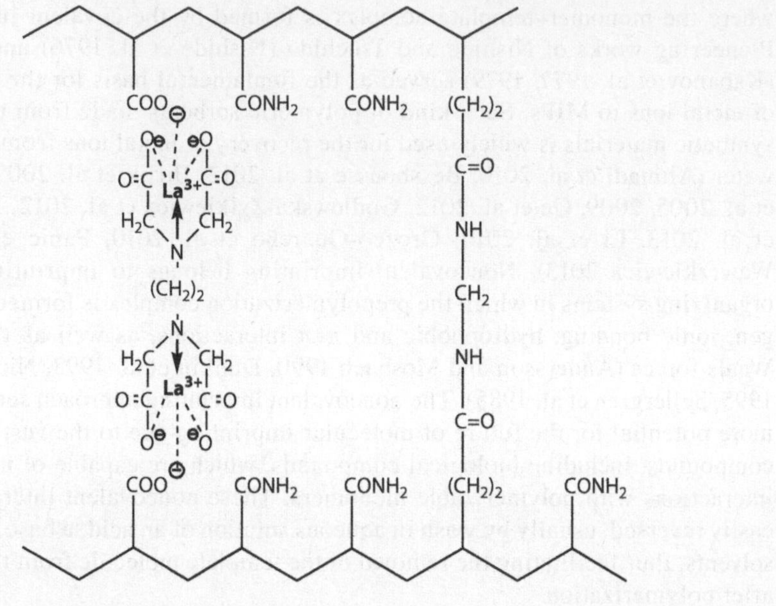

FIGURE 1.14 Scheme of immobilization of EDTA in AAm-AA hydrogel under *in situ* polymerization conditions.

(Rammika et al. 2011). The recovery of Ni(II) achieved using the Ni(II)-DMG imprinted nanofiber mats in water samples was found to range from 83% to 89%, while that of nonimprinted nanofiber mats was found to range from 59% to 65%, and that of polysulfone from 55% to 62%. The MIH was synthesized by immobilization of ethylenediaminetetraacetic acid–La(III) complex ([EDTA]:[La^{3+}] = 2:1 mol/mol) within AAm and AAc hydrogel matrix via *in situ* cross-linking polymerization (Bekturganov et al. 2010) (Figure 1.14).

It is expected that the EDTA–La(III) complex in hydrogel matrix is stabilized by electrostatic interaction between carboxylate anions and metal ions. After leaching out of La(III) ions by 0.1 N HCl, the MIH sample was used for recovery of trace concentration of rare-earth elements (REEs) from the real solution (Table 1.1).

Sorption of REE was also performed by commercially available Russian-made cation exchanger КУ-2-8н (Smirnov et al. 2002) (Table 1.2).

TABLE 1.1
Sorption of REE by MIH Sorbent from the Real Solution

Sorption	Initial Concentration of REE, mg/L							
	La	Ce	Pr	Nd	Y	Dy	Gd	Total
Stock solution	0.024	0.23	0.041	0.036	0.26	0.35	13.84	14.78
After sorption by MIH	0	0	0.04	$16 \cdot 10^{-3}$	0.028	$5 \cdot 10^{-4}$	13.17	13.25
Sorption degree, %	100	100	0	95.5	89.25	99.86	4.84	89.64

TABLE 1.2

Sorption of REE by КУ-2-8н Cation-Exchange Resin from the Real Solution

Sorption	Initial Concentration of REE, mg/L							
	La	Ce	Pr	Nd	Y	Dy	Gd	Total
Stock solution	—	0.46	0.065	0.061	0.23	0.20	6.54	7.556
After sorption by КУ-2-8н	—	0.29	0.065	0.021	0.032	0.13	4.75	5.291
Sorption degree, %	—	36.96	0	65.67	86.08	35.00	27.37	70.02

Comparison of the sorption effectiveness of REE by cation exchanger КУ-2-8н and MIH is in favor of the latter. Excepting for Pr and Gd, the EDTA-immobilized hydrogel sample adsorbs from 89% to 100% of REE during 20 min. Ammonium salt of EDTA was also used as an eluent in selective separation of REE (Lu, Sm, and Y) by ion-exchange resins based on iminodiacetic acid (Moore 2000). In spite of selective separation of REE by iminodiacetic resin in hydrogen form, the disadvantage of this process is the multistage character that consists of transferring of iminodiacetate resin at first to hydrogen form, then to ammonium form, saturation of iminodiacetate resin by REE solutions, and elution of REE by EDTA.

1.7 CONCLUDING REMARKS

Thus, the literature survey shows that the "smart" composite hydrogel materials are a fast developing and emerging field of polymer science. Synthetic and natural polymers including inorganic polymers, micro- and nanogels, metal nanoparticles, high- and low-molecular-weight ligands may be embedded into the hydrogel network, resulting in improvement of the mechanical properties and biocompatibility, making them as carriers for the controlled release of drugs and as catalysts, and providing stimuli-sensitive compositions. Structure, morphology, and physicochemical and physicomechanical properties of composite hydrogel materials are determined by both network structure and immobilized substances. The composite hydrogel materials can be applied in medicine, biotechnology, catalysis, environmental protection, and oil industry.

ABBREVIATIONS

AA	Allylamine
AAc	Acrylic acid
AAm	Acrylamide
AAm-AAc	Acrylamide and acrylic acid
AAm-SA	Poly(acrylamide-*co*-sodium acrylate)
AgNPs	Silver nanoparticles
AMPS	2-Acrylamido-2-methyl-1-propanesulfonic acid
ARPDs	Asphaltene–resin–paraffin depositions
AuNPs	Gold nanoparticles
BSA	Bovine serum albumin
CPAM	Cationic polyacrylamide

CTS	Chitosan
CTS-g-PAAc/MMT	Chitosan-g-poly(acrylic acid)/montmorillonite
Dex-MA	Methacrylated dextran
DLS	Dynamic light scattering
DMG	Dimethylglyoxime
EDTA	Ethylenediaminetetraacetic acid
FTIR	Fourier transform infrared spectroscopy
GE	Gelatin
IIP	Ion-imprinted polymers
IPN	Interpenetrating polymer network
MAA	Methacrylic acid
MB	Methylene blue
MBAA	Methylenebisacrylamide
MIH	Molecular imprinted hydrogels
MIP	Molecularly imprinted polymers
MIT	Molecular imprinted technology
MMT	Montmorillonite
MRT	Molecular recognition technology
NIPA	N-Isopropylacrylamide
NIPA-AAc	N-Isopropylacrylamide and acrylic acid
NIPA-IA	N-Isopropylacrylamide-itaconic acid
NIPA-HEMA	Poly(isopropylacrylamide-co-hydroxyethylmethacrylate)
PAAc-B-FeCo	Poly(acrylic acid)–bentonite–FeCo
PAAH	Poly(acrylamide) hydrogel
PdNPs	Palladium nanoparticles
PEI	Polyethyleneimine
PNIPA	Poly-N-isopropylacrylamide
PVP	Poly(N-vinyl-2-pyrrolidone)
PVP gel	Poly(N-vinyl-2-pyrrolidone)gel
PSA	Poly(sodium acrylate)
REE	Rare-earth elements
SA	Sodium alginate
SEM	Scanning electron microscopy
SS	Silk sericin
TON	Turnover numbers
VPTT	Volume-phase-transition temperature
w.r.	Weight ratio
XRD	X-ray diffraction

REFERENCES

Abdurrahmanoglu, S., V. Can, and O. Okay. 2008. Equilibrium swelling behavior and elastic properties of polymer–clay nanocomposite hydrogels. *J. Appl. Polym. Sci.* 109: 3714–3724.

Ahamed, M.E.H., X.Y. Mbianda, A.F. Mulaba-Bafubiandi, and L. Marjanovic. 2013. Selective extraction of gold(III) from metal chloride mixtures using ethylenediamine N-(2-(1-imidazolyl)ethyl) chitosan ion-imprinted polymer. *Hydrometallurgy* 140: 1–13.

Ahmadi, S.J., O. Noori-Kalkhoran, and S. Shirvani-Arani. 2010. Synthesis and characterization of new ion-imprinted polymer for separation and preconcentration of uranyl (UO_2^{2+}) ions. *J. Hazard. Mater.* 175: 193–197.

Akdemir, S. and N. Kayaman-Apohan. 2007. Investigation of swelling, drug release and diffusion behaviors of poly(*N*-isopropylacrylamide)/poly(*N*-vinylpyrrolidone) full-IPN hydrogels. *Polym. Adv. Technol.* 18: 932–939.

Ali, Sh.A. and Sh.A. Haladu. 2013. A novel cross-linked poly zwitterion/anion having pH-responsive carboxylate and sulfonate groups for the removal of Sr^{2+} from aqueous solution at low concentrations. *React. Funct. Polym.* 73: 796–804.

Ali, Sh.A. and O.Ch.S. Hamouz. 2012. Comparative solution properties of cyclocopolymers having cationic, anionic, zwitterionic and zwitterionic/anionic backbones of similar degree of polymerization. *Polymer* 53: 3368–3377.

Ali, Sh.A., O.Ch.S. Hamouz, and N.M. Hassan. 2013. Novel cross-linked polymers having pH-responsive amino acid residues for the removal of Cu^{2+} from aqueous solution at low concentrations. *J. Hazard. Mater.* 248–249: 47–58.

Al-Yaari, M. 2011. Paraffin wax deposition: Mitigation and removal techniques. *SPE Saudi Arabia Section Young Professionals Technical Symposium*, Dhahran, Saudi Arabia, March 14–16, 2011.

An, F., B. Gao, X. Huang, Y. Zhang, Y. Li, Y. Xu, Z. Zhang, J. Gao, and Z. Chen. 2013. Selectively removal of Al(III) from Pr(III) and Nd(III) rare earth solution using surface imprinted polymer. *React. Funct. Polym.* 73: 60–65.

Anderson, N.J., B.A. Bolto, R.J. Eldridge, and M.B. Jackson. 1993. Polyampholyts for water treatment with magnetic particles. *React. Polym.* 19: 87–95.

Andersson, L.I. and K. Mobach. 1990. Enantiomeric resolution on molecularly imprinted polymers prepared with only non-covalent and non-ionic interactions. *J. Chromatogr.* 516(2): 313–322.

Anil, K.A. 2007. Stimuli-induced pulsatile or triggered release delivery systems for bioactive compounds. Recent patents on endocrine. *Metab. Immun. Drug Discov.* 1: 83–90.

Anish, K.G. and W.S. Abdul. 2012. Environmental responsive hydrogels: A novel approach in drug delivery system. *J. Drug Deliv. Ther.* 2: 81–88.

Annenkov, V.V., E.N. Danilovtzeva, V.V. Saraev, and I.A. Alsarsur. 2000. Interaction of copolymer of acrylic acid and 1-vinylimidazole with copper(ll) ions in aqueous solution. *Izv. Russian Acad. Nauk Ser. Khim.* 12: 2047–2054.

Annenkov, V.V., E.N. Danilovtseva, V.V. Saraev, and A.I. Mikhaleva. 2003. Complexation of copper(II) ions with imidazole–carboxylic polymeric systems. *J. Polym. Sci.* 41: 2256–2263.

Arasawa, H., C. Odawara, R. Yokoyama, H. Saitoh, T. Yamauchi, and N. Tsubokawa. 2004. Grafting of zwitterion-type polymers on to silica gel surface and their properties. *React. Funct. Polym.* 61: 153–161.

Arica, M.Y., H.A. Öktem, Z. Öktem, and S.A. Tuncel. 1999. Immobilization of catalase in poly(isopropylacrylamide-*co*-hydroxyethylmethacrylate) thermally reversible hydrogels. *Polym. Int.* 48: 879–884.

Avvaru, N.R., N.R. de Tacconi, and K. Rajeshwar. 1998. Compositional analysis of organic-inorganic semiconductor composites. *Analyst* 123: 113–116.

Bajpai, A.K., S.K. Shukla, S. Bhanu, and S. Kankane. 2008. Responsive polymers in controlled drug delivery. *Prog. Polym. Sci.* 33: 1088–1118.

Balasubramanian, S.K., L. Yang, L.-Y.L. Yung, Ch.-N. Ong, W.-Y. Ong, and L.E. Yu. 2010. Characterization, purification, and stability of gold nanoparticles. *Biomaterials* 31(34): 9023–9030.

Behbahani, M., A. Bagheri, M. Taghizadeh, M. Salarian, O. Sadeghi, L. Adlnasab, and K. Jalali. 2013. Synthesis and characterisation of nano structure lead (II) ion-imprinted polymer as a new sorbent for selective extraction and preconcentration of ultra trace amounts of lead ions from vegetables, rice, and fish samples. *Food Chem.* 138: 2050–2056.

Bekturganov, N.S., N.K. Tusupbayev, S.E. Kudaibergenov, G.S. Tatykhanova, L.V. Semushkina, and Zh.E. Ibrayeva. 2010. Method of recovery of rare earth elements from solution. Innovation Patent of Kazakhstan No. 24563.

Bekturov, E.A. and S.E. Kudaibergenov. 1996. *Catalysis by Polymers*. Huthig & Wepf Verlag, Heidelberg, Germany, 153pp.

Bekturov, E.A., S.E. Kudaibergenov, R.M. Iskakov, A.K. Zharmagambetova, Zh.E. Ibraeva, and S. Shmakov. 2010. *Polymer-Protected Nanoparticles of Metals*. Print-S Almaty, 274pp. (in Russian).

Bekturov, E.A., S.E. Kudaibergenov, and V.B. Sigitov. 1986. Complexation of amphoteric copolymer of 2-methyl-5-vinylpyridine-acrylic acid with copper(II) ions and catalase like activity of polyampholyte-metal complexes. *Polymer* 27: 1269–1272.

Bergbreiter, D.E., B.L. Case, Y.-S. Liu, and J.W. Caraway. 1998. Poly(N-isopropylacrylamide) soluble polymer supports in catalysis and synthesis. *Macromolecules* 31: 6053–6062.

Bergmann, N.M. and A.P. Nicholas. 2008. Molecular imprinted polymers with specific recognition for macromolecules and proteins. *Prog. Polym. Sci.* 33(3): 271–288.

Bessbousse, H., J.-F. Verchere, and L. Lebrun. 2012. Characterisation of metal-complexing membranes prepared by the semi-interpenetrating polymer networks technique. Application to the removal of heavy metal ions from aqueous solutions. *Chem. Eng. J.* 187: 16–28.

Birlik, E., A. Ersoz, E. Acıkkalp, A. Denizli, and R. Say. 2007. Cr(III)-imprinted polymeric beads: Sorption and preconcentration studies. *J. Hazard. Mater.* 140: 110–116.

Buengera, D., F. Topuza, and J. Groll. 2012. Hydrogels in sensing applications. *Prog. Polym. Sci.* 37: 1678–1719.

Byrne, M.E., K. Park, and N.A. Peppas. 2002. Molecular imprinting within hydrogels. *Adv. Drug Deliv. Rev.* 54: 149–161.

Byrne, M.E. and V. Salian. 2008. Molecular imprinting within hydrogels II: Progress and analysis of the field. *Int. J. Pharm.* 364: 188–212.

Cameron, A., H.S. Andersson, L.I. Andersson, R.J. Ansell, N. Kirsch, I.A. Nicholls, J. O'Mahony, and M.J. Whitcombe. 2006. Molecular imprinting science and technology: A survey of the literature for the years up to and including 2003. *J. Mol. Recogn.* 19: 106–180.

Cândido, J., A.G.B. Pereira, A.R. Fajardo, M.P.S. Ricardo Nágila, P.A. Feitosa Judith, C.M. Edvani, and H.A.R. Francisco. 2013. Poly(acrylamide-*co*-acrylate)/rice husk ash hydrogel composites II. Temperature effect on rice husk ash obtention. *Compos. B Eng.* 51: 246–253.

Casoloro, M., F. Bignotti, L. Sartore, and M. Penco. 2001. The thermodynamics of basic and amphoteric poly(amidoamine)s containing peptide nitrogens as potential binding sites for metal ions. *Polymer* 42: 903–912.

Chan, W.C. and J.Y. Wu. 2001. Dynamic adsorption behaviors between Cu^{2+} ion and water-insoluble amphoteric starch in aqueous solutions. *J. Appl. Polym. Sci.* 81: 2849–2855.

Charles, O., S. Al Hamouz, and S.A. Ali. 2012. Removal of heavy metal ions using a novel cross-linked polyzwitterionic phosphonate. *Sep. Purif. Technol.* 98: 94–101.

Chauhan, G.S., B. Singh, and S. Kumar. 2005. Synthesis and characterization of N-vinyl pyrrolidone and cellulosics based functional graft copolymers for use as metal ions and ions and iodine sorbents. *J. Appl. Polym. Sci.* 98: 373–382.

Chauhan, K., G.S. Chauhan, and J.H. Ahn. 2009. Synthesis and characterization of novel guar gum hydrogels and their use as Cu^{2+} sorbents. *J. Biotech.* 100: 3599–3603.

Chen, H., D.M. Lentz, and R.C. Hedden. 2012a. Solution templating of Au and Ag nanoparticles by linear poly[2-(diethylamino)ethyl methacrylate]. *J. Nanopart. Res.* 14: 690–698.

Chen, H. and A. Wang. 2009. Adsorption characteristics of Cu(II) from aqueous solution onto poly(acrylamide)/attapulgite composite. *J. Hazard. Mater.* 165: 223–231.

Chen, S., H. Zhong, B. Gu, Y. Wang, X. Li, Zh. Cheng, L. Zhang, and Ch. Yao. 2012b. Thermosensitive phase behavior and drug release of in situ N-isopropylacrylamide copolymer. *Mater. Sci. Eng.* 32: 2199–2204.

Cheng, Y.-J., S. Zhou, and J.S. Gutmann. 2007. Morphology transition in ultrathin titania films: From pores to lamellae. *Macromol. Rapid Commun.* 28: 1392–1396.

Cho, E.Ch., J. Kim, A. Fernández-Nieves, and D.A. Weitz. 2008. Highly responsive hydrogel scaffolds formed by three-dimensional organization of microgel nanoparticles. *Nanoletters* 8: 168–172.

Chung, J.W., Y. Guo, S.-Y. Kwak, and R.D. Priestley. 2012. Understanding and controlling gold nanoparticle formation from a robust self-assembled cyclodextrin solid template. *J. Mater. Chem.* 22: 6017–6026.

Ciferri, A. and S.E. Kudaibergenov. 2007. Natural and synthetic polyampholytes. I. Theory and basic structures. *Makromol. Rapid Commun.* 28: 1953–1968.

Coughlan, D.C., F.P. Quilty, and O.I. Corrigan. 2004. Effect of drug physiochemical properties on swelling/deswelling kinetics and pulsatile drug release from thermoresponsive poly(N-isopropylacrylamide) hydrogels. *J. Control. Release* 98: 97–114.

Dai, J., P. Yao, N. Hua, P. Yang, and Y. Du. 2007. Preparation and characterization of polymer-protected Pt-Pt/Au core-shell nanoparticles. *J. Dispersion Sci. Technol.* 28: 872–875.

Dalaran, M., S. Emik, G. Güçlü, T.B. İyim, and S. Özgümüş. 2011. Study on a novel poly-ampholyte nanocomposite superabsorbent hydrogels: Synthesis, characterization and investigation of removal of indigo carmine from aqueous solution. *Desalination* 279: 170–182.

Dinu, M.V., M. Pradny, E.S. Dragan, and J. Michalek. 2013. Morphological and swelling properties of porous hydrogels based on poly(hydroxyethyl methacrylate) and chitosan modulated by ice-templating process and porogen leaching. *J. Polym. Res.* 20: 285.

Dolya, N., O. Rojas, S. Kosmella, B. Tiersch, J. Koetz, and S. Kudaibergenov. 2013. "One-pot" in situ formation of gold nanoparticles within poly(acrylamide) hydrogels. *Macromol. Chem. Phys.* 214: 114–121.

Dolya, N.A. 2009. Physico-chemical and catalytic properties of polymer-protected and gel-immobilized metals nanoparticles. PhD thesis, A. Bekturov Institute of Chemical Sciences, Almaty, Kazakhstan.

Dolya, N.A., Zh.E. Ibrayeva, E.A. Bekturov, and S.E. Kudaibergenov. 2009. Preparation, properties and catalytic activity of polymer-protected and gel-immobilized nanoparticles of gold, silver and palladium. *Bull. Natl. Acad. Sci. Republic of Kazakhstan* 4: 30–35.

Dolya, N.A., B.Kh. Musabayeva, M.G. Yashkrova, and S.E. Kudaibergenov. 2008a. Preparation, properties and catalytic activity of palladium nanoparticles immobilized within poly-N-isopropylacrylamide hydrogel matrix. *Chem. J. Kazakhstan* 1: 139–146.

Dolya, N.A., A.K. Zharmagambetova, B.Kh. Musabayeva, and S.E. Kudaibergenov. 2008b. Immobilization of polyethyleneimine-$[PdCl_4]^{-2}$ complexes into the matrix of poly-acrylamide hydrogels and study of hydrogenation of allyl alcohol with the help of gel-immobilized nanocatalyst. *Bull. Natl. Acad. Sci. Republic of Kazakhstan* 1: 55–59.

Dorris, A., S. Rucareanu, L. Reven, C.J. Barrett, and R. Bruce Lennox. 2008. Preparation and characterization of polyelectrolyte-coated gold nanoparticles. *Langmuir* 24(6): 2532–2538.

Dragan, E.S. and D.F. Apopei Loghin. 2013. Enhanced sorption of Methylene Blue from aqueous solutions by semi-IPN composite cryogels with anionically modified potato starch entrapped in PAAm matrix. *Chem. Eng. J.* 234: 211–222.

Dragan, E.S. and M.V. Dinu. 2013. Design, synthesis and interaction with Cu^{2+} ions of ice templated composite hydrogels. *Res. J. Chem. Environ.* 17: 4–10.

Dunkin, I.R., J. Lenfeld, and D.C. Sherrington. 1993. Molecular imprinting of flat polycon-densed aromatic-molecules in macroporous polymers. *Polymer* 34: 77–84.

Echeverria, C. and C. Mijangos. 2010. Effect of gold nanoparticles on the thermosensitivity, morphology, and optical properties of poly(acrylamide–acrylic acid) microgels. *Macromol. Rapid Commun.* 31: 54–58.

Eros, I., I. Csoka, E. Csanyi, and T.T. Wormsdorff. 2003. Examination of drug release from hydrogels. *Polym. Adv. Technol.* 14: 847–853.

Essawy, H. 2008. Poly(methyl methacrylate)-kaolinite nanocomposites prepared by interfacial polymerization with redox initiator system. *Colloid Polym. Sci.* 286: 795–803.

Fan, H.-T., J. Li, Z.-C. Li, and T. Sun. 2012. An ion-imprinted amino-functionalized silica gel sorbent prepared by hydrothermal assisted surface imprinting technique for selective removal of cadmium (II) from aqueous solution. *Appl. Surf. Sci.* 258: 3815–3822.

Feng, Q., F. Li, Q. Yan, Y.-C. Zhu, and C.-C. Ge. 2010. Frontal polymerization synthesis and drug delivery behavior of thermo-responsive poly(N-isopropylacrylamide) hydrogel. *Colloid Polym. Sci.* 288: 915–921.

Frimpong, R.A., S. Fraser, and J.Z. Hilt. 2006. Synthesis and temperature response analysis of magnetic-hydrogel nanocomposites. *J. Biomed. Mater. Res.* 80: 1–6.

Galaev, I.Y. and B. Mattiasson. 1999. 'Smart' polymers and what they could do in biotechnology and medicine. *Trends Biotechnol.* 17: 335–340.

Ge, F., M.-M. Li, H. Ye, and B.-X. Zhao. 2012. Effective removal of heavy metal ions Cd^{2+}, Zn^{2+}, Pb^{2+}, Cu^{2+} from aqueous solution by polymer-modified magnetic nanoparticles. *J. Hazard. Mater.* 211–212: 366–372.

Godlewska-Zyłkiewicz, B., E. Zambrzycka, B. Leśniewska, and A.Z. Wilczewska. 2012. Separation of ruthenium from environmental samples on polymeric sorbent based on imprinted Ru(III)-allyl acetoacetate complex. *Talanta* 89: 352–359.

Gong, J.P., Y. Katsuyama, T. Kurokawa, and Y. Osada. 2003. Double-network hydrogels with extremely high mechanical strength. *Adv. Mater.* 15: 1155–1158.

Gorner, T., R. Gref, D. Michenot, F. Sommerb, M.N. Tranc, and E. Dellacherie. 1999. Lidocaine-loaded biodegradable nanospheres. Optimization of the drug incorporation into the polymer matrix. *J. Control. Release* 57: 259–268.

Hagiwara, H., Y. Shimizu, T. Hoshi, T. Suzuki, M. Ando, K. Ohkubo, and C. Yokoyama. 2001. Heterogeneous Heck reaction catalyzed by Pd/C in ionic liquid. *Tetrahedron Lett.* 42: 4349–4351.

Haraguchi, K., R. Farnworth, A. Ohbayashi, and T. Takehisa. 2003. Compositional effects on mechanical properties of nanocomposite hydrogels composed of poly(N,N-dimethylacrylamide) and clay. *Macromolecules* 36: 5732–5741.

Haraguchi, K. and H.J. Li. 2006. Mechanical properties and structure of polymer-clay nanocomposite gels with high clay content. *Macromolecules* 39: 1898–1905.

Haraguchi, K., K. Murata, and T. Takehisa. 2013. Stimuli-responsive properties of nanocomposite gels comprising (2-methoxyethylacrylate-co-N,N-dimethylacrylamide) copolymer-clay networks. *Macromol. Symp.* 329: 150–161.

Haraguchi, K. and T. Takehisa. 2002. Nanocomposite hydrogels: A unique organic–inorganic network structure with extraordinary mechanical, optical, and swelling/deswelling properties. *Adv. Mater.* 14: 1120–1124.

Haraguchi, K., T. Takehisa, and S. Fan. 2002. Effects of clay content on the properties of nanocomposite hydrogels composed of poly(N-isopropylacrylamide) and clay. *Macromolecules* 35: 10162–10171.

Haruta, M. 1997. Novel catalysis of gold deposited on metal oxides. *Catal. Surveys Japan* 1: 61–73.

Haruta, M. and M. Daté. 2001. Advances in the catalysis of Au nanoparticles (review). *Appl. Catal. A: General* 222: 427–437.

Hendrickson, O.D., A.V. Zherdev, and B.B. Dzantiyev. 2006. Molecularly imprinted polymers and their application in biochemical analysis. *Achievements Biol. Chem.* 46: 149–192.

Hoare, T.R. and D.S. Kohane 2008. Hydrogels in drug delivery: Progress and challenges. *Polymer* 49: 1993–2007.

Hoffman, A. and P.S. Stayton. 2004. Bioconjugates of smart polymers and proteins: Synthesis and application. *Macromol. Symp.* 207: 139–151.

Holgado, M.A., J.L. Arias, M.J. Cózar, J. Alvarez-Fuentes, A.M. Gañán-Calvo, and M. Fernández-Arévalo. 2008. Synthesis of lidocaine-loaded PLGA microparticles by flow focusing: Effects on drug loading and release properties. *Int. J. Pharm.* 358: 27–35.

Holtz, J.H. and S.A. Asher. 1997. Polymerized colloidal crystal hydrogel films as intelligent chemical sensing materials. *Nature* 389: 829–832.

Hoppe, C.E., M. Lazzari, I. Pardiñas-Blanco, and M.A. López-Quintela. 2006. One-step synthesis of gold and silver hydrosols using poly(*N*-vinyl-2-pyrrolidone) as a reducing agent. *Langmuir* 22: 7027–7034.

Ibrayeva, Zh.E. 2010. Composite polymer hydrogels. *Chem. J. Kazakhstan* 2: 165–175.

Ibrayeva, Zh.E., S.E. Kudaibergenov, and E.A. Bekturov. 2013. *Stabilization of Metal Nanoparticles by Hydrophilic Polymers* (in Russian). LAP Lambert Academic Publishing, Saarbrücken, Germany, 376pp.

Isobe, N., D. Lee, Y. Kwon, S. Kimura, Sh. Kuga, M. Wada, and U. Kim. 2011. Immobilization of protein on cellulose hydrogel. *Cellulose* 18: 1251–1256.

Jaggard, W.S. and A. Allen. 1977. Gel-like composition for use as a pig in a pipeline, US Patent No. 4003393.

Jagur-Grodzinski, J. 2010. Polymeric gels and hydrogels for biomedical and pharmaceutical applications. *Polym. Adv. Technol.* 21: 27–47.

James, D., G. Venkateswaran, and T. Prasada Rao. 2009. Removal of uranium from mining industry feed simulant solutions using trapped amidoxime functionality within a mesoporous imprinted polymer material. *Microporous Mesoporous Mater.* 119: 165–170.

Ji, X.Z., H.J. Liu, L.L. Wang, Y.K. Sun, and Y.W. Wu. 2013. Study on adsorption of Th(IV) using surface modified dibenzoylmethane molecular imprinted polymer. *J. Radioanal. Nucl. Chem.* 295: 265–270.

Jia, X., Y. Lia, B. Zhanga, Q. Chenga, and Sh. Zhanga. 2008. Preparation of poly(vinyl alcohol)/kaolinite nanocomposites via in situ polymerization. *Mater. Res. Bull.* 43: 611–617.

Jiajia G., C. Jibao, and S. Qingde. 2009. Ion imprinted polymer particles of neodymium: Synthesis, characterization and selective recognition. *J. Rare Earths* 27: 22.

Jiang, B. and Y. Zhang. 1993. Immobilization of catalase on crosslinked polymeric hydrogels-effect of anion on the activity of immobilized enzyme. *Eur. Polym. J.* 29: 1251–1254.

Jiang, H.Q., S. Manolache, A.C.L. Wong, and F.S. Denes. 2004. Plasma-enhanced deposition of silver nanoparticles onto polymer and metal surfaces for the generation of antimicrobial characteristics. *J. Appl. Polym. Sci.* 93: 1411–1422.

Jiang, W. and K. Irgum. 1999. Covalently bonded polymeric zwitterionic stationary phase for simultaneous separation of inorganic cations and anions. *Anal. Chem.* 71: 333–344.

Kabanov, V.A., A.A. Efendiev, and D.D. Orujev. 1977. Obtaining of complex-forming polymer sorbent with location of macromolecules "tuned" to the sorbed ion. *Vysokomol. Soedin. Ser. B* 19: 91–92.

Kabanov, V.A., A.A. Efendiev, and D.D. Orujev. 1979. Complex-forming polymeric sorbents with macromolecular arrangement favorable for ion sorption. *J. Appl. Polym. Sci.* 24: 259–267.

Kabiri, K. and M.J. Zohuriaan-Mehr. 2003. Superabsorbent hydrogel composites. *Polym. Adv. Technol.* 14: 438–444.

Kalfus, J., N. Singh, and A.J. Lesser. 2012. Reinforcement in nano-filled PAA hydrogels. *Polymer* 53: 2544–2547.

Kevadiya, B.D., G.V. Joshi, H.M. Mody, and H.C. Bajaj. 2011. Biopolymer–clay hydrogel composites as drug carrier: Host–guest intercalation and in vitro release study of lidocaine hydrochloride. *Appl. Clay Sci.* 52: 364–367.

Khvan, A.M., V.V. Chupov, O.V. Noa, and N.A. Plate. 1985. Experimental study of intramolecular crosslinking of poly-*N*-methacryloyl-L-lysine by copper(II) ions. *Vysokomol. Soedin. Ser. A* 27: 1243–1248.

Kidambi, S., J.H. Dai, J. Li, and M.L. Bruening. 2004. Selective hydrogenation by Pd nanoparticles embedded in polyelectrolyte multilayers. *J. Am. Chem. Soc.* 126: 2658–2659.

Kim, J.-H. and T.R. Lee. 2007. Hydrogel-templated growth of large gold nanoparticles: Synthesis of thermally responsive hydrogel-nanoparticle composites. *Langmuir* 23: 6504–6509.

Kirsch, N., C. Alexander, M. Lubke, M.J. Whitcombe, and E.N. Vulfson. 2000. Enhancement of selectivity of imprinted polymers via post-imprinting modification of recognition sites. *Polymer* 41: 5583–5590.

Klinger, D. and K. Landfester. 2012. Stimuli-responsive microgels for the loading and release of functional compounds: Fundamental concepts and applications. *Polymer* 53: 5209–5231.

Kohler, K., R.G. Heidenreich, J.G.E. Krauter, and M. Pietsch. 2001. Highly active palladium/ activated carbon catalysts for heck reactions: Correlation of activity, catalyst properties, and Pd leaching. *Chem. Eur. J.* 8: 622–631.

Kowalczyk, M., Z. Hubicki, and D. Kołodynska. 2013. Modern hybrid sorbents—New ways of heavy metal removal from waters. *Chem. Eng. Proc.* 70: 55–65.

Kudaibergenov, S.E. 2002. *Polyampholytes: Synthesis, Characterization and Application.* Kluwer Academic/Plenum Publishers, New York, 220pp.

Kudaibergenov, S.E. 2008. Polyampholytes. In: *Encyclopedia of Polymer Material and Technology.* John Wiley & Sons. Inc., pp. 1–30.

Kudaibergenov, S.E., Zh. Adilov, D. Berillo, G. Tatykhanova, Zh. Sadakbaeva, Kh. Abdullin, and I. Galaev. 2012b. Novel macroporous amphoteric gels: Preparation and characterization. *Express Polym. Lett.* 6: 346–353.

Kudaibergenov, S.E., L.A. Bimendina, and M.G. Yashkarova. 2007. Preparation and characterization of novel polymeric betaines based on aminocrotonates. *J. Macromol. Sci. A: Pure Appl. Chem.* 44: 899–912.

Kudaibergenov, S.E. and A. Ciferri. 2007. Natural and synthetic polyampholytes, functions and applications. *Makromol. Rapid. Commun.* 28: 1969–1986.

Kudaibergenov, S.E., N. Dolya, G. Tatykhanova, Zh. Ibrayeva, B. Musabayeva, M. Yashkarova, and L. Bimendina. 2007. Semi-interpenetrating polymer networks of polyelectrolytes. *Eurasian Chem. Technol. J.* 9: 177–192.

Kudaibergenov, S.E., Zh.E. Ibraeva, N.A. Dolya, B.Kh. Musabayeva, A.K. Zharmagambetova, and J. Koetz. 2008. Semi-interpenetrating hydrogels of polyelectrolytes, polymer-metal complexes and polymer-protected palladium nanoparticles. *Macromol. Symp.* 274: 11–21.

Kudaibergenov, S.E., W. Jaeger, and A. Laschewsky. 2006. Polymeric betaines: Synthesis, characterization and application. *Adv. Polym. Sci.* 201: 157–224.

Kudaibergenov, S.E., N. Nueraje, and V. Khutoryanskiy. 2012a. Amphoteric nano-, micro-, and macrogels, membranes, and thin films. *Soft Matter* 8: 9302–9321.

Kudaibergenov, S.E., G. Tatykhanova, and Zh. Ibraeva. 2011. Immobilization and controlled release of bioactive substances from stimuli-responsive hydrogels. *Biodefence.* NATO Science for Peace and Security Series-A: Chemistry and Biology. Springer, Dordrecht, The Netherlands, Chapter 19, pp. 79–188.

Kumar, A., A. Srivastava, I. Galaev, and B. Mattiasson. 2007. Smart polymers: Physical forms and bioengineering applications. *Prog. Polym. Sci.* 32: 1205–1237.

Kundakci, S., Ö.B. Üzüm, and E. Karadağ. 2008. Swelling and dye sorption studies of acrylamide/2-acrylamido-2-methyl-1-propanesulfonic acid/bentonite highly swollen composite hydrogels. *React. Funct. Polym.* 68: 458–473.

Kurokawa, Y. and M. Sasaki. 1982. Complexation between polyions and hydrous inorganic oxides and adsorption properties of complex. *Makromol. Chem.* 183: 679–685.

Lao, L.L. and R.V. Ramanujan. 2004. Magnetic and hydrogel composite materials for hyperthermia applications. *J. Mater. Sci. Mater. Med.* 15: 1061–1064.

Lázaro Martínez, J.M., M.F. Leal Denis, V. Campo Dall'Orto, and G.Y. Buldain. 2008a. Synthesis, FTIR, solid-state NMR and SEM studies of novel polyampholytes or polyelectrolytes obtained from EGDE, MAA and imidazoles. *Eur. Polym. J.* 44: 392–407.

Lázaro Martínez, J.M., M.F. Leal Denis, L.L. Piehl, E. Rubín de Celis, G.Y. Buldain, and V.C. Dall'Orto. 2008b. Studies on the activation of hydrogen peroxide for color removal in the presence of a new Cu(II)-polyampholyte heterogeneous catalyst. *Appl. Catal. Environ.* 82: 273–283.

Lázaro Martínez, J.M., E. Rodríguez-Castellón, R.M. Torres Sánchez, L.R. Denaday, G.Y. Buldain, and V.C. Dall'Orto. 2011. XPS studies on the Cu(I,II)–polyampholyte heterogeneous catalyst: An insight into its structure and mechanism. *J. Mol. Catal.* 339: 43–51.

Leadbeater, N.E. and M. Marco. 2002. Ligand-free palladium catalysis of the Suzuki reaction in water using microwave heating. *Org. Lett.* 4(17): 2973–2976.

Lee, K.Y. and S.H. Yuk. 2007. Polymeric protein delivery systems. *Prog. Polym. Sci.* 32: 669–697.

Lee, W.F. and L.G. Yang. 2004. Superabsorbent polymeric materials. XII. Effect of montmorillonite on water absorbency for poly(sodium acrylate) and montmorillonite nanocomposite superabsorbents. *J. Appl. Polym. Sci.* 92: 3422–3429.

Lee, Y.-J. and P.V. Braun. 2003. Tunable inverse opal hydrogel pH sensors. *Adv. Mater.* 15: 563–566.

Li, Q., H. Liua, T. Liu, M. Guo, B. Qinga, X. Ye, and Z. Wu. 2010. Strontium and calcium ion adsorption by molecularly imprinted hybrid gel. *Chem. Eng. J.* 157: 401–407.

Li, S., Y. Wu, J. Wang, Q. Zhang, Y. Kou, and S. Zhang. 2010. Double-responsive polyampholyte as a nanoparticle stabilizer: Application to reversible dispersion of gold nanoparticles. *J. Mater. Chem.* 20: 4379–4384.

Li, Z.-C., H.-T. Fan, Yi. Zhang, M.-X. Chen, Z.-Y. Yu, X.-Q. Cao, and T. Sun. 2011. Cd(II)-imprinted polymer sorbents prepared by combination of surface imprinting technique with hydrothermal assisted sol–gel process for selective removal of cadmium(II) from aqueous solution. *Chem. Eng. J.* 171: 703–710.

Lianga, R. and M. Liu. 2007. Preparation of poly(acrylic acid-*co*-acrylamide)/kaolin and release kinetics of urea from it. *J. Appl. Polym. Sci.* 106: 3007–3015.

Lianga, R., M. Liu, and L. Wu. 2007. Controlled release NPK compound fertilizer with the function of water retention. *React. Funct. Polym.* 67: 769–779.

Lin, J., J. Wu, Z. Yang, and M. Pu. 2001. Synthesis and properties of poly(acrylic acid)/mica superabsorbent nanocomposite. *Macromol. Rapid Commun.* 22: 422–424.

Liu, D.Z., M.T. Sheu, C. Chen, Y.R. Yang, and H.O. Ho 2007. Release characteristics of lidocaine from local implant of polyanionic and polycationic hydrogels. *J. Control. Release* 118: 333–339.

Liusheng, Z., B. Brittany, and A. Frank. 2011. Stimuli responsive nanogels for drug delivery. *Soft Matter* 7: 5908–5916.

Lombardo Lupano, L.V., J.M. Lázaro Martínez, L.L. Piehl, E.R. de Celis, and V. Campo Dall'Orto. 2013. Activation of H_2O_2 and superoxide production using a novel cobalt complex based on a polyampholyte. *Appl. Catal. General* 467: 342–354.

Lu, F., J. Ruiz, and D. Astruc. 2004. Palladium-dodecanethiolate nanoparticles as stable and recyclable catalysts for the Suzuki-Miyaura reaction of aryl halides under ambient conditions. *Tetrahedron Lett.* 45: 9443–9445.

Lu, Zh., G. Liu, and S. Duncan. 2003. Poly(2-hydroxyethyl acrylate-*co*-methyl acrylate)/SiO_2/TiO_2 hybrid membranes. *J. Membr. Sci.* 221: 113–122.

Mahltig, B., N. Cheval, J.-F. Gohy, and A. Fahmi. 2010. Preparation of gold nanoparticles under presence of the diblock polyampholyte PMAA-b-PDMAEMA. *J. Polym. Res.* 17: 579–588.

Makysh, G.Sh., L.A. Bimendina, and S.E. Kudaibergenov. 2002. Interaction of richlocaine with some linear and crosslinked polymers. *Polymer* 43: 4349–4353.

Makysh, G.Sh., L.A. Bimendina, K.B. Murzagulova, and S.E. Kudaibergenov. 2003. Interaction of a new anesthetic drug richlocain with linear and weakly crosslinked poly-*n*-vinylpyrrolidone. *J. Appl. Polym. Sci.* 89: 2977–2981.

Manpreet, K., Rajnibala, and A. Sandeep. 2013. Stimuli responsive polymers and their applications in drug delivery. *Int. J. Adv. Pharm. Sci.* 4: 477–495.

Marcin, K., J. Romanski, K. Michniewicz, J. Jurczak, and Z. Stojek. 2010. Influence of polymer network-metal ion complexation on the swelling behaviour of new gels with incorporated α-amino acid groups. *Soft Matter* 6: 1336–1342.

Martínez, J.M.L., A.K. Chattah, G.A. Monti, M.F.L. Denis, G.Y. Buldain, and V. Campo Dall'
    Orto. 2008. New copper(II) complexes of polyampholyte and polyelectrolyte polymers:
    Solid-state NMR, FTIR, XRPD and thermal analyses. *Polymer.* 49: 5482–5489.
Martínez-Martínez, D., C. López-Cartes, A. Fernández, and J.C. Sánchez-López. 2008.
    Comparative performance of nanocomposite coatings of TiC or TiN dispersed in a-C
    matrixes. *Surf. Coat. Technol.* 203: 756–760.
Mattiasson, B., A. Kumar, and I. Yu Galaev. 2010. *Macroporous Polymers: Applications,
    Production, Properties and Biotechnological/Biomedical Applications.* CRC Press,
    Boca Raton, FL, 513pp.
Mayes, A.G. and M.J. Whitcombe. 2005. Synthetic strategies for the generation of molecu-
    larly imprinted organic polymers. *Adv. Drug Deliv. Rev.* 57: 1742–1778.
Metin, O., S. Sahin, and S. Ozkar. 2009. Water-soluble poly(4-styrenesulfonic acid-*co*-maleic
    acid) stabilized ruthenium(0) and palladium(0) nanoclusters as highly active catalysts in
    hydrogen generation from the hydrolysis of ammonia-borane. *Int. J. Hydrogen Energy*
    34: 6304–6313.
Milasinovic, N., M.K. Krusic, Z. Knezevic-Jugovic, and J. Filipovic. 2010. Hydrogels of
    N-isopropylacrylamide copolymers with controlled release of a model protein. *Int.
    J. Pharm.* 383: 53–61.
Moore, B.M. 2000. Selective separation of rare earth elements by ion exchange in an iminodi-
    acetic resin, US Patent No. 6093376.
Morrow, B.J., E. Matijević, and D.V. Goia. 2009. Preparation and stabilization of monodisperse
    colloidal gold by reduction with aminodextran. *J. Colloid Interface Sci.* 335: 62–69.
Mosbach, K. and O. Ramstrom. 1996. The emerging technique of molecular imprinting and its
    future impact on biotechnology. *J. Biotechnol.* 14: 163–170.
Motoyuki, I. and K. Hidehiro. 2009. Surface modification for improving the stability of
    nanoparticles in liquid media. *Kona Powder Part. J.* 27: 119–129.
Nakamura, T. and M. Ogawa. 2013. Adsorption of cationic dyes within spherical particles of
    poly(*N*-isopropylacrylamide) hydrogel containing smectite. *Appl. Clay Sci.* 83–84: 469–473.
Nakayama, A., A. Kakugo, J.P. Gong, Y. Osada, M. Takai, T. Erata, and S. Kawano. 2004.
    High mechanical strength double-network hydrogel with bacterial cellulose. *Adv. Funct.
    Mater.* 14: 1124–1128.
Nicholls, I.A., O. Ramstrom, and K. Mosbach. 1995. Insights into the role of the hydrogen-
    bond and hydrophobic effect on recognition in molecularly imprinted polymer synthetic
    peptide receptor mimics. *J. Chromatogr.* 691: 349–353.
Nishide, H., J. Deguchi, and E. Tsuchida. 1976. Selective adsorption of metal-ions on cross-
    linked poly(vinylpyridine) resin prepared with a metal-ion as a template. *Chem. Lett.* 5:
    169–174.
Note, C., J. Koetz, L. Wattebled, and A. Laschewsky. 2007. Effect of a new hydrophobically
    modified polyampholyte on the formation of inverse microemulsions and the prepara-
    tion of gold nanoparticles. *J. Colloid Interface Sci.* 308: 162–169.
Orozco-Guareño, E., F. Santiago-Gutiérrez, J.L. Morán-Quiroz, S.L. Hernandez-Olmos,
    V. Soto, W. de la Cruz, R. Manríquez, and S. Gomez-Salazar. 2010. Removal of Cu(II)
    ions from aqueous streams using poly(acrylic acid-co-acrylamide) hydrogels. *J. Colloid
    Interface Sci.* 349: 583–593.
Panic, V.V., Z.P. Madzarevic, T. Volkov-Husovic, and S.J. Velickovic. 2013. Poly(methacrylic
    acid) based hydrogels as sorbents for removal of cationic dye basic yellow 28: Kinetics,
    equilibrium study and image analysis. *Chem. Eng. J.* 217: 192–204.
Pardo-Yissar, V., R. Gabai, A.N. Shipway, T. Bourenko, and I. Willner. 2001. Gold nanopar-
    ticle/hydrogel composites with solvent-switchable electronic properties. *Adv. Mater.* 13:
    1320–1323.
Park, T.G. 1993. Stabilization of enzyme immobilized in temperature-sensitive hydrogels.
    Biotechol. Lett. 15: 57–60.

Pavlyuchenko, V.N. and S.S. Ivanchev. 2009. Composite polymer hydrogels. *Vysokomol. Soedin. Ser. A* 51: 1075–1095.

Peppas, N.A., J.Z. Hilt, A. Khademhosseini, and R. Langer. 2006. Hydrogels in biology and medicine: From molecular principles to bionanotechnology. *Adv. Mater.* 18: 1345–1360.

Phan, N.T.S., D.H. Brown, and P. Styring. 2004. A polymer-supported salen-type palladium complex as a catalyst for the Suzuki-Miyaura cross-coupling reaction. *Tetrahedron Lett.* 45: 7915–7919.

Polakovič, M., T. Görner, R. Gref, and E. Dellacherie.1999. Lidocaine loaded biodegradable nanospheres: II. Modelling of drug release. *J. Control. Release* 60: 169–177.

Polyakov, M.V. 1931. Adsorption properties and structure of silica gel. *Zh. Fiz. Khim.* 2: 799–805.

Popa, M., N. Bajan, A.A. Popa, and A. Verestiuc. 2006. The preparation, characterization and properties of catalase immobilized on crosslinked gellan. *J. Macromol. Sci. A: Pure Appl. Chem.* 43: 355–367.

Qiu, Y. and K. Park. 2012. Environment-sensitive hydrogels for drug delivery. *Adv. Drug Deliv. Rev.* 64: 49–60.

Ram, S., L. Agrawal, A. Mishra, and S.K. Roy. 2011. Synthesis and optical properties of surface stabilized gold nanoparticles with poly(N-vinylpyrrolidone). Polymer molecules of a nanofluid. *Adv. Sci. Lett.* 4: 3431–3438.

Rammika, M., G. Darko, and N. Torto. 2011. Incorporation of Ni(II)-dimethylglyoxime ion-imprinted polymer into electrospun polysulphone nanofibre for the determination of Ni(II) ions from aqueous samples. *Water SA* 37: 539–546.

Ray, S.S. and M. Okamoto. 2003. Polymer/layered silicate nanocomposite: A review from preparation to processing. *Prog. Polym. Sci.* 28: 1539–1641.

Rivas, B.L., S. Villegas, and B. Ruf. 2006. Water-insoluble polymers containing amine, sulfonic acid, and carboxylic acid groups: Synthesis, characterization, and metal-ion-retention properties. *J. Appl. Polym. Sci.* 99: 3266–3274.

Romana, S., R.K. Ning, and N.Z. Richard. 2012. Surface-imprinted polymers in microfluidic devices. *Sci. China Chem.* 55: 469–483.

Rutkevičius, M., S.K. Munusami, Z. Watson, A.D. Field, M. Salt, S.D. Stoyanov, J. Petkov, G.H. Mehl, and V.N. Paunov. 2012. Fabrication of novel lightweight composites by a hydrogel templating technique. *Mater. Res. Bull.* 47: 980–986.

Rychkov, V.N. 2003. Uranium sorption from sulfate solutions with polyampholytes. *Radiochemistry* (in Russian) 45: 56–60.

Rzaev, Z.M., S. Dinçer, and E. Pişkin. 2007. Functional copolymers of N-isopropylacrylamide for bioengineering applications. *Prog. Polym. Sci.* 32: 534–595.

Saber-Samandari, S. and M. Gazi. 2013. Cellulose-graft-polyacrylamide/hydroxyapatite composite hydrogel with possible application in removal of Cu (II) ions. *React. Funct. Polym.* 73: 1523–1530.

Sahiner, N. 2004. In situ metal particle preparation in cross-linked poly(2-acrylamido-2-methyl-1-propansulfonic acid) hydrogel networks. *Colloid Polym. Sci.* 285: 283–292.

Sahiner, N. 2013. Soft and flexible hydrogel templates of different sizes and various functionalities for metal nanoparticle preparation and their use in catalysis. *Prog. Polym. Sci.* 38: 1329–1356.

Saraji, M. and H. Yousefi. 2009. Selective solid-phase extraction of Ni(II) by an ion-imprinted polymer from water samples. *J. Hazard. Mater.* 167: 1152–1157.

Sellergren, B. 1997. Noncovalent molecular imprinting: Antibody-like molecular recognition in polymeric network materials. *Trends Anal. Chem.* 16: 310–320.

Sellergren, B., B. Ekberg, and K. Mosbach. 1985. Molecular imprinting of amino acid derivatives in macroporous polymers. Demonstration of substrate and enantioselectivity by chromatographic resolution of racemic mixtures of amino acid derivatives. *J. Chromatogr. A* 347: 1–10.

Sershen, S.R., G.A. Mensing, M. Ng, N.J. Halas, D.J. Beebe, and J.L. West. 2005. Independent optical control of microfluidic valves formed from optomechanically responsive nanocomposite hydrogels. *Adv. Mater.* 17: 1366–1368.

Sershen, S.R., S.L. Westcott, N.J. Halas, and J.L. West. 2000. Temperature-sensitive polymer–nanoshell composites for photothermally modulated drug delivery. *J. Biomed. Mater. Res.* 51: 293–298.

Sershen, S.R., S.L. Westcott, J.L. West, and N.J. Halas. 2001. Anopto-mechanical nanoshell-polymer composite. *Appl. Phys.* 73: 379–381.

Shamsipur, M., A. Besharati-Seidani, J. Fasihi, and H. Sharghi. 2010. Synthesis and characterization of novel ion-imprinted polymeric nanoparticles for very fast and highly selective recognition of copper(II) ions. *Talanta* 83: 674–681.

Shan, J. and H. Tenhu. 2007. Recent advances in polymer protected gold nanoparticles: Synthesis, properties and applications (review). *Chem. Commun.* 44: 4580–4598.

Sharifkanov, A.Sh., Sh.S. Akhmedova, K.B. Murzagulova, and P.A. Galenko-Yaroshevskii. 2011. Richlokain—Dermatoprotecting pharmacological agent, Russian Patent No. 2261710.

Sheeney-Hai-Ichia, L., G. Sharabi, and I. Willner. 2002. Control of the electronic properties of thermosensitive poly(*N*-isopropylacrylamide) and Au-nano-particle/poly(*N*-isopropylacrylamide) composite hydrogels upon phase transition. *Adv. Funct. Mater.* 12: 27–32.

Shi, L., S. Khondee, T.H. Linz, and C. Berkland. 2008. Poly(*N*-vinylformamide) nanogels capable of pH-sensitive protein release. *Macromolecules* 41: 6546–6554.

Shiju, N.R. and V.V. Guliants. 2009. Recent developments in catalysis using nanostructured materials (review). *Appl. Catal. A: General* 356: 1–17.

Shikanov, A., A.J. Domb, and C.F. Weiniger. 2007. Long acting local anesthetic–polymer formulation to prolong the effect of analgesia. *J. Control. Release* 117: 97–103.

Shirsath, S.R., A.P. Hage, M. Zhou, S.H. Sonawane, and M. Ashokkumar. 2011. Bentonite nanoclay-FeCo nanocomposite hybrid hydrogel: A potential responsive sorbent for removal of organic pollutant from water. *Desalination* 281: 429–437.

Shirsath, S.R., A.P. Patil, R. Patil, J.B. Naik, P.R. Gogate, and Sh.H. Sonawane. 2013. Removal of Brilliant Green from wastewater using conventional and ultrasonically prepared poly(acrylic acid) hydrogel loaded with kaolin clay: A comparative study. *Ultrason. Sonochem.* 20: 914–923.

Sigitov, V.B., S.E. Kudaibergenov, and E.A. Bekturov. 1987. Complexation of copper(II) with polyampholyte 2-methyl-5-vinylpyridine-acrylic acid in aqueous solution. *Koord. Khim.* 13: 600–604.

Singh, D.K. and S. Mishra. 2009. Synthesis, characterization and removal of Cd(II) using Cd(II)-ion imprinted polymer. *J. Hazard. Mater.* 164: 1547–1551.

Sivudu, K.S., N.M. Reddy, M.N. Prasad, K. Mohana Raju, Y. Murali Mohan, J.S. Yadav, G. Sabitha, and D. Shailaja. 2008. Highly efficient and reusable hydrogel-supported nano-palladium catalyst: Evaluation for Suzuki-Miyaura reaction in water. *J. Mol. Catal. A: Chem.* 295: 10–17.

Smirnov, D.I., T.V. Molchanova, L.I. Vodolazov, and V.A. Peganov. 2002. The sorption recovery of rare earth elements, yttrium and aluminum from the red mud. *Non-ferrous Metals* 8: 64–69.

Starodoubtsev, S.G., N.A. Churochkina, and A.R. Khokhlov. 2000. Hydrogel composites of neutral and slightly charged poly(acrylamide) gels with incorporated bentonite interaction with salt and ionic surfactants. *Langmuir* 16: 1529–1534.

Stein, D.B. 2009. *Handbook of Hydrogels: Properties, Preparation & Applications*. Nova Science Publishers, Inc., New York, 750pp.

Svetlichnyy, D.S., N.A. Dolya, Zh.E. Ibrayeva, and S.E. Kudaibergenov. 2009a. Immobilization of TiO_2 nanoparticles within poly(acrylamide) hydrogel matrix and evaluation of swelling behavior, thermodynamic parameters and mechanical properties of composite networks. *Materials of Russia-Kazakhstan-Japan Conference "Perspective technologies, equipment and analytical systems for material science and nanomaterials,"* Volgograd, Russia, pp. 126–136.

Svetlichnyy, D.S., N.A. Dolya, Zh.E. Ibraeva, and S.E. Kudaibergenov. 2009b. Swelling behavior and mechanical properties of composite materials derived from poly(acrylamide) hydrogel and kaolin microparticles. *Bull. Kazakh Natl. Techn. Univ.* 4: 154–162.

Tanaka, Y., J.P. Gong, and Y. Osada. 2005. Novel hydrogels with excellent mechanical performance. *Prog. Polym. Sci.* 30: 1–9.

Taşdelen, B., N. Kayaman-Apohan, O. Güven, and B.M. Baysal. 2004. Preparation of poly(*N*-isopropylacrylamide/itaconic acid) copolymeric hydrogels and their drug release behavior. *Int. J. Pharm.* 78: 343–351.

Tatykhanova, G.S. 2009. Immobilization of biological active substances within the matrix of pH- and thermosensitive polymeric hydrogels. PhD thesis, A. Bekturov Institute of Chemical Sciences, Almaty, Kazakhstan, 109pp.

Tatykhanova, G., Zh. Sadakbayeva, D. Berillo, I. Galaev, Kh. Abdullin, Zh. Adilov, and S. Kudaibergenov. 2012. Metal complexes of amphoteric cryogels based on allylamine and methacrylic acid. *Macromol. Symp.* 317: 7–17.

Terlemezian, E., S. Veleva, and A. Arsov. 1990. Thermodynamic investigation of the sorption of Fe^{3+} and Cu^{2+} ions by a fibrous polyampholyte. *Acta Polym.* 40: 42–45.

Tokonami, S., H. Shiigi, and T. Nagaoka. 2009. Review: Micro- and nanosized molecularly imprinted polymers for high-throughput analytical applications. *Anal. Chim. Acta* 641: 7–13.

Uzu, O., R. Napier, and K. Ngwuobia. 2000. Gel technology applications in pipeline servicing. *Nigerian Annual International Conference and Exhibition*, Abuja, Nigeria, August 7–9, 2000.

Vasapollo, G., R. Del Sole, L. Mergola, M.R. Lazzoi, A. Scardino, S. Scorrano, and G. Mele. 2011. Molecularly imprinted polymers: Present and future prospective. *J. Mol. Sci.* 12: 5908–5945.

Verheyen, E., S. van der Wal, H. Deschout, K. Braeckmans, S. de Smedt, A. Barendregt, W.E. Hennink, and C.F. van Nostrum. 2011. Protein macromonomers containing reduction-sensitive linkers for covalent immobilization and glutathione triggered release from dextran hydrogels. *J. Control. Release* 156: 329–336.

Vesna, V.P., Z.P. Madzarevic, T. Volkov-Husovic, and S.J. Velickovic. 2013. Poly(methacrylic acid) based hydrogels as sorbents for removal of cationic dye basic yellow 28: Kinetics, equilibrium study and image analysis. *Chem. Eng. J.* 217: 192–204.

Wang, C., N.T. Flynn, and R. Langer. 2004a. Morphologically well-defined gold nanoparticles embedded in thermo-responsive hydrogel matrices. *Mater. Res. Soc. Symp. Proc.* 820: R2.2.1–R2.2.6.

Wang, C., N.T. Flynn, and R. Langer. 2004b. Controlled structure and properties of thermoresponsive nanoparticle–hydrogel composites. *Adv. Mater.* 16: 1074–1079.

Wang, G., K. Kuroda, T. Enoki, A. Grosberg, S. Masamune, T. Oya, Y. Takeoka, and T. Tanaka. 2000. Gel catalysts that switch on and off. *Proc. Natl. Acad. Sci. USA* 97: 9861–9864.

Wang, L., J. Zhang, and A. Wang. 2008. Removal of methylene blue from aqueous solution using chitosan-g-poly (acrylic acid)/montmorillonite superadsorbent nanocomposite. *Colloids Surf. A* 322: 47–53.

Wawrzkiewicz, M. 2013. Removal of C.I. Basic Blue 3 dye by sorption onto cation exchange resin, functionalized and non-functionalized polymeric sorbents from aqueous solutions and wastewaters. *Chem. Eng. J.* 217: 414–425.

Weissman, J.M., H.B. Sunkara, A.S. Tse, and S.A. Asher. 1996. Thermally switchable periodicities from novel mesocopically ordered materials. *Science* 274: 959–960.

Wen, W., W. Dongsheng, and L. Yuan. 2013. Controlled release of bovine serum albumin from stimuli-sensitive silk sericin based interpenetrating polymer network hydrogels. *Polym. Int.* 62: 1257–1262.

Whitcombe, M.J., M.E. Rodriguez, P. Villar, and E.N. Vulfson. 1995. A new method for the introduction of recognition site functionality into polymers prepared by molecular imprinting: Synthesis and characterization of polymeric receptors for cholesterol. *J. Am. Chem. Soc.* 117: 7105–7111.

Wizeman, W. and P. Kofinas. 2001. Molecularly imprinted polymer hydrogels displaying isomerically resolved glucose binding. *Biomaterials* 22: 1485–1491.

Wu, H., L. Wang, J. Zhang, Z. Shen, and J. Zhao. 2011. Catalytic oxidation of benzene, toluene and *p*-xylene over colloidal gold supported on zinc oxide catalyst. *Catal. Commun.* 12: 859–865.

Wu, W., W. Li, L.Q. Wang, K. Tu, and W. Sun. 2006. Synthesis and characterization of pH- and temperature-sensitive silk sericin/poly(*N*-isopropylacrylamide) interpenetrating polymer networks. *Polym. Int.* 55: 513–519.

Wulff, G. 1995. Molecular imprinting in cross-linked materials with the aid of molecular templates—A way towards artificial antibodies. *Angew. Chem. Int. Ed. Engl.* 34: 1812–1832.

Wunder, S., Y. Lu, M. Albrecht, and M. Ballauff. 2011. Catalytic activity of faceted gold nanoparticles studied by a model reaction: Evidence for substrate-induced surface restructuring. *ACS Catal.* 1: 908–916.

Xi, X., L. Yi, J. Shi, and S. Cao. 2003. Palladium complex of poly(4-vinylpyridine-*co*-acrylic acid) for homogeneous hydrogenation of aromatic nitro compounds. *J. Mol. Catal.: Chem.* 192: 1–7.

Xiang, Y., Zh. Peng, and D. Chen. 2006. A new polymer/clay nano-composite hydrogel with improved response rate and tensile mechanical properties. *Eur. Polym. J.* 42: 2125–2132.

Xie, F., G. Liu, F. Wu, G. Guo, and G. Li. 2012. Selective adsorption and separation of trace dissolved Fe(III) from natural water samples by double template imprinted sorbent with chelating diamines. *Chem. Eng. J.* 183: 372–380.

Xu, Sh., Sh. Zhang, and J. Yang. 2008. An amphoteric semi-IPN nanocomposite hydrogels based on intercalation of cationic polyacrylamide into bentonite. *Mater. Lett.* 62: 3999–4002.

Xu, S.M., S.F. Zhang, R.W. Lu, J.Z. Yang, and C.X. Cui. 2003. Study on adsorption behavior between Cr(VI) and crosslinked amphoteric starch. *J. Appl. Polym. Sci.* 89: 262–267.

Yang, X. and L. Ni. 2012. Synthesis of hybrid hydrogel of poly(AM-*co*-DADMAC)/silica sol and removal of methyl orange from aqueous solutions. *Chem. Eng. J.* 209: 194–200.

Yesmurzayeva, N., B. Selenova, and S. Kudaibergenov. 2013. Preparation and catalytic activity of gold nanoparticles stabilized by poly(*n*-vinylpyrrolidone) and deposited onto aluminum oxide. *Am. J. Nanomater.* 1: 1–4.

Yoshida, T., T.C. Lai, G.S. Kwon, and K. Sako. 2013. pH- and ion-sensitive polymers for drug delivery. *Expert Opin. Drug Deliv.* 10: 1497–1513.

Zhang, H., Y. Lu, G. Zhang, Sh. Gao, D. Sun, and Y. Zhong. 2008. Bupivacaine-loaded biodegradable poly(lactic-*co*-glycolic) acid microspheres: I. Optimization of the drug incorporation into the polymer matrix and modelling of drug release. *Int. J. Pharm.* 351: 244–249.

Zhang, J., B. Zhao, L. Meng, H. Wu, X. Wang, and Ch. Li. 2007. Controlled synthesis of gold nanospheres and single crystals in hydrogel. *J. Nanopart. Res.* 9: 1167–1171.

Zhang, Q., X. Li, Y. Zhao, and L. Chen. 2009. Preparation and performance of nanocomposite hydrogels based on different clay. *Appl. Clay Sci.* 46: 346–350.

Zhang, Y., R.W. Cattrall, I.D. McKelvie, and S.D. Kolev. 2011. Gold, an alternative to platinum group metals in automobile catalytic converters (review). *Gold Bull.* 44(3): 145–153.

Zhang, Y.X., F.P. Wu, M.Zh. Li, and E.J. Wang. 2005. pH switching "on-off" semi-IPN hydrogel based on cross-linked poly(acrylamide-*co*-acrylic acid) and linear polyallylamine. *Polymer* 46: 7695–7700.

Zharmagambetova, A.K., N.A. Dolya, Zh.E. Ibrayeva, B.K. Dyussenalin, and S.E. Kudaibergenov. 2010. Physico-chemical and catalytic properties of polymer-protected and gel-immobilized palladium nanoparticles in hydrogenation of 2-propene-1-ol. *Bull. Acad. Sci. Repub. Kazakhstan, Ser. Chem.* 3: 6–9.

Zheksembayeva, N.A., G.S. Tatykhanova, D.K. Sabitova, V.B. Sigitov, and S.E. Kudaibergenov. 2012. Study of new composite hydrogel pigs for pipeline cleaning. *Intern. J. Transport&Logistic.* 12: 241–247.

Zhou, J., J. Ralston, R. Sedev, and D.A. Beattie. 2009. Functionalized gold nanoparticles: Synthesis, structure and colloid stability. *J. Colloid Interface Sci.* 331: 251–262.

Zhumaly, A.A., E.Yu. Blagikh, Zh.E. Ibrayeva, and S.E. Kudaibergenov. 2013. Preparation and properties of composite materials based on poly(acrylamide) hydrogel and clay minerals. *Bull. Kazakh Natl. Techn. Univ.* 5: 234–241.

2 Cryogels for Affinity Chromatography

Nilay Bereli, Deniz Türkmen, Handan Yavuz,
Igor Yuri Galaev, and Adil Denizli

CONTENTS

2.1 INTRODUCTION

Affinity chromatography is a method in which specific and reversible interactions are used for the isolation, separation, and purification of biomolecules from crude samples (Wilchek 2004). In affinity chromatography, a ligand molecule having specific recognition ability is bound on a matrix. The molecule to be purified is selectively captured by the ligand bound on the solid matrix by simply passing the solution containing the target through the chromatographic column under favorable conditions. The target molecules are then eluted by using proper elutants under conditions favoring elution, by adjusting the pH, ionic strength, or temperature, using specific solvents or competitive free ligands, so that the interactions between the ligand and the target are broken and the target molecules are obtained in a purified form. The affinity-based approach was introduced in 1968 by Cuatrecasas et al. (1968) to purify proteins, and today, it still represents one of the most powerful techniques available for purification of bioactive materials (Matejtschuk 1997). The general affinity-based approach

TABLE 2.1
Various Methods Derived from Affinity Chromatography

Affinity capillary electrophoresis	Immunoaffinity chromatography
Affinity electrophoresis	Lectin affinity chromatography
Affinity partitioning	Library-derived affinity ligands
Affinity precipitation	Membrane affinity chromatography
Affinity repulsion chromatography	Metal-chelate affinity chromatography
Affinity tag chromatography	Molecular imprinting technique
Avidin–biotin-immobilized system	Perfusion affinity chromatography
Covalent affinity chromatography	Protein A affinity chromatography
Dye affinity chromatography	Receptor affinity chromatography
High-performance affinity chromatography	Tandem affinity purification
Histidine affinity chromatography	Thiophilic chromatography
Hydrophobic interaction chromatography	Weak affinity chromatography

was subsequently adopted for a variety of other affinity methods, such as affinity capillary electrophoresis, immunoaffinity chromatography, lectin affinity chromatography, affinity tag chromatography, histidine affinity chromatography, and thiophilic affinity chromatography (Wilchek 2004) (Table 2.1).

Chromatography, both analytical and large scale, is the predominant technology in downstream separations. Although conventional packed-bed columns have been used for many applications, they have some important drawbacks such as the slow diffusional mass transfer and the large void volume between the beads (McCoy et al. 1996). New stationary phases such as the nonporous polymeric beads and perfusion chromatography packing are designed to resolve these problems, but these matrices are not sufficient to resolve these limitations in essence. Recently, cryogels are considered as a novel generation of stationary phases in the separation science owing to their easy preparations, excellent flow properties, and high performances compared to conventional columns for the separation of biomolecules. Several potential advantages of cryogels are large pores, short diffusion path, low pressure drop, and very short residence time for both adsorption and elution (Arvidsson et al. 2003, Babaç et al. 2006, Dragan and Dinu 2013, Lozinsky et al. 2001, Yilmaz et al. 2009).

2.2 CHROMATOGRAPHY OF ANTIBODIES USING MACROPOROUS CRYOGELS

Antibodies represent glycoproteins having carbohydrate-recognition motifs, and they play an important role in biochemistry, biotechnology, and biotherapeutics (Labib et al. 2009). Antibody-based biotherapeutics and in vivo diagnostics are gaining wider approval from many health authorities all over the world (Reichert et al. 2005). Monoclonal antibodies represent the fastest growing biopharmaceutical market segment with a potential to reach total global sales of 50 billion USD. Approximately 30% of biotherapeutics today in clinical trials are antibodies and Fc fusion proteins (Rosa et al. 2010). Currently, 24 monoclonal antibodies, 2 monoclonal antibody

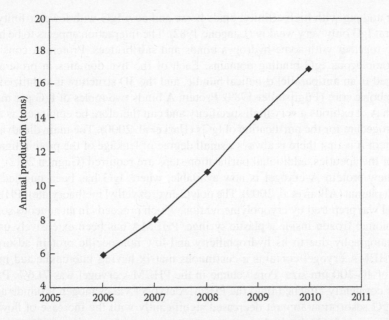

FIGURE 2.1 Annual production of monoclonal antibodies. (From Denizli, A., *Hacettepe J. Biol. Chem.*, 39, 1, 2011.)

fragments, and 4 Fc fusion proteins have been approved for therapeutic use, and over 150 are under clinical research (Carter 2006). Immunoglobulin G (IgG) purified from human serum is frequently used to treat a variety of disorders such as primary and secondary immunodeficiencies, infections, and inflammatory and autoimmune diseases (Shukla and Thömmes 2010). The administration of IgG products has not only improved the quality of life but also saved lives in a variety of indications. The large number of IgG products in development certainly supports the case for a standardized approach (Low et al. 2007). But the large quantities in which some of them will be required put considerable economic pressure on both the current methods and the facilities required (Figure 2.1). The worldwide consumption of human IgG is nearly tripled between 1992 and 2003 from 19.4 to 52.6 tons (Denizli 2011).

2.2.1 PROTEIN A AFFINITY CHROMATOGRAPHY

Staphylococcal protein A is one of the first discovered IgG-binding molecule and has been extensively used as a biospecific ligand during the past decade. At a large scale, protein A chromatography still persists as the most common technique for antibody purification. There are a number of naturally IgG-binding proteins that have been described. Of these, the most significant is protein A, which is a cell wall–associated protein domain exposed on the surface of the gram-positive bacterium *Staphylococcus aureus* (Gagnon 2013). Protein A consists of a single polypeptide chain, with a molecular weight of 42,000 Da, and does not contain significant amounts of carbohydrate. It does, however, bind with different affinities to the Fc region of IgGs from a variety of sources, for example, it binds to IgG from human, rabbit, and pig with high affinity; binds horse and cow IgG with lower affinity; and binds rat IgG only very weakly (Langone 1982). The interaction appears to be hydrophobic together with some hydrogen bonds and salt bridges. Protein A consists of five homologous IgG-binding domains. Each of the five domains in protein A is arranged in an antiparallel α-helical bundle, and the 3D structure is stabilized via a hydrophobic core (Füglistaller 1989). Protein A binds two moles of IgG per mole of protein A. It exhibits a very high specificity and can therefore be employed as a one-step procedure for the purification of IgGs (Hari et al. 2000). The main disadvantage of protein A is that there is always a small degree of leakage of the protein ligand so that for therapeutics, additional purification steps are required (Gagnon 2012).

A new protein A cryogel is now available, where IgG has been purified from human plasma (Alkan et al. 2009). The poly(2-hydroxyethyl methacrylate) (PHEMA) cryogel was prepared by cryopolymerization, which proceeds in an aqueous solution of monomer frozen inside a plastic syringe. PHEMA has been extensively used in chromatography due to its hydrophilicity and low nonspecific protein adsorption. The PHEMA cryogel contains a continuous matrix having interconnected macropores of 10–200 μm size. Pore volume in the PHEMA cryogel was 71.6%. Protein A was covalently attached onto the PHEMA cryogel via cyanogen bromide activation. IgG adsorption amount decreased significantly with the increase of flow rate. The maximum IgG adsorbed amount was 83.2 mg/g with a purity of 85%. Cleaning regimes consist of alternating cycles of low concentrations of NaOH and salt, resulting in the cryogel maintaining good yield out to 10 cycles, a significant improvement over the earlier PHEMA protein A where the yield of IgG purification dropped to 5% after 10 cycles.

2.2.2 HISTIDINE AFFINITY CHROMATOGRAPHY

Pseudospecific ligands such as histidine, tryptophan, and phenylalanine can be used to purify a wide range of biomolecules (Altıntaş and Denizli 2009). They are small molecules with higher chemical and physical stability and lower cost. Hydrophobic amino acids tryptophan and phenylalanine with their predominant aromatic stacking properties were shown to bind selectively proteins rich in aromatic residues (Uygun et al. 2010). Histidine was used as an affinity ligand for the purification of IgG (Vijayalakshmi 1989). The adsorption mechanism between histidine and IgG is based on several molecular interactions such as hydrogen bonding, electrostatic, and mild hydrophobicity. The different physicochemical properties of histidine are because of the nonsymmetric arrangement of its carbon atoms and its broad pK_a range. Histidine is also characterized by its hydrophobicity and its capacity to transfer charge because of its imidazole ring. Owing to these properties, in addition to its role in the acid–base system, histidine interacts with its microenvironment by different mechanisms under different pH, ionic strength, and temperatures (Vijayalakshmi 1996).

Theoretical considerations and understanding of the interactions between protein molecules and the histidine-coupled matrix have shown that the mechanism would be water mediated. This would involve changes in dielectric constant at the adsorption interface owing to the combined electrostatic, hydrophobic, and charge-transfer

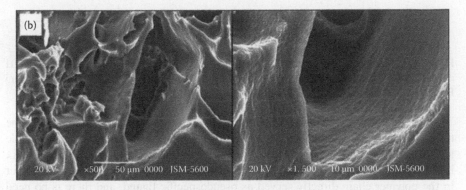

FIGURE 2.2 (a) The molecular structure of PHEMAH and (b) SEM images of PHEMAH cryogel. (From Bereli, N. et al., *Sep. Sci. Technol.*, 47, 1813, 2012.)

interactions between histidine and the specific amino acid residues available on the protein surface (Bhattacharyya et al. 2003).

Denizli's research group has developed a new approach to obtain an efficient and cost-effective purification of IgG from human plasma (Bereli et al. 2012). Macroporous cryogel was obtained by the cryopolymerization of 2-hydroxyethyl methacrylate (HEMA) and *N*-methacryloyl-(L)-histidine-methyl ester (MAH) in an ice bath. Molecular structure and SEM images of poly(hydroxyethyl methacrylate-*co-N*-methacryloyl-(L)-histidine) (PHEMAH) cryogel were given in Figure 2.2a and b.

There are many macropores whose diameter is over 50 μm on the bulk structure of the cryogel. SEM images also showed that the macropores were uniformly distributed into the polymeric network. The macropores reduce mass transfer resistance and facilitate convective transport. So the PHEMAH cryogel had good flow properties. IgG adsorption amount from human plasma was 97.3 mg/g with a purity of 94.6%. The selectivity of the PHEMAH cryogel was confirmed via fast protein liquid chromatography (FPLC). As shown in Figure 2.3, albumin (67 kDa) and IgG (150 kDa) separation from human plasma was observed at 2.4 and 9.3 min.

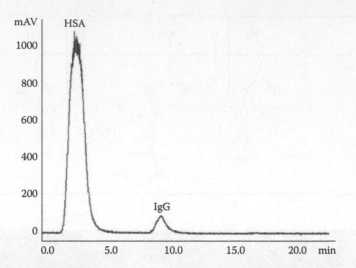

FIGURE 2.3 FPLC separation of albumin and IgG from human plasma on a PHEMAH cryogel column: flow rate was 2.0 mL/min, and detection was performed at 280 nm. (From Bereli, N. et al., *Sep. Sci. Technol.*, 47, 1813, 2012.)

PHEMAH cryogel column has many advantages over conventional columns. The time consuming and high cost of ligand binding step has inspired a search for suitable low-cost matrices. An expensive and critical step in the preparation of affinity matrix is the binding of an affinity ligand to the matrix. In this work, the MAH comonomer in the polymer chain directly served as pseudospecific ligand, and there is no need to activate the matrix for the ligand binding. Another problem is that of slow release of the covalently bonded ligands off the matrix. Ligand release is a general problem encountered in any affinity adsorption technique, which caused a decrease in binding capacity. Ligand leakage also causes contamination that will interfere with the analysis of the purified biomolecule. Ligand binding step was eliminated in this approach. MAH was polymerized with HEMA and there was no ligand leakage.

2.2.3 THIOPHILIC AFFINITY CHROMATOGRAPHY

Thiophilic adsorption chromatography was developed by Porath et al. in the 1980s. Hutchens and Porath have shown thiophilic adsorption of proteins on both functional groups present in the ligand structure, thioether sulfur, and the adjacent sulfone group, in a cooperative manner (Hutchens and Porath 1986). Recently, Bakhspour et al. prepared thiophilic matrices to isolate IgG from human serum based on the PHEMA cryogel (Bakhspour et al. 2014). After the thiophilic ligand of 2-mercaptoethanol was attached on the surface, the cryogel exhibited a strong specificity toward IgG in a salt-independent manner. The maximum adsorption capacity of the thiophilic cryogel was 74.8 mg IgG per gram of cryogel from human serum. Since the selective recognition between the thiophilic ligand and IgG is attributed to electron donor–acceptor interaction, the temperature effect on adsorption was not greatly pronounced. The purity of the isolated IgG exceeded 89% with a recovery of about 81%. Prominent advantages of this method, such as strong specificity, rapid processing, mild conditions, conventional equipment, and excellent reusability, make this chromatographic technology embody great potentialities to isolate the antibodies on a large scale.

Agarose beads embedded cryogels showed the fascinating characteristics of high stability, biological inertness, and facile activation (Sun et al. 2012a). Sun et al. prepared continuous supermacroporous agarose beads embedded in agarose–chitosan composite cryogels by cryopolymerization. After binding 2-mercaptopyridine onto divinyl sulfone–activated matrix, the composite cryogels were used for the purification of IgG. The obtained cryogels possess interconnected pores of 10–100 μm size. The specific surface area was 350 m²/g with maximum adsorption capacity of IgG 71.4 mg/g. The cryogels can be reused at least 15 times without significant loss in adsorption capacity.

2.3 METAL-CHELATE AFFINITY CHROMATOGRAPHY USING MACROPOROUS CRYOGELS

Immobilized metal-chelate affinity chromatography (IMAC) of proteins, with metal chelate linked to Sepharose, was introduced by Porath et al. in 1975, and since then, it has been adopted for the purification of many therapeutic proteins, peptides, nucleic acids, hormones, and enzymes (Altıntaş et al. 2007). IMAC introduces a new approach for selectively interacting materials on the basis of their affinities for chelated metal ions. The separation is based on the interaction of a Lewis acid (electron pair acceptor), that is, a chelated metal ion, with electron donor atoms (N, O, and S) on the surface of the protein. Proteins are assumed to interact mainly through the imidazole group of histidine and, to a lesser extent, the indoyl group of tryptophan and the thiol group of cysteine. Cooperation between neighboring amino acid side chains and local conformations play important roles in protein binding. Aromatic amino acids and the amino-terminal of the peptides also have some contributions (Emir et al. 2004).

Jack bean (*Canavalia ensiformis*) is the source of interesting proteins that contribute to modern biochemistry, and urease is the primary of these proteins. Recently, Tekiner et al. developed a novel IMAC matrix (Tekiner et al. 2012). While IMAC conventionally uses a metal-ion-loaded chelating group, it has been shown that Cu^{2+} ions directly loaded PHEMAH (MAH consisting of imidazole groups) and this worked quite well as an IMAC medium (Figure 2.4). This obviates the need for attachment of any chelating groups like iminodiacetic acid (IDA). They showed that

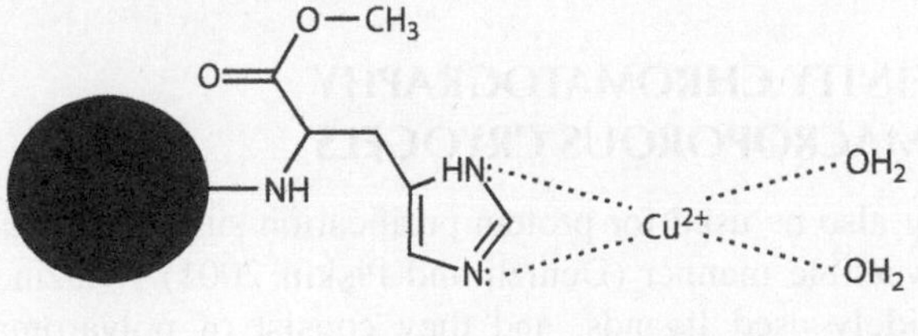

FIGURE 2.4 Cu^{2+} chelation through the PHEMAH cryogel.

high urease adsorption capacity was obtained from jack bean (*C. ensiformis*) (up to 67.8 mg/g) with a recovery of 88.8%. They also reported that jack bean urease was purified 162-fold with PHEMAH–Cu^{2+} cryogel.

Akduman et al. successfully showed that Zn^{2+}-loaded PHEMA–glycidyl methacrylate (GMA) cryogel is a potential alternative for the purification of yeast alcohol dehydrogenase (Akduman et al. 2013). Dynamic binding capacity at flow rate of 4 mL/min was estimated to be 4.6 mg/g of support with Zn^{2+}-chelated PHEMA–GMA cryogel. They reported that alcohol dehydrogenase from baker's yeast was purified 71.2-fold. Since cryogel is made from one piece of polymer resin, there is no need for column packing step. As a result of convective flow, dynamic binding capacities are higher even at higher flow rates. The main advantage of the convective-based system over the conventional gels is that they can be operated at high flow rates without compromising on the purity and the binding capacity of the product. Thus, combining the advantages of cryogel with chelated metal ions, alcohol dehydrogenase from baker's yeast can be efficiently purified in a single step.

Recently, one interesting example for metal affinity chromatography devices is the particle-embedded cryogels (Çimen and Denizli 2012). Cryogels are good alternatives to separation with many advantages (Andaç et al. 2008). However, the low binding capacities of cryogel columns due to low specific surface area often limit their practical applications (Bereli et al. 2008). In actual separation processes, it is of great importance to improve the binding capacity of macroporous cryogel. Therefore, particle embedding would be a useful improvement mode to use in the preparation of novel composite cryogels for increasing surface area (Yao et al. 2006, 2007, Le Noir et al. 2007). This approach makes use of a combinatorial selection strategy to enhance adsorption capacity. Denizli and coworkers reported the use of selective IgG adsorption with PHEMA cryogel having embedded PGMA–IDA–Cu^{2+} particles (Bereli et al. 2010). The scanning electron images of PGMA particles, PHEMA, and PGMA–IDA–Cu^{2+} embedded PHEMA cryogels were given in Figure 2.5. The presence of embedded beads can be seen clearly. IgG adsorption capacity of the PGMA–IDA–Cu^{2+} embedded PHEMA cryogel from human serum was 257 mg/g.

Bibi et al. prepared megaporous cryogels with metal-ion affinity functionality, which possess enhanced protein-binding ability (Bibi et al. 2013). These highly porous materials (pore size up to 100 μm) allowed the direct capture of recombinant His6-tagged protein from a partially clarified extract. They demonstrated that IDA-functionalized poly(methacrylic acid)-based cryogels charged with Cu^{2+} ions can be tailored to allow for increased and highly specific binding capacities of NAD(P)H-dependent 2-cyclohexen-1-one-reductase by exploiting IMAC.

2.4 DYE AFFINITY CHROMATOGRAPHY USING MACROPOROUS CRYOGELS

Textile dyes may also be used for protein purification since they bind proteins in a selective and reversible manner (Denizli and Pişkin 2001). Triazine reactive dyes are the most widely used ligands, and they consist of polyaromatic sulfonated compounds containing a triazine reactive group, which facilitates their binding to

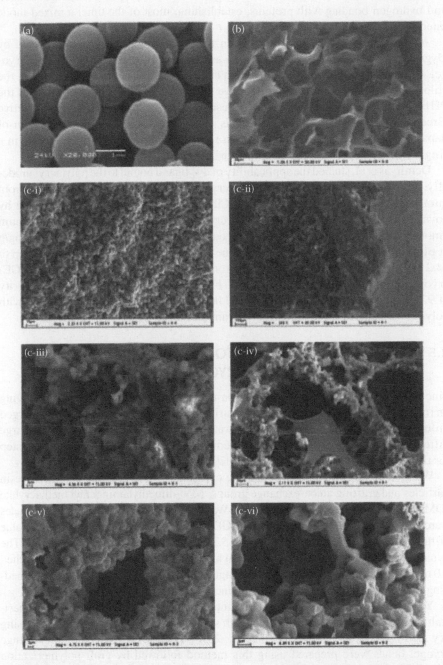

FIGURE 2.5 SEM images of (a) PGMA beads, (b) PHEMA, and (c) PGMA–IDA–Cu^{2+} embedded PHEMA cryogels. (From Bereli, N. et al., *Mater. Sci. Eng.*, 30, 323, 2010.)

insoluble matrices. Dye ligands can engage in ionic, hydrophobic, charge-transfer, and hydrogen bonding with proteins, establishing, most of the time, a mixed-mode interaction.

An interesting dye affinity application for papain purification was achieved by Uygun et al. (2012). Reactive Green 5 was attached to PHEMA monolithic cryogel. Under the experimental conditions, a chemical reaction took place between the chlorine-containing group of the Reactive Green 5 and the hydroxyl groups of the PHEMA, with the elimination of NaCl, resulting in covalent attachment of Reactive Green 5 onto the PHEMA cryogel. This matrix was used for the purification of papain from *Carica papaya* latex. Papain from *C. papaya* was purified 42-fold in a single step. The specific activity of the purified papain was 3.45 U/mg.

Demiryas et al. studied the applicability of dye ligand bound to the poly(acrylamide-allyl glycidyl ether) (poly[AAm-AGE]) cryogel for the albumin purification from human plasma (Demiryas et al. 2007). Cibacron Blue F3GA was immobilized by covalent binding onto poly(AAm-AGE) cryogel via epoxy groups. The maximum amount of albumin adsorption from aqueous solution in acetate buffer was 27 mg/g at pH 5.0. Higher albumin adsorption value was obtained from human plasma (up to 74.2 mg/g). Albumin elution from Cibacron Blue F3GA–attached poly(AAm-AGE) cryogel was achieved using 0.1 M Tris/HCl buffer containing 0.5 M NaCl with a purity of 92%. They reported that albumin could be repeatedly adsorbed and desorbed with poly(AAm-AGE) cryogel without significant loss in the adsorption capacity.

2.5 ION-EXCHANGE CHROMATOGRAPHY USING MACROPOROUS CRYOGELS

Since its first introduction in the 1960s, ion-exchange chromatography is still playing an important role for the separation and purification of biomolecules bearing charged groups. This technique is capable of separating molecules having only minor charge differences and can be used as an intermediate purification step as well as final step for the polishing of bioproduct (Heftman 2004).

Wang et al. fabricated novel composite cryogels by incorporating polymeric resin particles and grafting anion-exchange groups, N,N'-dimethylaminoethyl methacrylate (DMAEMA), on the pore wall surfaces (Wang et al. 2013). The embedded particles were prepared by grinding poly(GMA-EGDMA) monoliths. The surface area value of the composite cryogel increased from 5.3 to 14.8 m^2/g with particle loading. The dynamic binding capacity of bovine serum albumin on the composite cryogel reached 6.0 mg/mL column, which was 2.8 times higher than a cryogel column without embedding the ground resin particles due to the increase in the specific surface area.

Yun et al. described stable and continuous microchannel flow focusing by peristaltic pump and a rapid freezing approach suitable for scale-up preparation by using dry ice for cryopolymerization (Yun et al. 2013). Poly(AAm)-based anion-exchange cryogel beads were prepared using this method followed by graft polymerization with DMAEMA. The performance of a supermacroporous packed bed with these cryogel beads was determined for bovine γ-globulin. The mean diameter of cryogel beads was about 1110 μm. The chromatographic capacity of γ-globulin in the cryogel bead-packed bed is around 2 mg/mL bed even at high velocity.

2.6 CHROMATOGRAPHY OF DNA USING MACROPOROUS CRYOGELS

DNA-based therapies such as gene therapy and genetic vaccination using plasmid DNAs are gaining wider attention in recent years. For the safer application that meets the regulatory issues, high purity of the product is required and it should be manufactured at the industrial scale. Chromatographic methods such as size exclusion, ion-exchange, hydrophobic interaction, and affinity-based methods can be used for the large-scale purification of DNA (Urthaler et al. 2005).

The DNA affinity cryogel matrix having nanospines for a hydrophobic affinity chromatography can be synthesized using novel cryogelation method (Üzek et al. 2013). The main disadvantage of cryogel form is the low surface area and thus low ligand density and low adsorption capacity. Methods such as particle embedding, freeze-drying, and synthesis of cryogel using nanoparticles can be performed to increase the specific surface area of cryogels to be used (Hajizadeh et al. 2013). Üzek et al. reported that the inclusion of freeze-drying stage to conventional cryogelation process drastically increased the specific surface area of the cryogels without affecting the macroporosity. The specific surface area values of cryogels were 17.6 m^2/g for conventional cryogelation and 36.0 m^2/g for freeze-dried cryogelation. PHEMA cryogel containing phenylalanine as hydrophobic ligand can be used for plasmid DNA purification from *Escherichia coli* lysate. The specific selectivity for plasmid DNA was 237.5-folds greater than all impurities.

Odabaşı et al. prepared Zn^{2+}-chelated supermacroporous PHEMA-MAH monolithic cryogels for DNA adsorption. They obtained 32.93 mg/g DNA adsorption amount with 45.8 μmol/g MAH content and 49.2 μmol/g Zn^{2+} loading. Without Zn^{2+} loading, the DNA adsorbed amount was significantly lower, indicating the importance of functionalization of cryogels to control the adsorption property (Odabaşı et al. 2011).

Recently, Perçin et al. prepared HEMA-based cryogels containing MAH as a ligand and reported the adsorption capacity as 13.5 mg/g for plasmid DNA (Perçin et al. 2011).

2.7 CHROMATOGRAPHY OF LIVING CELLS USING MACROPOROUS CRYOGELS

One of the main focuses of biochemical research has been to better understand individual molecules and their roles in the living process. Great efforts have been made to obtain a better understanding on how these biomolecules are organized in more complex structures and also how these structures function in the living cell. Extensive experience of working with individual biomolecules has been based on the existence of highly efficient techniques for isolation and purification of molecular entities with molecular masses less than 10^6 Da. However, purification of larger objects, often combined under the name of biological nanoparticles, like plasmids, cell organelles, viruses, protein inclusion bodies, macromolecular assemblies, as well as the separation of cells of different kinds, still remains a challenge. Macroporous cryogels provide an elegant solution to this challenge as they have a highly interconnected system of open macropores. *E. coli* cells with a size of 1 × 3 μm (Madigan et al. 2000) were expected to pass rather easily through the pores of 10–100 μm in size. In fact, no *E. coli* cells were retained on plain acrylamide-based cryogels or on anion-exchange cryogels in the presence of 1 M NaCl, sufficient to suppress electrostatic interactions. The surface of *E. coli* cells is negatively charged, and they bound efficiently to anion-exchange cryogel column at low ionic strength and eluted with 70%–80% efficiency at high ionic strength of 1 M NaCl (Arvidsson et al. 2002). When cells are retained to the column, an important question is whether the binding takes place in the whole volume of the column or the cells are accumulated at certain areas, for example, at the column top. One column was sacrificed after binding of ampicillin-resistant *E. coli* cells. The matrix was taken out from the column, and a few small pieces were taken from the central part of the column and placed on agar plate containing ampicillin. A similar growth around all pieces indicated homogeneous binding of *E. coli* cells to the cryogel matrix (Figure 2.6), which is also confirmed by SEM (Figure 2.7).

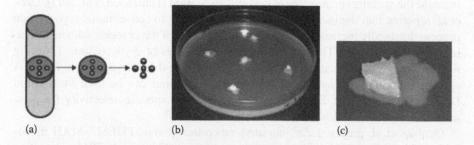

(a)   (b)   (c)

FIGURE 2.6 Growth from pieces of anion-exchange cryogel matrix on ampicillin containing agar plates after binding ampicillin-resistant *E. coli* cells: (a) scheme for generating pieces of cryogel matrix, (b) agar plate with pieces of the cryogel matrix with bound cells, and (c) enlarged view of one of the pieces.

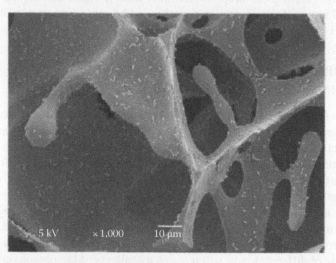

FIGURE 2.7 SEM images of supermacroporous anion-exchange matrix with bound *E. coli* cells. (Reproduced from Arvidsson, P. et al., *J. Chromatogr. A*, 977, 27, 2002.)

The developed cryogel matrices allowed for binding and elution of bound cells. Hence, when different types of cells have different affinities to the matrix, it was possible to separate them in chromatographic mode. Separation of different cell types was demonstrated for two model systems: the mixtures of wild-type *E. coli* (w.t. *E. coli*) and recombinant *E. coli* cells displaying poly-His peptides (His-tagged *E. coli*) and of w.t. *E. coli* and *Bacillus halodurans* cells. W.t. *E. coli* and His-tagged *E. coli* were quantitatively captured from the feedstock containing equal amounts of both cell types and recovered by selective elution with imidazole and EDTA, with yields of 80% and 77%, respectively. The peak obtained after EDTA elution was eightfold enriched with His-tagged *E. coli* cells as compared with the peak from imidazole elution, which contained mainly weakly bound w.t. *E. coli* cells. Haloalkalophilic *B. halodurans* cells had low affinity to the Cu^{2+}–IDA cryogel column and could be efficiently separated from a mixture with w.t. *E. coli* cells, which were retained and recovered in high yields from the column with the gradient of imidazole solution. All the cells maintained their viability after the chromatographic separation (Dainiak et al. 2005).

Cell chromatography was further advanced by introducing immune-affinity interactions in combination with capturing antibody-labeled cells on macroporous cryogel with immobilized protein A (Kumar et al. 2003). The macropores in cryogels are big enough to allow for passing not only small microbial cells but also much larger mammalian cells like red blood cells (Figure 2.8) (Noppe et al. 2007).

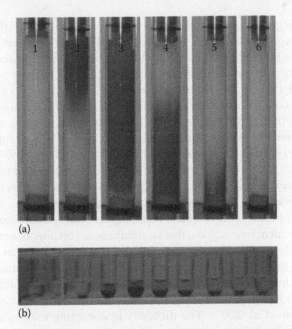

(a)

(b)

FIGURE 2.8 (a) Flow pattern of whole blood through a naked cryogel column. One milliliter blood was applied to the naked cryogel column at a flow rate of 0.5 mL/min in isotonic buffer solution. Column 1, column before application; columns 2–5, column during the run; and column 6, column after the flow of blood sample. (b) Flow-through fractions of the column was collected and left to stand. No red blood cell lysis was observed in the fractions. (Reproduced from Noppe, W. et al., *J. Biotechnol.*, 131, 293, 2007.)

Binding of antibody-treated cells to protein A-cryogel
(a)

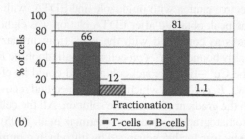

(b)

FIGURE 2.9 (a) Schematic presentation of the binding of goat antihuman IgG-treated human blood lymphocytes on a supermacroporous monolithic cryogel column with immobilized protein A. (b) Fractionation of B- and T-lymphocytes. B-lymphocytes: binding 91%, recovery 70%, viability 80%. T-lymphocytes: breakthrough 81%, viability >90%.

After treating lymphocytes with goat antihuman IgG (H+L), the IgG-positive B-lymphocytes were efficiently separated from T-lymphocytes on cryogel matrix with covalently immobilized protein A. IgG-labeled B-lymphocytes were retained by the column, whereas T-lymphocytes, which do not interact with these antibodies, passed through the column. More than 90% of the B-lymphocytes were retained in the column, while the cells in the breakthrough fraction were enriched in T-lymphocytes (81%). The viability of isolated T-lymphocytes was above 90%. The bound B-lymphocytes were released from the column by displacement with cheap human or dog IgG recovered under mild conditions with the 60%–70% yield and without significantly impairing the cell viability (Figure 2.9). The technique can be applied in general to cell separation systems where IgG antibodies against specific cell surface markers are available (Kumar et al. 2003).

Adsorption of cells to any surface, including chromatographic matrix, is characterized by polyvalent interactions, that is, simultaneous binding of multiple receptors on the cell surface to multiple ligands on another surface, and can be collectively much stronger than corresponding monovalent interactions (Mammen et al. 1998). Under typical chromatographic conditions (10^{10} to 10^{12} ligands and receptors per cm^2 and 10^{-10} to 10^{-8} cm^2 contact area), the number of interactions can be between 1 and 10,000 (Cao et al. 2002). The difficulty in disrupting multivalent interactions is one of the main challenges in using affinity chromatography for cell separation. An external force affecting the entire cell or the matrix in an integral way is often required to simultaneously disrupt multiple interactions and detach cell from the matrix under eluting conditions. Usually, such integral force is generated by the passage of air–liquid interfaces (Gomez-Suarez et al. 2001) or by using flow-induced shear forces (Ming et al. 1998).

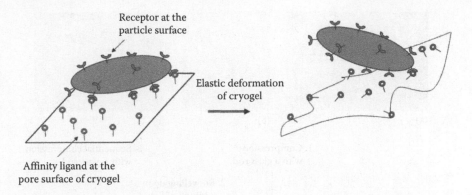

FIGURE 2.10 Schematic illustration of the mechanism of detachment of affinity-bound cell induced by compression of cryogel matrix.

In biological systems, cells often come into contact with soft surfaces, for example, tissue or an extracellular matrix, which can undergo changes in elasticity (e.g., wound healing). In vivo, the primary step of any bacterial or virus infection is the adsorption of the pathogen onto an elastic, hydrophilic matrix, namely, the tissue of a living organism. Only recently have systematic studies of the effect of substrate mechanics on cell adhesion been carried out, and softness and elasticity of the surface were shown to be important for cell–surface interactions (Engler et al. 2004, Wong et al. 2004). These trends are independent of the adhesive ligand. An important implication of these findings is that the use of soft materials in cell–affinity separation may decrease nonspecific cell–surface interactions and allows for the introduction of integral disruptive force via elastic deformation of the matrix. The cryogel matrices with their elasticity look as ideally suited for such applications (Figure 2.10).

The effect of mechanical compression of the cryogel matrix on elution of affinity-bound cells has been studied in 96-well microplate system (Figure 2.11). Elastic cryogel monoliths have properties highly favorable for using them in 96-well systems. First, the elasticity of monoliths allows them to be slightly compressed and be easily introduced into the wells and retained therein with no risk of leakage in between the monolith and the walls of the well. Second, the capillary forces keep the liquid inside the pores of the gel, making the monolith column drainage protected. On the one hand, the application of a certain volume of liquid on top of a cryogel monolith column results in the displacement of exactly the same volume of liquid from the column, while on the other, it stays filled with the liquid all the time (Galaev et al. 2005). Finally, cryogel monoliths could be easily compressed by a glass rod within their respective wells, and hence, the effect of compression on elution of bound cells could be studied.

Compression, even in the absence of the eluent in the equilibration buffer, detached 40%–80% of bound cells. Manual compression with a glass rod takes about 2 s and results in the expression of about 370 μL of liquid out of cryogel monolith. Thus, detached cells enter the flow at a high velocity, which prevent most of the cells from readsorbing. The rate of cell release is faster than the kinetics of

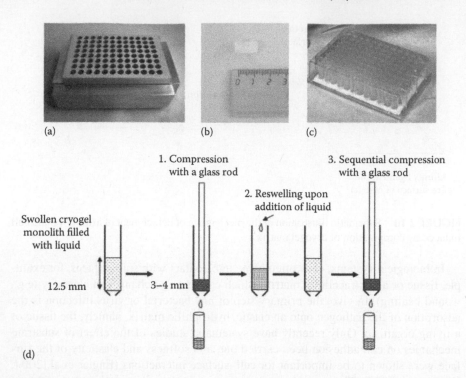

FIGURE 2.11 A 96-well microplate system with cryogel monoliths. (a) Metal mold for cryogel production, (b) individual cryogel monolith, (c) 96-well microplate with monolithic cryogel columns, and (d) schematic presentation of elastic deformation of cryogel monolith in the respective well.

bond formation between the particles and the affinity matrix. Because immobilized affinity ligands are not damaged by compression, it is probable that the bonds may reform when the cells are left in contact with the affinity cryogel once compression is complete. The amount of cells released increased with increasing concentration of the specific eluent in the running buffer. The effect of compression was especially pronounced in the case of soft cryogel monoliths, which have higher porosity and elasticity than dense cryogels. Recombinant *E. coli* K12 strain pop 6510 (*thr, leu, tonB, th*i, *lacY1, recA, dex5, metA, supE,* and *dex5*) with plasmid pLH2 encoding the hybrid LamB-His (two 6× His) monomers (Sousa et al. 1996) bound efficiently to Ni^{2+}-loaded iminodiacetate cryogel monoliths via histidine tails available at the cell surface. Due to carbohydrate residues available at the surface of yeast cells, they bound specifically to cryogel monolith with immobilized concanavalin A (Con A). Practically all bound yeast cells and recombinant *E. coli* cells were released by compressing soft Con A- and Ni^{2+}-loaded iminodiacetate cryogel column in the presence of 10 mM α-D-mannopyranoside and 3 mM EDTA, respectively. Nearly quantitative release of yeast cells by compressing the matrix was achieved in the presence of 40–60 mM glucose. Glucose is an eluent with a lower affinity for Con A ligands; therefore, higher concentrations were required than in experiments with α-D-mannopyranoside. Thus, the recovery of affinity-bound cells depends on both the concentration and the nature of the specific eluent. Quantitative recovery of yeast cells by compressing dense Con A cryogel monoliths was observed at 0.5 M α-D-mannopyranoside, and 80% was recovered in the presence of 0.7 M glucose. Recombinant *E. coli* cells recovered by squeezing retained their viability (Dainiak et al. 2006).

These results illustrate the interaction of cells with an elastic affinity surface. Affinity-mediated binding and recovery of cells under various conditions mimic the interactions of bacteria, viruses, macrophages, etc., with elastic tissues in biological systems. Indeed, it was possible to simulate the contraction of biological tissue triggered by conformational transitions of polymers forming the tissue matrix. When cryogel monoliths were prepared from a so-called thermoresponsive polymer, poly(*N*-isopropyl acrylamide) (polyNIPAAm), they shrink dramatically when the temperature is increased above a certain critical transition temperature T_c, which is about 32°C, depending on the conditions and the structure of the gel. The gel shrinking takes place due to the coil-to-globule transition of polyNIPAAm macromolecules (Shibayama and Tanaka 1993). PolyNIPAAm-based cryogel monoliths shrank in a way similar to ordinary polyNIPAAm gels when the temperature was increased above Tc. When bearing Cu^{2+}-loaded iminodiacetate ligands, polyNIPAAm cryogel monoliths bound *E. coli* cells in a way similar to that of polyacrylamide Cu^{2+}-loaded iminodiacetate cryogel monoliths. Compression of the polyNIPAAm-based cryogel monoliths at 25°C, in the absence of imidazole, detached 56% of the bound cells. The bound cells were eluted with 65% efficiency by using 0.2 M imidazole buffer at 25°C, at the temperature below Tc. However, when elution was carried out with the same buffer at 40°C, that is, above Tc, the polyNIPAAm-based cryogel monolith shrank almost instantaneously upon contact with the "warm" buffer, resulting in the release of 85% of the bound cells. The cells eluted at the elevated temperature retained their viability. Hence, the improved elution of bound cells when the matrix underwent elastic deformation has been demonstrated for different types of deformation as a result of either external forces (mechanical compression) or internal forces (shrinking of the matrix because of the coil-to-globule transition of the polymer forming the matrix). The phenomenon of detachment of specifically bound cells from the matrix when undergoing elastic deformation is of generic character, because it was demonstrated for a variety of bioparticles of different sizes (inclusion bodies, microbial cells, yeasts, and mammalian cells) as well as for synthetic particles, for different ligand–receptor pairs (IgG–protein A, sugar–Con A, metal ion–chelating ligand), and when the deformation was caused by either external or internal forces. From the practical point of view, the release of specifically bound cells, especially fragile mammalian cells, by mechanical compression of elastic macroporous cryogels is a unique and efficient method of cell detachment. Mechanical compression may be scaled up by using simple mechanical devices. However, homogeneous increase of temperature in a large volume to recover cells from temperature-sensitive materials may be difficult to realize. The elasticity of cryogels ensures high recovery of bound cells under mild conditions, thus ensuring retained viability. This detachment strategy employing continuous porous structure makes cryogels very attractive for applications in affinity–cell separations.

2.8   CHROMATOGRAPHY OF PROTEINS USING PROTEIN-IMPRINTED MACROPOROUS CRYOGELS

Increasing reports regarding the isolation or purification of proteins for therapeutic purpose using the molecular imprinting technology have been presented in recent years. Molecular imprinting technology is an attractive biomimetic approach that creates specific recognition sites to the shape and functional group arrangement to template biomolecules. Molecular imprinting has been used successfully for imprinting of small molecules such as peptides, amino acids, bilirubin, creatinine, and metal ions (Alexander et al. 2006). The difficulties involved with using the technique for the template imprinting of proteins are their large molecular sizes, the fragility, and the complexity of the molecules. In spite of these difficulties, there is still a strong incentive to prepare bioimprinted polymers, and attempts have been made to prepare protein-imprinted polymers via different strategies (Takeuchi and Hishiya 2008).

Supermacroporous cryogels are hydrogels prepared under freezing point of the diluents. Thanks to the two important properties of cryogel preparation, they are promising candidates for protein imprinting. First, in the preparation of the cryogels, the water-soluble monomers are generally used because of unique features of solid crystals of water molecules, which is also a well-matched selection for the imprinting of biomolecules. Second, the polymerization occurs under freezing point of water, especially at between $-12°C$ and $-20°C$. This restricts molecular motions in biomolecules, which allows imprinting them more easily and specifically, meanwhile preventing them to denature (Bereli et al. 2008). Several recent reports demonstrate the combination of biorecognition and adsorption discussed in the following paragraphs by using molecularly imprinted cryogels to be eventually used for the separation of proteins. A general consensus for an ideal matrix is that it should exhibit high adsorption capacity.

Lysozyme is an important protein, which has a potential use as an anticancer drug (Ghosh et al. 2000) and also frequently used as cell-disrupting agent for bacterial cells, as an antibacterial agent in ophthalmologic preparations, as a food additive in milk products, and as a drug for the treatment of ulcers and infections (Das et al. 1992). Thus, efficient, simple, and cost-effective production and purification of lysozyme is required. Bereli et al. designed lysozyme-imprinted macroporous cryogels for one-step and easy-to-apply lysozyme purification from egg white. They imprinted lysozyme through the reversible metal-chelate coordination interactions of lysozyme with Cu^{2+} ions bound through the MAH comonomer of the PHEMA-based cryogel (PHEMA-MAH) (Figure 2.12). The binding capacity of the lysozyme-imprinted cryogels was 22.9 mg/g polymer, which is a higher amount than those of the nonimprinted cryogels (3.5 mg/g) (Figure 2.13). The relative selectivity of the imprinted cryogel for lysozyme with respect to the bovine serum albumin and cytochrome c was 4.6 and 3.2 times greater than the nonimprinted cryogels, respectively. The reported recovery of lysozyme from the egg white was about 86% with a 94% purity (Bereli et al. 2008).

In another study, Tamahkar and coworkers prepared cytochrome c–imprinted supermacroporous cryogels. They also used metal coordination interactions of cytochrome c with Cu^{2+} ions chelated through the MAH comonomer of PHEMA-based

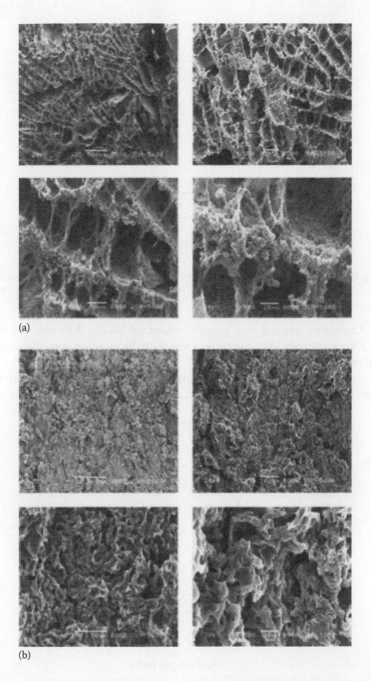

FIGURE 2.12 SEM images of the inner parts of the supermacroporous (a) Lyz-MIP and (b) NIP cryogels. (From Bereli, N. et al., *J. Chromatogr. A*, 1190, 18, 2008.)

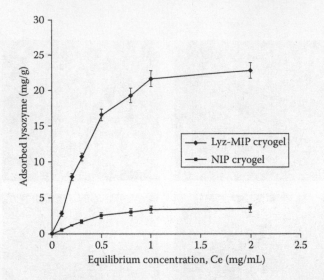

FIGURE 2.13 Effect of equilibrium lysozyme concentration on adsorption amount. Flow rate, 1.0 mL/min; adsorbing buffer, 20 mM HEPES; pH, 7.0; T, 25°C. (From Bereli, N. et al., *J. Chromatogr. A*, 1190, 18, 2008.)

cryogel. Adsorption amount was greatly influenced by the flow rate; the binding capacity decreased from 35 to 7 mg/g polymer with the increase in the flow rate from 0.5 to 3.0 mL/min (Figure 2.14).

They obtained 126 mg/g polymer cytochrome *c* binding amount. Selective binding studies were performed with lysozyme and bovine serum albumin. The relative

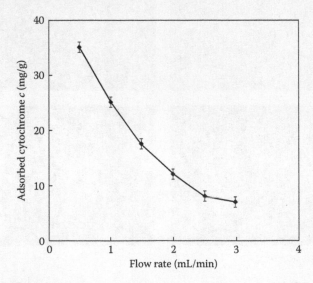

FIGURE 2.14 Effect of flow rate on cytochrome *c* binding: cytochrome *c* concentration, 0.2 mg/mL; adsorbing buffer, 0.1 M carbonate; pH, 9.0; T, 25°C. (From Tamahkar, E. et al., *J. Sep. Sci.*, 34, 3433, 2011.)

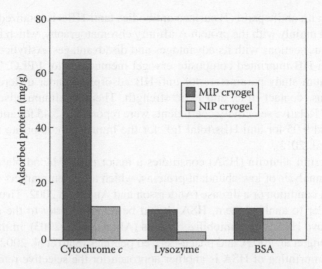

FIGURE 2.15 Binding of cytochrome c and competitive proteins both in MIP and NIP cryogel: flow rate, 1.0 mL/min; adsorbing buffer, 0.1 M carbonate; pH, 9.0; T, 25°C. (From Tamahkar, E. et al., *J. Sep. Sci.,* 34, 3433, 2011.)

selectivity coefficients were 1.7 and 5.2 times greater for lysozyme and bovine serum albumin than those of the nonimprinted PHEMA cryogel, respectively (Figure 2.15). They confirmed the selectivity of MIP cryogel with FPLC as well (Tamahkar et al. 2011).

As indicated earlier, imprinting of the proteins is a difficult process because of the fragility, denaturation problems, and conformation dependency of proteins for a proper application. In order to overcome these drawbacks, epitope imprinting, in which a small portion of the protein that is adequate for the capture of whole protein is used as a template, seems to be a reasonable solution. In such a study, Aslıyüce et al. imprinted Fab fragments of IgG for its specific recognition. In this study, they prepared Fab fragments by papain digestion of IgG. Digestion mixtures were analyzed and separated by FPLC system equipped with HiTrap rProtein A FF column. Unbound Fab and bound Fc fragments were collected, and obtained Fab fragments were imprinted by two approaches: the first one is directly through the interaction with MAH molecules (MIPDirect) and the second is through metal-coordinated interaction with Cu^{2+} ions chelated onto the MAH molecules (MIP Cu^{2+} assisted). In both cases, PHEMA was the base matrix. They investigated the selectivity of the Fab-imprinted cryogel with respect to IgG, Fc fragment, and albumin. They reported relative selectivity constants as 1.47, 2.64, and 3.89 for MIPDirect and 2.90, 8.98, and 11.51 for MIPCu^{2+} assisted for Fab/IgG, Fab/Fc, and Fab/albumin as biomolecule pairs, respectively. They commented that the metal-ion assistance improved the selectivity features of the Fab-imprinted cryogels and allowed to study under milder conditions with enhanced adsorptive properties (Aslıyüce et al. 2013).

The hepatitis B virus (HBV) infection is a worldwide problem; 75% of the world's population is living in the high infection region and 350 million are chronic carriers (Carman 1997, Hsu et al. 1999). For the prevention of the disease, extensive vaccination with antihepatitis B surface antibodies (anti-HBs) is required. Anti-HBs are produced mainly with the protein A affinity chromatography, which is discussed in the previous sections with its advantages and disadvantages. Aslıyüce et al. have prepared anti-HB-imprinted composite cryogel membranes for FPLC. They determined optimum study parameters for anti-HB adsorption under different anti-HB concentrations, contact time, and ionic strength. Their maximum adsorption was 701.4 mIU/g. Relative selectivity coefficients were reported as 5.45 for anti-HBs/total anti-HAV and 9.05 for anti-HBs/total IgE for the imprinted composite membranes (Aslıyüce et al. 2012).

Human serum albumin (HSA) constitutes a major part of blood plasma; thus, it prevents the analysis of low-abundant proteins, which are often markers of a certain physiological condition or a disease (Andersson and Anderson 2002, Tirumalai et al. 2003). In order to analyze them, HSA should be removed prior to the analysis. In order to remove HSA, dye-immobilized resins (Ahmed et al. 2003), immunoaffinity sorbents (Wang et al. 2003), and phage-derived peptides (Sato et al. 2002) are generally in use. Imprinting of HSA is another approach for the selective removal of the albumin from human plasma. For the selective removal of albumin, Andaç et al. have prepared albumin-imprinted PHEMAPA cryogels. They imprint HSA through the hydrophobic interactions of protein with MAPA functional monomer. The optimum adsorption parameters were first investigated in the aqueous media, and the maximum HSA adsorption amount was reported as 25.9 mg/g under these conditions. The selectivity of the imprinted cryogels was investigated with respect to the myoglobin and human transferrin and reported as 56.6 for transferrin and 18.5 for myoglobin. The depletion efficiency of albumin from human serum was shown with the FPLC instrument and SDS-PAGE electrophoresis. Due to their results, they concluded that the PHEMAPA-HSA cryogel selectively removed the HSA (Andaç et al. 2013).

Andaç et al. also prepared HSA-imprinted composite cryogels for the high-capacity removal of the albumin for the proteomic applications. They embedded HSA-imprinted PGMA beads into the PHEMA cryogel. The HSA-imprinted composite cryogels exhibited a high binding capacity and selectivity for HSA in the presence of human transferrin and myoglobin. The competitive adsorption amount for HSA in MIP composite cryogel was 722.1 mg/dL in the presence of competitive proteins. Their depletion ratio from human serum was highly increased by embedding beads into cryogel (85%) (Andaç et al. 2012).

As in all the proteomic studies of the blood plasma, high-abundant proteins in erythrocytes, that is, hemoglobin and carbonic anhydrase, also interfere with the analysis of low-abundant and disease-related proteins. Hemoglobin constitutes 95% of the total cytosolic proteins of erythrocytes and masks observation and the detection of the low-abundant proteins in 2D gel electrophoresis. Thus, it is essential to remove these proteins before a detailed analysis of cytosolic proteins of erythrocytes. Derazshamshir et al. prepared hemoglobin-imprinted PHEMAH cryogels for hemoglobin depletion by using MAH as a functional monomer with 90% gelation yield. They have investigated the adsorption efficiency of the cryogel by circulating the hemolysate in one-step affinity column system, and they reported the hemoglobin adsorption capacity of 167.4 mg/g from the 1:4 diluted hemolysate with high selectivity (Derazshamshir et al. 2010).

2.9 OTHER CHROMATOGRAPHIC APPLICATIONS USING MACROPOROUS CRYOGELS

Recently, composite cryogels containing porous particles were prepared under cryogelation conditions (Hajizadeh et al. 2012). The composite cryogels with immobilized Con A were used for capturing glycoproteins. Hajizadeh et al. reported that the plain polyvinyl alcohol (PVA) cryogel allowed higher flow through and had better mechanical stability than the gels containing porous polymer particles. Increasing particle concentration in composite cryogels increased ligand density, which enhanced the amount of bound horseradish peroxidase from 0.98 to 2.9 mg/mL of gel.

Mannose-specific lectin Con A was purified from *C. ensiformis* seeds by Perçin et al. (2012). Mannose is used as the affinity ligand and was covalently attached onto the PHEMA cryogel via carbodiimide activation. Mannose-carrying PHEMA cryogel provides one-step purification of Con A by biospecific affinity chromatography. They showed that the Con A binding capacity of the mannose-attached PHEMA cryogel decreased significantly with the increasing flow rate from 0.5 to 2.0 mL/min. Unfortunately, the binding behavior was not flow independent and thus led to low binding capacities at reasonable flow rates.

Sun et al. have prepared macroporous agarose/chitosan composite monolithic cryogels carrying 2-aminophenylboronic acid for affinity purification of the major egg-white glycoproteins, ovalbumin and ovotransferrin (Sun et al. 2012b). The composite cryogels contained a continuous interpenetrating polymer network matrix with interconnected pores of 10–100 μm in size. The composite cryogels offered high mechanical stability and had specific recognition for glycoproteins. The binding capacity was 55.6 mg/g for ovalbumin.

Avcibaşi et al. prepared PHEMA–[*N,N*-bis(2,6-diisopropylphenyl)-perylene-3,4,9,10-tetracarboxylic diimide] (DIPPER) monolithic cryogel by radical cryopolymerization and used it for capturing of albumin (Avcibaşi et al. 2010). The chemical structure and internal image of PHEMA–DIPPER were shown in Figure 2.16. Because of the highly hydrophobic DIPPER comonomer, the dominant interaction between cryogel and albumin should be hydrophobic. They reported that the maximum adsorption capacity was 40.9 mg/g polymer.

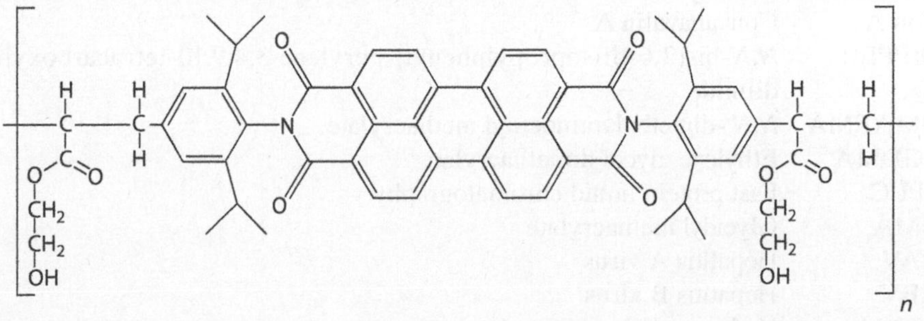

FIGURE 2.16 Chemical structure of PHEMA–DIPPER cryogel. (From Avcibaşi, N. et al., *Appl. Biochem. Biotechnol.*, 162, 2232, 2010.)

Fibronectin is a multifunctional glycoprotein, which is an important protein in many biological processes such as cell proliferation and differentiation, embryogenesis, hemostasis, and wound healing (Pelta et al. 2000). Perçin et al. prepared gelatin-attached PHEMA cryogel to purify fibronectin from human plasma. They chose gelatin as a specific ligand due to the high affinity of fibronectin to gelatin. Optimum amount of gelatin attached by carbodiimide activation on 1 g of PHEMA cryogel was determined as 21 mg. Adsorption capacity of the gelatin-attached PHEMA cryogel was 38 mg/mL. According to their results, fibronectin was purified from human plasma with the recovery of 64%, and the purity of fibronectin was shown as a single band on polyacrylamide gel electrophoresis. The selectivity of the gelatin-attached PHEMA cryogel was also tested on FPLC and the FPLC system allowed separation of fibronectin in a short time (Perçin et al. 2013).

2.10 CONCLUSIONS

New-generation stationary phases, cryogels, have found increasing use in the separation science due to their easy preparations, excellent flow properties, and high performances compared with the conventional matrices. Cryogels are controllable megaporous 3D polymeric gel networks formed under freezing conditions. The unique properties of cryogels like osmotic, chemical, and mechanical stability, large pores, short diffusion path, low pressure drop, and short residence time for both binding and elution stages make them attractive matrices for affinity chromatography of large molecules such as proteins, plasmids, and even whole cells, as well as small molecules. Therefore, cryogels can be used in the various affinity chromatography applications in downstream processes. A combination of the indicated advantages of cryogels and efforts that made them more effective by increasing their surface area with the development of more selective, efficient, easy-to-use, cheap, and stable affinity ligands will enable the design of unique and convenient affinity adsorbents.

ABBREVIATIONS

AAm	Acrylamide
AGE	Allyl glycidyl ether
anti-HBs	Antihepatitis B surface antibodies
Con A	Concanavalin A
DIPPER	N,N-bis(2,6-diisopropylphenyl)-perylene-3,4,9,10-tetracarboxylic diimide
DMAEMA	N,N'-dimethylaminoethyl methacrylate
EGDMA	Ethylene glycol dimethacrylate
FPLC	Fast protein liquid chromatography
GMA	Glycidyl methacrylate
HAV	Hepatitis A virus
HBV	Hepatitis B virus
HEMA	Hydroxyethyl methacrylate
HSA	Human serum albumin
IDA	Iminodiacetic acid

IgG   Immunoglobulin G
IMAC   Immobilized metal-chelate affinity chromatography
MAH   N-methacryloyl-(L)-histidine-methyl ester
MAPA   N-methacryloyl-(L)-phenyl alanine-methyl ester
MIP   Molecular-imprinted polymer
NIPAAm   N-isopropyl acrylamide
PHEMAH   Poly(hydroxyethyl methacrylate-co-N-methacryloyl-(L)-histidine)
PHEMAPA   Poly(hydroxyethyl methacrylate-co-N-methacryloyl-(L)-phenyl alanine)
PVA   Polyvinyl alcohol

REFERENCES

Ahmed, N., G. Barker, K. Oliva, D. Garfin, G. Rice. 2003. An approach to remove albumin for the proteomic analysis of low abundance biomarkers in human serum. *Proteomics* 3: 1980–1987.

Akduman, B., M. Uygun, D.A. Uygun, S. Akgöl, A. Denizli. 2013. Purification of yeast alcohol dehydrogenase by using immobilized metal affinity cryogels. *Mater. Sci. Eng. C* 33: 4842–4848.

Alexander, C., H. Andersson, L. Andersson, R. Ansell, N. Kirsch, I.A. Nicholls, J. Mahony, M.J. Whitcombe. 2006. Molecular imprinting science and technology: A survey of the literature for the years up to and including 2003. *J. Mol. Recognit.* 19: 106–180.

Alkan, H., N. Bereli, Z. Baysal, A. Denizli. 2009. Antibody purification with protein A attached supermacroporous PHEMA cryogel. *Biochem. Eng. J.* 45: 201–208.

Altıntaş, E.B., A. Denizli. 2009. Monosize magnetic hydrophobic beads for lysozyme purification under magnetic field. *Mater. Sci. Eng. C* 29: 1627–1634.

Altıntaş, E.B., N. Tüzmen, L. Uzun, A. Denizli. 2007. Immobilized metal affinity adsorption for antibody depletion from human serum with monosize beads. *Ind. Eng. Chem. Res.* 46: 7802–7810.

Andaç, M., G. Baydemir, H. Yavuz, A. Denizli. 2012. Molecularly imprinted composite cryogel for albumin depletion from human serum. *J. Mol. Recognit.* 25: 555–563.

Andaç, M., I.Y. Galaev, A. Denizli. 2013. Molecularly imprinted poly(hydroxyethyl methacrylate) based cryogel for albumin depletion from human serum. *Colloids Surf. B* 109: 259–265.

Andaç, M., F.M. Plieva, A. Denizli, I.Y. Galaev, B. Mattiasson. 2008. Poly(hydroxyethyl methacrylate)-based macroporous hydrogels with disulfide cross-linker. *Macromol. Chem. Phys.* 209: 577–584.

Andersson, N.L., N.G. Anderson. 2002. The human plasma proteome. *Mol. Cell. Proteomics* 11: 845–867.

Arvidsson, P., F.M. Plieva, V.I. Lozinsky, I.Y. Galaev, B. Mattiasson. 2003. Direct chromatographic capture of enzyme from crude homogenate using immobilized metal affinity chromatography on a continuous supermacroporous adsorbent. *J. Chromatogr. A* 986: 275–290.

Arvidsson, P., F.M. Plieva, I.N. Savina, V.I. Lozinsky, S. Fexby, L. Bülow, I. Yu. Galaev, B. Mattiasson. 2002. Chromatography of microbial cells using a continuous supermacroporous affinity and ion-exchange column. *J. Chromatogr. A* 977: 27–38.

Aslıyüce, S., L. Uzun, A.Y. Rad, S. Ünal, R. Say, A. Denizli. 2012. Molecular imprinting based composite cryogel membranes for purification of anti-hepatitis B surface antibody by fast protein liquid chromatography. *J. Chromatogr. B* 889–890: 95–102.

Aslıyüce, S., L. Uzun, R. Say, A. Denizli. 2013. Immunoglobulin G recognition with Fab fragments imprinted monolithic cryogels: Evaluation of the effects of metal-ion assisted-coordination of template molecule. *React. Funct. Polym.* 73: 813–820.

Avcibaşi, N., M. Uygun, M.E. Çorman, S. Akgöl, A. Denizli. 2010. Application of super-macroporous monolithic hydrophobic cryogel in capturing of albumin. *Appl. Biochem. Biotechnol.* 162: 2232–2243.

Babaç, C., H. Yavuz, I.Y. Galaev, E. Pişkin, A. Denizli. 2006. Binding of antibodies to concanavalin A modified monolithic cryogel. *React. Funct. Polym.* 66: 1263–1271.

Bakhspour, M., N. Bereli, S. Şenel. 2014. Preparation and characterization of thiophilic cryogels with 2-mercaptoethanol as the ligand for IgG purification. *Colloids Surf. B* 113: 261–268.

Bereli, N., M. Andac, G. Baydemir, R. Say, I.Y. Galaev, A. Denizli. 2008. Protein recognition via ion coordinated molecularly imprinted supermacroporous cryogels. *J. Chromatogr. A* 1190: 18–26.

Bereli, N., G. Ertürk, A. Denizli. 2012. Histidine containing macroporous affinity cryogels for immunoglobulin G purification. *Sep. Sci. Technol.* 47: 1813–1820.

Bereli, N., G. Şener, E.B. Altintaş, H. Yavuz, A. Denizli. 2010. Poly(glycidyl methacrylate) beads embedded cryogels for pseudo-specific affinity depletion of albumin and immunoglobulin G. *Mater. Sci. Eng.* 30: 323–329.

Bhattacharyya, R., R.P. Saha, U. Samana, P. Chakrabarti. 2003. Geometry of interaction of the histidine ring with other planar and basic residues. *J. Proteome Res.* 2: 255–263.

Bibi, N.S., N.K. Singh, R.N. Dsouza, M. Aasim, M. Fernandez-Lahore. 2013. Synthesis and performance of megaporous immobilized metal-ion affinity cryogels for recombinant protein capture and purification. *J. Chromatogr. A* 1272: 145–149.

Cao, X., R. Eisenthal, J. Hubble. 2002. Detachment strategies for affinity adsorbed-cells. *Enzyme Microb. Technol.* 31: 153–160.

Carman, W.F. 1997. The clinical significance of surface antigen variants of hepatitis B virus. *J. Viral Hepat.* 4: 11–20.

Carter, P.J. 2006. Potent antibody therapeutics by design. *Nat. Rev. Immunol.* 6: 343–357.

Çimen, D., A. Denizli. 2012. Immobilized metal affinity monolithic cryogels for cytochrome c purification. *Colloids Surf. B* 93: 29–35.

Cuatrecasas, P., M. Wilchek, C.B. Anfinsen. 1968. Selective enzyme purification by affinity chromatography. *Proc. Natl. Acad. Sci. USA* 61: 636–643.

Dainiak, M.B., A. Kumar, I.Y. Galaev, B. Mattiasson. 2006. Detachment of affinity-captured bioparticles by deformation of a macroporous hydrogel. *Proc. Natl. Acad. Sci. USA* 103: 849–854.

Dainiak, M.B., F.M. Plieva, I.Y. Galaev, R. Hatti-Kaul, B. Mattiasson. 2005. Cell chromatography: Separation of different microbial cells using IMAC supermacroporous monolithic columns. *Biotechnol. Prog.* 21: 644–649.

Das, S., S. Banerjee, J. Dasgupta. 1992. Experimental evaluation of preventive and therapeutic potentials of lysozyme. *Chemotherapy* 38: 350–357.

Demiryas, N., N. Tuzmen, I.Y. Galaev, E. Piskin, A. Denizli. 2007. Poly(acrylamide-allyl glycidyl ether) cryogel as a stationary phase in dye affinity chromatography. *J. Appl. Polym. Sci.* 105: 1808–1816.

Denizli, A. 2011. Purification of antibodies by affinity chromatography. *Hacettepe J. Biol. Chem.* 39: 1–18.

Denizli, A., E. Pişkin. 2001. Dye-ligand affinity systems. *J. Biochem. Biophys. Methods* 49: 391–416.

Derazshamshir, A., G. Baydemir, M. Andaç, R. Say, I.Y. Galaev, A. Denizli. 2010. Molecularly imprinted PHEMA-based cryogel for depletion of hemoglobin from human blood. *Macromol. Chem. Phys.* 211: 657–668.

Dragan, E.S, M.V. Dinu. 2013. Design, synthesis and interaction with Cu^{2+} ions of ice templated composite hydrogels. *Res. J. Chem. Environ.* 17: 4–10.

Emir, S., R. Say, H. Yavuz, A. Denizli. 2004. A new metal chelate affinity adsorbent for cytochrome c. *Biotechnol. Prog.* 20: 223–228.

Engler, A.J., L. Richert, J.Y. Wong, C. Picart, D.E. Discher. 2004. Surface probe measurements of the elasticity of sectioned tissue, thin gels and polyelectrolyte multilayer films: Correlations between substrate stiffness and cell adhesion. *Surf. Sci.* 570: 142–154.

Füglistaller, P. 1989. Comparison of immunoglobulin binding capacities and ligand leakage using eight different protein A affinity matrices. *J. Immunol. Methods* 124: 171–177.

Gagnon, P. 2012. Technology trends in antibody purification. *J. Chromatogr. A* 1221: 57–70.

Gagnon, P. 2013. Emerging challenges to protein A: Chromatin-directed clarification enables new purification options. *Bioprocess Int.* 11: 44–52.

Galaev, I.Y., M.B. Dainiak, F.M. Plieva, R. Hatti-Kaul, B. Mattiasson. 2005. High throughput processing of particulate-containing samples using supermacroporous elastic monoliths in microtiter (multiwall) plate format. *J. Chromatogr. A* 1065: 169–175.

Ghosh, R., S.S. Silva, Z.F. Cui. 2000. Lysozyme separation by hollow fibre ultrafiltration. *Biochem. Eng. J.* 6: 19–24.

Gomez-Suarez, C., H.J. Busscher, H.C. van der Mei. 2001. Analysis of bacterial detachment from substrate surfaces by the passage of air-liquid interface. *Appl. Environ. Microbiol.* 67: 2531–2537.

Hajizadeh, S., H. Kirsebom, A. Leistner, B. Mattiasson. 2012. Composite cryogel with immobilized concanavalin A for affinity chromatography of glycoproteins. *J. Sep. Sci.* 35: 2978–2985.

Hajizadeh, S., C. Xu, H. Kirsebom, L. Ye, B. Mattiasson. 2013. Cryogelation of molecularly imprinted nanoparticles: A macroporous structure as affinity chromatography column for removal of b-blockers from complex samples. *J. Chromatogr. A* 1274: 6–12.

Hari, P.R., W. Paul, C.P. Sharma. 2000. Adsorption of human IgG on Cu(II)-immobilized cellulose affinity membrane: Preliminary study. *J. Biomed. Mater. Res.* 50: 110–113.

Heftman, E., ed. 2004. *Chromatography: Fundamentals and Applications of Chromatography and Related Differential Migration Methods, Part B*, 6th edn. Elsevier, Amsterdam, the Netherlands.

Hsu, H.Y., M.H. Chang, S.H. Liaw, Y.H. Ni, H.L. Chen. 1999. Changes of hepatitis B surface antigen variants in carrier children before and after universal vaccination. *Hepatology* 30: 1312–1317.

Hutchens, T.W., J. Porath. 1986. Thiophilic adsorption of immunoglobulins-analysis of conditions optimal for selective immobilization and purification. *Anal. Biochem.* 159: 217–226.

Kumar, A., F.M. Plieva, I.Y. Galaev, B. Mattiasson. 2003. Affinity fractionation of lymphocytes using a monolithic cryogel. *J. Immunol. Methods* 283: 185–194.

Labib, M., M. Hedström, M. Amin, B. Mattiasson. 2009. A multipurpose capacitive biosensor for assay and quality control of human IgG. *Biotechnol. Bioeng.* 104: 312–320.

Langone, J.J. 1982. Applications of immobilized protein A in immunochemical techniques. *J. Immunol. Methods* 55: 277–296.

Le Noir, M., F. Plieva, T. Hey, B. Guieyse, B. Mattiasson. 2007. Macroporous molecularly imprinted polymer/cryogel composite systems for the removal of endocrine disrupting trace contaminants. *J. Chromatogr. A* 1154: 158–164.

Low, D., R. O'Leary, N.S. Pujar. 2007. Future of antibody purification. *J. Chromatogr. B* 848: 48–63.

Lozinsky, V.I., F.M. Plieva, I.Y. Galaev, B. Mattiasson. 2001. The potential of polymeric cryogels in bioseparation. *Bioseparation* 10: 163–188.

Madigan, M.T., J.M. Martinko, J. Parker. 2000. *Brock Biology of Microorganisms*. Prentice-Hall, Upper Saddle River, NJ.

Mammen, M., S.K. Choi, G.M. Whitesides. 1998. Polyvalent interactions in biological systems: Implications for design and use of multivalent ligands and inhibitors. *Angew. Chem. Int. Ed.* 37: 2754–2794.

Matejtschuk, P., ed. 1997. *Affinity Separations: A Practical Approach.* IRL Press, Oxford, U.K.

McCoy, M., K. Kalghatgi, F.E. Regnier, N. Afeyan. 1996. Perfusion chromatography—Characterization of column packings for chromatography. *J. Chromatogr. A* 743: 221–229.

Ming, F., W.J.D. Whish, J. Hubble. 1998. Estimation of parameters for cell-surface interactions-maximum binding force and detachment constant. *Enzyme Microb. Technol.* 22: 94–99.

Noppe, W., F.M. Plieva, K. Vanhoorelbeke, H. Deckmyn, M. Tuncel, A. Tuncel, I.Y. Galaev, B. Mattiasson. 2007. Macroporous monolithic gels, cryogels, with immobilized phages from phage-display library as a new platform for fast development of affinity adsorbent capable of target capture from crude feeds. *J. Biotechnol.* 131: 293–299.

Odabaşı, M., G. Baydemir, M. Karataş, A. Derazshamshir. 2011. Preparation and characterization of metal-chelated poly(HEMA-MAH) monolithic cryogels and their use for DNA adsorption. *J. Appl. Polym. Sci.* 116: 1306–1312.

Pelta, J., H. Berry, G.C. Fadda, E. Pauthe, D. Lairez. 2000. Statistical conformation of human plasma fibronectin. *Biochemistry* 39: 5146–5154.

Perçin, I., E. Aksoz, A. Denizli. 2013. Gelatin-immobilised poly(hydroxyethyl methacrylate) cryogel for affinity purification of fibronectin. *Appl. Biochem. Biotechnol.* 171: 352–365.

Perçin, I., E. Sağlar, H. Yavuz, E. Aksöz, A. Denizli. 2011. Poly(hydroxyethyl methacrylate) based affinity cryogel for plasmid DNA purification. *Int. J. Biol. Macromol.* 48: 577–582.

Perçin, I., H. Yavuz, E. Aksöz, A. Denizli. 2012. Mannose-specific lectin isolation from *Canavalia ensiformis* seeds by PHEMA-based cryogel. *Biotechnol. Prog.* 28: 756–761.

Reichert, J.M., C.J. Rosesweig, L.B. Faden, M.C. Dewitz. 2005. Monoclonal antibody successes in the clinic. *Nat. Biotechnol.* 23: 1073–1078.

Rosa, P.A.J., I.F. Ferreira, A.M. Azevedo, M.R. Aires-Barros. 2010. Aqueous two-phase systems: A viable-platform in the manufacturing of biopharmaceuticals. *J. Chromatogr. A* 1217: 2296–2305.

Sato, A.K., D.J. Sexton, L.A. Morganelli, E.H. Cohen, Q.L. Wu, G.P. Conley, Z. Streltsova et al. 2002. Development of mammalian serum albumin purification media by peptide phage display. *Biotechnol. Prog.* 18: 182–192.

Shibayama, M., T. Tanaka. 1993. Volume phase transition and related phenomena of polymer gels. In *Responsive Gels: Volume Transitions I*, Dusek, K. (ed.). Springer, Berlin, Germany, pp. 1–62.

Shukla, A.A., J. Thömmes. 2010. Recent advances in large-scale production of monoclonal antibodies and related proteins. *Trends Biotechnol.* 28: 253–261.

Sousa, C., A. Cebolla, V. DeLorenzo. 1996. Enhanced metalloadsorption of bacterial cells displaying poly-His peptides. *Nat. Biotechnol.* 14: 1017–1020.

Sun, S., Y. Tang, Q. Fu, X. Liu, M. Guo, Y. Zhao, C. Chang. 2012a. Monolithic cryogels made of agarose chitosan composite and loaded with agarose beads for purification of immunoglobulin G. *Int. J. Biol. Macromol.* 50: 1002–1007.

Sun, S., Y. Yang, Q. Fu, X. Liu, W. Du, K. Guo, Y. Zhao. 2012b. Preparation of agarose/chitosan composite cryogels for affinity purification of glycoproteins. *J. Sep. Sci.* 35: 893–900.

Takeuchi, T., T. Hishiya. 2008. Molecular imprinting of proteins emerging as a tool for protein recognition. *Org. Biomol. Chem.* 6: 2459–2467.

Tamahkar, E., N. Bereli, R. Say, A. Denizli. 2011. Molecularly imprinted supermacroporous cryogels for cytochrome c recognition. *J. Sep. Sci.* 34: 3433–3440.

Tekiner, P., I. Perçin, B. Ergün, H. Yavuz, E. Aksöz. 2012. Purification of urease from jack bean with copper(II) chelated PHEMAH cryogels. *J. Mol. Recognit.* 25: 549–554.

Tirumalai, R.S., K.C. Chan, D.A. Prieto, H.J. Issaq, T.P. Conrads, T.D. Veenstra. 2003. Characterization of the low molecular weight human serum proteome. *Mol. Cell. Proteomics* 2: 1096–1103.

Urthaler, J., R. Schlegl, A. Podgornik, A. Strancar, A. Jungbauer, R. Necina. 2005. Application of monoliths for plasmid DNA purification, development and transfer to production. *J. Chromatogr. A* 1065: 93–106.

Uygun, D.A., B. Akduman, M. Uygun, S. Akgöl, A. Denizli. 2012. Purification of papain using Reactive Green 5 attached supermacroporous monolithic cryogel. *Appl. Biochem. Biotechnol.* 167: 552–563.

Uygun, D.A., M.E. Çorman, N. Öztürk, S. Akgöl, A. Denizli. 2010. Poly(hydroxyethyl methacrylate-methacryloyl amino tryptophan) nanospheres and their utilization as affinity adsorbents for lipase adsorption. *Mater. Sci. Eng. C* 30: 1285–1290.

Üzek, R., L. Uzun, S. Şenel, A. Denizli. 2013. Nanospines incorporation into the structure of the hydrophobic cryogels via novel cryogelation method: An alternative sorbent for plasmid DNA purification. *Colloids Surf. B* 102: 243–250.

Vijayalakshmi, M.A. 1989. Pseudobiospecific ligand affinity chromatography. *Trends Biotechnol.* 7: 71–76.

Vijayalakshmi, M.A. 1996. Histidine ligand affinity chromatography. *Mol. Biotechnol.* 6: 347–357.

Wang, C., X.Y. Dong, Z. Jiang, Y. Sun. 2013. Enhanced adsorption capacity of cryogel bed by incorporating polymeric resin particles. *J. Chromatogr. A* 1272: 20–25.

Wang, Y.Y., P. Cheng, D.W. Chan. 2003. A simple affinity spin tube filter method for removing high-abundant common proteins on enriching low-abundant biomarkers for serum proteomic analysis. *Proteomics* 3: 243–248.

Wilchek, M. 2004. My life with affinity. *Protein Sci.* 13: 3066–3070.

Wong, J.Y., J.B. Leach, X.Q. Brown. 2004. Balance of chemistry, topography, and mechanics at the cell-biomaterial interface: Issues and challenges for assessing the role of substrate mechanics on cell response. *Surf. Sci.* 570: 119–133.

Yao, K., S. Shen, J. Yun, L. Wang, F. Chen, X. Yu. 2007. Protein adsorption in supermacroporous cryogels with embedded nanoparticles. *Biochem. Eng. J.* 36: 139–146.

Yao, K., J. Yun, S. Shen, L. Wang, X. He, X. Yu. 2006. Characterization of a novel continuous supermacroporous monolithic cryogel embedded with nanoparticles for protein chromatography. *J. Chromatogr. A* 1109: 103–110.

Yilmaz, F., N. Bereli, H. Yavuz, A. Denizli. 2009. Supermacroporous hydrophobic affinity cryogels for protein chromatography. *Biochem. Eng. J.* 43: 272–279.

Yun, J., J.T. Dafoe, E. Peterson, L. Xu, S.J. Yao, A.J. Daugulis. 2013. Rapid freezing cryo-polymerization and microchannel liquid-flow focusing for cryogel beads: Adsorbent preparation and characterization of supermacroporous bead-packed bed. *J. Chromatogr. A* 1284: 148–154.

3 Particulate/Cell Separations Using Macroporous Monolithic Matrices

Akshay Srivastava, Akhilesh Kumar Shakya, and Ashok Kumar

CONTENTS

3.1   INTRODUCTION

Particle-based stationary phases have been used in chromatographic techniques such as high-pressure liquid chromatography (HPLC), anion-exchange chromatography, and capillary liquid chromatography (CLC). Despite high stability and load-bearing capacity, a large void volume between the particles and high mass transfer resistance in flow of large macromolecules makes them inappropriate in separation processes (Bisjak et al. 2005, Podgornik and Strancar 2005, Unger et al. 2008). These limitations can be overcome by the use of monolithic stationary phases based on silica or polymers. Polymeric monoliths are generally made of synthetic, natural organic polymers, or a combination of natural and synthetic polymers. Polymer macroporous monoliths can be easily prepared by *in situ* polymerization of monomers with suitable cross-linkers directly in chromatographic units. The porosity in these monoliths can be obtained by the use of porogens and precise control of cross-linker or other reaction conditions. The liquid is forced to flow through these monolith pores with minimum mass transfer resistance. The convective flow between these pores enables fast separation of large molecules unlike high mass transfer resistance in particulate stationary phases (Holdsvendova et al. 2007). Ease in synthesis, fast separation, and efficient performance of these monolith matrices make them better than conventional stationary phases. Moreover, surface modifications and coupling of ligands are other advantages over conventional matrices.

3.1.1 MONOLITHS

Monoliths have shown their importance in bioseparation and their advantages have been compared with conventional chromatography procedures. Monoliths are more advantageous because of their good mass transfer, easy scale up, low back pressure, and high flow rates than conventional columns used in bioseparation. Such macroporous matrices have their importance in the separation of large bioparticles such as plasmid DNA, viruses, bacteria, and large mammalian cells. Industrially, monoliths are mainly manufactured for analytical applications but some larger prototype monoliths are commercially available up to 8 L of capacity (Jungbauer and Hahn 2004). This chapter focuses on the progress in macroporous monoliths for the separation of particulate and cell separation. It also appears that the scientific community has a keen interest in understanding the separation potential of these matrices. Theoretically, monolithic column is composed of a single continuous structure that fills the entire column without any void. Interconnected pores form a continuous skeleton that ultimately forms flow channels across the length of the matrix (Miller 2004). Such porous network channels in the monolithic matrices provide high axial permeability and large internal pore surface area that leads to less back pressure compared to conventional separation columns. The channels also allow better interaction between the analyte and the binding sites of the stationary phase (Miller 2004, Svec and Huber 2006, Vlakh and Tennikova 2009). Additionally, in particle-based columns the pores are partially used and cause major limitation to diffusion, whereas the interphase mass transfer is governed by convection in monoliths and the total pore volume is used. There are various macroporous monolithic columns developed and are briefly discussed later.

Generally, silica-based monoliths are promising monoliths in separation and analysis of proteins or peptides. These monoliths have mesopores for efficient separation of biomolecules. The small silica monolith columns with high pores have shown higher permeability than conventional columns. These columns enable high speed separation of biomolecules with high efficiency. Recently, new monolithic silica columns were developed with bimodal pores with diameter of around 2 μm. These macropores enhance rapid flow and transportation of analytes on active silica gel under low pressures. The large surface area inside the pores facilitates fast adsorption and desorption of analytes. These columns have significantly high porosity in comparison to conventional particle-based columns. Due to high surface area, low back pressure, fast and efficient separation of analytes can also be made possible by connecting many silica monoliths in series (Josic and Clifton 2007). Silica-based monolithic columns are prepared from alkoxysilane. These monoliths have relatively large flow-through pores that allow higher permeability than the conventional particle-packed columns, which are advantageous for rapid separation at a high flow rate. Moreover, high-efficiency separation can be achieved by connecting it to octadecylsilylated monolithic silica columns (Minakuchi et al. 1996). Large porous monolithic silica columns were prepared with unique characteristics and high-efficiency chromatographic performance (Motokawa et al. 2002, Miyamoto et al. 2008). Monolithic columns prepared from an alkoxysilane and polymer produce hierarchical macro- and mesoporosity that has potential for high-efficiency separations (Minakuchi et al. 1996, 1998, Tanaka et al. 2001, Ikegami 2004).

Metal oxide-based monoliths have also been prepared for different applications (Smått et al. 2006, 2012). Generally, metal oxide monoliths are composed of different oxides of metals such as aluminum, zirconium, and titanium. Surface properties of these metallic oxides can easily modify according to the need of application through different chemical reactions. Unlike silica monoliths, covalent attachment of different chemicals or functional groups is not easy with these monoliths, while coating of polymer can be easily done with oxide monoliths. Moreover, the surface of metal oxides can be easily modified with Lewis bases. These monoliths are mechanically stable and can withstand harsh conditions of temperature, pressure, and strong acids and bases (Nawrocki et al. 2004a,b). Synthesis of aluminum, zirconium, and hafnium metal oxide monoliths were reported by Hoth et al. (2005). In synthesis, first silica tubes with inner diameter of 50 μm were filled by a solution of hafnium chloride, propylene oxide, and methylformamide. These compounds condense inside silica tubes and form the monoliths. Microscopy study revealed the micron-size pores with interconnectivity in the monolith. A different approach was adopted in the synthesis of zirconium monoliths (Hoth et al. 2005).

Carbon-based monoliths are other monoliths that were recently prepared and demonstrated in proteomics. These monoliths exhibit hierarchical interconnected porosity (Liang et al. 2003, Taguchi et al. 2003). In the synthesis of these monoliths, first, silica beads were socked in carbon precursor and then pyrolyzed at high temperature. Later, silica particles were removed by using hydrofluoric acid that leaves porous interconnected monolith.

Monolithic polymers were first introduced by Svec and Fréchet (1992) and seen as a new monolith technology for separation science. The best advantages provided are rigid structure having high permeability and sufficiently large pores for high liquid flow rates without causing column back pressure. Polymer-based monoliths provide unique characteristics largely dependent on the type of polymer, cross-linker, porogen, and synthesis conditions. Polymer monoliths have many advantages over other conventional stationary phases and they can be synthesized in a single step by copolymerization of reactive monomers with appropriate cross-linkers in the presence of porogens (Bisjak et al. 2005, Lin et al. 2012). Good porosity and surface chemistry are major factors to achieve efficient separation of biomolecules. Alteration in porosity depends on the type and amount of porogens used in the reaction, which affect the salvation of polymers at the beginning of reaction, while other parameters like monomers, cross-linker, and the type of initiator affect porosity and surface chemistry of monoliths. Like the monomers, cross-linkers become the part of monoliths after polymerization, while porogens can be washed out after the reaction. Thus, the properties of monoliths depend on the chemistry of monomers and cross-linkers. Despite these factors, reaction conditions such as temperature and pH also affect monolith properties. Monomers such as silica (Minakuchi et al. 1996), acrylamide (AAm) (Freitag and Vogt 2000), acrylates (Luo et al. 2002), styrene (Wang et al. 1993), and their variants have been widely used in the synthesis of CLC. Methacrylates with different alkyl substituents were reported to affect the surface properties of monoliths. In synthesis of these monoliths, ethylene glycol dimethacrylate (EGDMA) is a popular cross-linker. The possibility of chemical modification, biofunctionalization, and grafting on polymer monoliths are other advantages of polymeric monoliths. Functionalization of pore surface with polymeric nanoparticles is a new approach to make monolith with specific chemical property and nanostructuring. For example, Xu and colleagues have recently used this concept in the synthesis of polymer monolith containing gold nanoparticles. This monolith enables the separation of thiol-containing peptides from the mixture of different peptides (Xu et al. 2010). Due to their porous properties, these polymeric materials have been extensively investigated for different chromatographic separations (Nordborg et al. 2011, Xu and Oleschuk 2013), sample pretreatment (Lin et al. 2011), microfluidic devices (Satterfield et al. 2007), immobilized enzyme reactors (Ma et al. 2007), and solid-phase microextraction (Xu et al. 2011, Tong et al. 2012).

3.1.2 CRYOGELS

Cryogels are chemically/physically cross-linked hydrogels prepared at subzero temperature. The cryogenic treatment (freezing, storage in the frozen state for a definite time, and defrosting) of low or high molecular weight polymeric precursors capable of gelation and/or polymerization can produce cryogels. Cryotropic gelation (or cryostructuration) takes place via cryogenic treatment of the systems containing aqueous solvent and potential monomer or polymeric precursors capable of polymerization or gelation, respectively. The crystallization of the solvent is the essential feature of cryogelation, which distinguishes cryogelation from chilling-induced gelation when the gelation takes place on decreasing temperature (e.g., gelation of agarose or gelatin solutions on cooling, which proceeds without any phase transition of the solvent). Production of cryogels without the use of organic solvents, effective control over the pore size, and architecture make this approach favorable compared to other techniques such as solvent casting/particulate leaching or phase separation combined with freeze-drying, which are used for the production of macroporous scaffolds.

The unique features of cryogelation processes (Figure 3.1) are as follows: (1) Cryotropic polymerization/gelation that proceeds in a nonfrozen liquid microphase (NFLMP) existing in a macroscopically frozen sample. At moderately low temperatures below the freezing point of solvent (often water), some part of the solvent remains nonfrozen. All soluble solutes are concentrated in this nonfrozen part (so-called NFLMP) where the chemical reaction or processes of physical gelation proceed with time. (2) Crystals of the frozen solvent grow until they merge and after melting leave behind the interconnected pores, thus playing a role of porogen. As the size and alignment of solvent crystals are varied to a large extent, the pores in the prepared cryogels can be in the range of 1–200 μm. (3) Typically, the critical

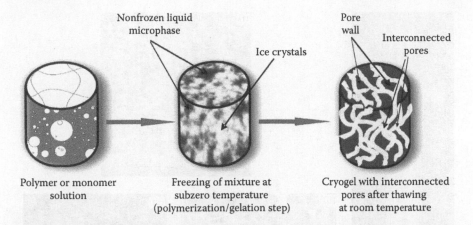

Nonfrozen liquid microphase

Ice crystals

Pore wall

Interconnected pores

Polymer or monomer solution

Freezing of mixture at subzero temperature (polymerization/gelation step)

Cryogel with interconnected pores after thawing at room temperature

FIGURE 3.1 Cryogelation process: polymer or monomer mixed in aqueous solvent and then the whole system incubated at subzero temperature for cryopolymerization and/or gelation along with the ice crystal formation that acts as a porogen. After complete gelation, the ice crystals melt and leave behind large interconnected pores when incubated at room temperature and washed with water to remove unreacted monomers or polymeric precursors. (From Kumar, A. and Srivastava, A., *Nat. Protoc.*, 5, 1737, 2010.)

concentration of gelation (CCG) is decreased for cryogels, as compared to the conventional gels prepared at room temperature, due to the concentration of reagents in NFLMP (so-called cryoconcentration).

The shape and size of the crystals formed and surface tension of solvent along with the pore wall determine the shape (mainly circular) and size of the pores formed after defrosting the sample. In general, the size of ice crystals depends on how fast the system is frozen, provided other parameters (e.g., concentration of the dissolved substances, volume, and geometrical shape of the sample) remain the same (Lozinsky 2002). The pore size depends on the initial concentration of precursors in a solution, their physicochemical properties, and the freezing conditions (Plieva et al. 2006, 2007). Contrary to conventional gels (which are homophase systems where solvent is bound to the polymer network), the cryogels are heterophase systems where solvent (water) is both present inside the interconnected pores and also bound to the polymer network. Depending on the gel precursors and chemical reaction used, the micro- and macroporous structure of cryogels can be varied to a large extent.

The cryogels can be prepared in different shapes and sizes. The most common format of cryogel is monolith for cell chromatography applications (Figure 3.2a).

The porous nature of cryogel was well demonstrated by its high flow rate and scanning electron microscopic images (Figure 3.2b, i and ii). Also, micro-CT analysis clearly demonstrates that the cryogel material is a highly porous system and the pores are interconnected (Figure 3.2b, iii) (Plieva et al. 2005). In particular, it has been demonstrated that cryogels are highly suitable as matrices for cell separation (Lozinsky et al. 2003, Kumar et al. 2005). Cryogels derivatized with affinity ligands have been developed for chromatographic cell separation. They could be prepared in different formats, such as monolithic columns, sheets, and beads,

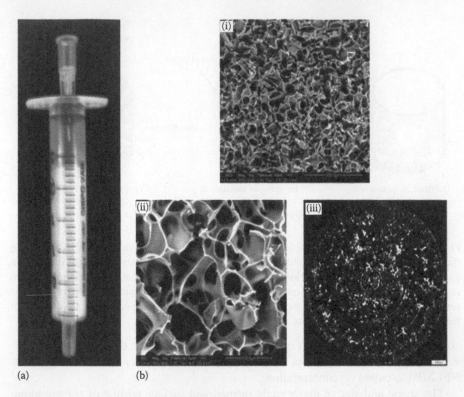

(a)   (b)

FIGURE 3.2 Physical characteristics of cryogels. (a) Cryogel monolith chromatography column, (b) (i, ii) scanning electron microscopy images of AAm cryogel and (iii) MicroCT image of cryogel showing interconnected pores. (Adapted and modified from Kumar, A. and Srivastava, A., *Nat. Protoc.*, 5, 1737, 2010.)

from any hydrophilic or hydrophobic polymer that can form a gel. These continuous macroporous gels offer advantages for cell separation by providing solutions to some of the major limitations of cell chromatography. The gel phase that is the polymer with tightly bound water comprises only 10% of the total cryogel volume, and most of the monolithic column (90%) is an interconnected system of supermacropores filled with water. Compared to the bead-packed columns, in cryogel monoliths, pores are uniformly distributed throughout the whole column volume. Cryogels are particularly designed for the chromatography of biological particles and are characterized by large (up to 200 μm), highly interconnected pores with 90% average porosity, high elasticity, convective flow, and good ligand immobilization properties. All these properties provide the cryogels with appropriate matrix characteristics for cell separations. These monolithic matrices have shown a lot of promise in their application to the separation of proteins, oligonucleotides, plasmids, viruses, nanoparticles, cell organelles, and whole cells (Dainiak et al. 2004, 2007a, 2008, Ahlqvist et al. 2006). Due to the better mass transfer and open structure, the chromatography columns based on continuous beds (monolithic supports) provide high purity and rapid separation of large molecules. The most important feature of these chromatographic columns is that they are independent of the flow rates, thus rapid mass transfer as well as high efficiency can be obtained even at high flow rates. Also, separations can be fast, which increases the productivity of chromatographic processes as compared to traditional-packed bed chromatographic columns (Annaka et al. 2003, Arvidsson et al. 2003, Kumar et al. 2005, Dainiak et al. 2007a). Another important advantage of the cryogel system is that it provides a closed system that can be operated aseptically for biological applications. Preparative cell separation technique is an important tool to provide a highly purified population of specific cell type for diagnostic, biotechnological, and biomedical applications. Importance of the area of cell separation can be judged by its application for human health care and medical product development (Kumar et al. 2007).

3.2 SEPARATION OF NUCLEIC ACIDS

Monoliths have enormous importance in molecular biology for the isolation of nucleic acids and nucleotides/oligonucleotides (Bisjak et al. 2005) (Table 3.1).

From the past few decades, oligonucleotides have imperative importance in molecular biology. As primers and probes, unmodified oligonucleotides have been used in different molecular biology techniques such as polymer chain reaction (PCR) and Western blotting (Padmapriya et al. 1994, Lesignoli et al. 2001), while modified

TABLE 3.1
Monolith Columns for Separation of Nucleic Acids

Monolith Matrices	Nucleic Acids/Nucleotides	Mode of Separation
Polyhydroxymethyl methacrylate (Holdsvendova et al. 2007)	Oligonucleotides	Hydrophilic interaction capillary liquid chromatography
CIM-DEAE (Forcic et al. 2005)	Genomic DNA	Electrostatic interactions
CIM (Strancar et al. 2002)	Plasmid DNA	Anion-exchange chromatography
CIM-DEAE (Perica et al. 2009)	Viral double-stranded RNA	Anion-exchange chromatography
Polyhydroxyethyl methacrylate-co-vinylphenylboronate p(HEMA-co-VPBA) (Srivastava et al. 2012)	Yeast and bacterial RNA	Affinity interaction
Porous polymer monolith (Chatterjee et al. 2010)	Mammalian cell RNA	Solid phase extraction system
Polycation grafted monoliths (Hanora et al. 2006)	Plasmid DNA	Electrostatic interactions
DEAE monolith (Krajacic et al. 2007)	Plant RNA	Electrostatic interactions
Microchip based on silica monolith (Wu et al. 2006)	Human genomic DNA	Micro solid phase extraction
Methacrylate-based porous polymer monolith (Satterfield et al. 2007)	Messenger RNA	Microfludic purification

Note: CIM, commercially available monolith based on polymethacrylate.

oligonucleotides can be used for therapeutic purposes (Crooke 1998, Malvy et al. 1999). Therefore, it is extremely important to study and analysis of purified oligonucleotides. Majorly two techniques, such as electrophoresis and HPLC, can be used frequently for the separation of oligonucleotides from their impurities. HPLC is an appropriate method for purification at large scale (Gilar and Bouvier 2000). Polymer monolith based on hydroxyethyl methacrylate was demonstrated for oligonucleotide separation by CLC (Holdsvendova et al. 2007). The ability of monolith columns has also been demonstrated for the separation and purification of plasmid DNA (Urthaler et al. 2005). Unlike proteins, plasmids have different physiological properties such as hydrodynamic diameters of plasmids are higher than the proteins, while diffusion is slower than the proteins (Zochling et al. 2004). Theoretically, on anion-exchange columns, plasmids can bind at multiple points with average 50 sites of the matrix in comparison to proteins that could bind with 3–10 sites only (Yamamoto et al. 2007). Polymer-based monoliths are efficient for plasmid binding and purification. For example, methacrylate-based convective interaction media (CIM) disk- and tube-based polymer monolith could bind 15 mg plasmid DNA and this capacity is similar to commercially available monolith columns. In this direction, a monolith matrix based on poly(3-diethylamino-2-hydroxypropylmethacrylate-divinylbenzene) has been synthesized for oligonucleotide separation through anion-exchange chromatography (Bisjak et al. 2005).

Monolith columns have also been developed for fast and efficient separation of genomic DNA. Chromatographic methods such as anion exchange (Teeters et al. 2003), affinity (Costioli et al. 2003), ion pair reversed phase (Oberacher et al. 2000), and size exclusion (Huber 1998) have been widely used in DNA purification. Since DNA is negatively charged due to phosphate groups, therefore it shows binding over positively charged matrix. Diethylaminoethanol (DEAE) and quaternary amines are widely applied in matrix modification for getting positive charge in the matrix (Huber 1998). Recently, CIM-based methacrylate polymer monoliths have been developed for DNA separation (Branovic et al. 2004). They can even separate different forms of DNA like open, linear, supercoiled, and plasmid DNA in short time. Some findings have also reported high binding capacity of plasmid DNA (8 mg/mL) over methacrylate monoliths (Strancar et al. 2002). In another study, a CIM-based DEAE monolith column has been developed for the separation of prokaryotic and eukaryotic DNA. Both types of DNAs are successfully bound to CIM column and specifically eluted through increasing salt concentrations. The quality and integrity of bound DNA was also best demonstrated by polymerase chain reactions (PCRs). Moreover, these columns are regenerative and can be active after each usage (Forcic et al. 2005).

Monoliths have been also developed for RNA extraction and purification. Conventionally, DEAE-based monoliths have been used for RNA purification (Krajacic et al. 2007), while the process required several steps to get the final purified product. In order to get efficient and fast purification of RNA, recently, cryogel-based monolith demonstrated RNA purification in single step (Srivastava et al. 2012). Polymer cryogel monolith based on poly(hydroxyethyl methacrylate-*co*-vinylphenylboronate) (poly(HEMA-*co*-VPBA)) has demonstrated for direct purification of RNA from cell lysate. Due to the presence of macropores, unhindered flow was observed by flowing cell lysate through the column. Based on affinity interaction, RNA molecules bind to the column from the mixture of RNA and DNA. Recently, microfludic technology has been tried for the extraction of RNA from microbes for diagnostic purposes. A micro solid-phase extraction method was designed for viral RNA extraction in plastic microfludic device (Bhattacharyya and Klapperich 2008).

3.3 SEPARATION OF LARGE PROTEINS

Both polymer- and silica-based monoliths have been explored in proteomics as a stationary phase in different forms of chromatography (Table 3.2).

Monoliths based on reversed-phase concept are one of the most commonly used materials in protein separation and their analysis. This type of support is available in capillary form and they can be made of either silica or polymers. Various silica- and polymer-based monoliths have been demonstrated for the separation of tryptic peptides before their analysis by mass spectrometry (MS) (Cabrera 2004, Luo et al. 2005). In this course, polystyrene-*co*-divinylbenzene monolith microcolumn was designed and compared with C18 silica microparticles column for peptide separation from mixtures of peptides. The monolith was more efficient and reproducible for peptide separation than microparticle-based columns (Toll et al. 2005). Later, Tanaka and colleagues designed silica-based reversed-phase micro- and nano-HPLC monoliths for the separation of tryptic peptides after peptide digestion (Minakuchi et al. 1996). In some aspects, silica-based monoliths have advantage over polymer monoliths in terms of available large surface area, which is around twofolds more than polymer monoliths (Leinweber and Tallarek 2003).

TABLE 3.2
Different Polymer Monoliths for Preparative Protein Separation

Monolith Composition	Purified Proteins	Purified Proteins
CIM (Strancar et al. 1997)	Blood coagulation factor VIII	Ion exchange
CIM (Brne et al. 2007)	IgM	Ion exchange
CIM (Ostryanina et al. 2002)	IgG	Protein A affinity
Poly(glycidyl methacrylate-*co*-ethylene dimethacrylate) (Luo et al. 2002)	IgG	Protein A affinity
PAAm (Freitag and Vogt 2000)	α-lactalbumin	Ion-exchange displacement
CIM (Isobe and Kawakami 2007)	Alcohol dehydrogenase, alcohol oxidase	Hydrophobic interaction
CIM (Peterka et al. 2006)	Tumor necrosis factor-α	Metal affinity
Poly(glycidyl methacrylate/ divinylbenzene)-iminodiacetic acid (Aprilita et al. 2005)	Phosphopeptides	Metal affinity
Cu(II)-IDA-polyacrylamide-*co*-allylglycidyl ether (Kumar et al. 2006)	Urokinase	Metal affinity

Note: CIM, commercially available monolith based on polymethacrylate.

Another mode of protein separation is based on ion-exchange principles utilizing the potential of monoliths (Li and Lee 2009) and such monolith was first based on compressed polyacrylamide (PAAm) (Liao et al. 1991). Due to low mechanical stability, this column cannot sustain for long term. Column was further improved and designed in the form of monolith with 10–25 μm capillaries for protein separation (Li et al. 1994). Despite the AAm-based monoliths, methacrylate-based ion-exchange CIM monolith discs have also been developed for fast separation of proteins. CIM columns efficiently fractionated serum and membrane proteins from organs such as the liver (Rucevic et al. 2006). The CIM columns have also demonstrated fast and efficient separation of IgM from the mixture of IgGs (Brne et al. 2007). Moreover, CIM monoliths were developed for selective separation of low-abundance membrane proteins from large biological mixtures (Rucevic et al. 2006). This separation was based on affinity interaction between protein A in CIM and membrane proteins.

The monoliths based on immobilized metal affinity have been extensively used in the proteomics for separation of His-tag containing proteins, phosphoproteins, and glycoproteins. Immobilized metal affinity chromatography (IMAC) monoliths are based on coordination chemistry where immobilized metal binds to the chelator of protein. For example, poly(glycidyl methacrylate/divinylbenzene)-iminodiacetic acid monolith immobilized with Fe^{+3} was successfully used for the separation of phosphopeptides (Aprilita et al. 2005), while IMAC with Cu^{+2} was designed for the separation and purification of His-tag containing proteins (Kumar et al. 2006). Diethylene glycol methacrylate and glycidyl methacrylate get polymerized and then functionalized with iminodiacetic acid (IDA) for the synthesis of IMAC monolith. After immobilization of copper ion, monolith can be used for binding of His-tag proteins/peptides. Since this interaction is temporary, it can break by the use of molecule that competitively binds to metal ions. Therefore, bound His-tag proteins can elute from the matrix in higher concentration of imidazole. Monoliths based on lectin and glycoprotein affinity have also been explored in proteomics. For example, different lectins immobilized to monoliths were employed for the purification of glycoproteins before analyzing them by MS (Qiu and Regnier 2005, Wuhrer et al. 2005). A chromatography with different immobilized lectins in series can offer a solution of fast high-throughput screening of protein glycosylation (Josic and Clifton 2007).

In recent trends, monoliths can be synthesized in the form of miniature columns for high-throughput screening of proteins. Recently, poly(lauryl methacrylate-co-ethylene dimethacrylate)- and poly(styrene-co-divinylbenzene)-based monoliths were designed in the form of microchips and demonstrated for the separation of a mixture of proteins such as ribonuclease A, myoglobulin, and cytochrome C (Levkin et al. 2008).

3.4   SEPARATION OF PROKARYOTIC CELLS

The separation of prokaryotic cells through monolithic columns is only possible with cryogel-based monoliths. Cryogel monoliths have recently been developed both in the column format (Kumar et al. 2003) and in 96-minicolumn plate format (Dainiak et al. 2007b) to be used for the application in cell chromatography.

The pore size range of 10–100 μm (and even larger) in cryogels makes it possible to use these matrices for the chromatography of cells having sizes up to 1–10 μm or above, without being mechanically entrapped in the column. Even cells as big as red blood cells (7 μm) are convectionally transported by the liquid flow through the monolithic plain cryogel column, without being mechanically trapped. Cells, while passing through the cryogel column, do not experience large shear forces because of the liquid flow to be laminar in the interconnected pores of 10–100 μm in the monolithic cryogel. Cells bind to the monolithic cryogel column when there is a possibility for cells to interact with some specific groups (charges, hydrophobic moieties, or affinity ligands) introduced at the surface of pores in the cryogel column. For example, *E. coli* cells were bound to an ion-exchange monolithic cryogel column at low ionic strength and were eluted with 70%–80% recovery at NaCl concentrations of 0.35–0.4 M, while the same cells could bind to a cryogel column bearing Cu(II)-loaded iminodiacetate (Cu(II)-IDA) ligands and were eluted with around 80% recovery using either 10 mM imidazole or 20 mM ethylenediamine tetraacetic acid (EDTA) (Arvidsson et al. 2002). One could expect different microbial cells to have different cell surface properties with different chemical groups exposed to the outer medium. Exploiting these differences allows for the separation of specific cells from the mixed population. This was demonstrated on two model systems where the mixtures of wild-type *E. coli* and recombinant *E. coli* cells displaying poly-His peptides (His-tagged *E. coli*). Wild-type *E. coli* and *Bacillus halodurans* cells were separated from each other (Dainiak et al. 2005). This approach was also tested when a different ligand concanavalin A (ConA) was immobilized on the cryogel monolith for the separation of yeast and *E. coli* cells (Dainiak et al. 2006a). A nearly base-line chromatographic separation was achieved under optimized conditions. Apart from microbial cells, other bioparticles like inclusion bodies (Ahlqvist et al. 2006), mitochondria (Teilum et al. 2006), and viruses (Williams et al. 2005) were specifically captured and separated using monolithic cryogel columns.

Beside the protein ligands, chemical ligands such as boronate compounds have also been utilized in affinity interaction for the separation of cells. In this interaction, boronate interacts with *cis*-diol groups containing compounds and involves ester formation between boronate and *cis*-diol group. This interaction is reversible and depends on the pH of the reaction medium. Phenylboronic acid is a well-known boronate compound that was explored in the form of a ligand in boronate affinity chromatography (Srivastava et al. 2012). Previously, boronate affinity matrices were developed for the separation of *cis*-diol containing compounds (Weith et al. 1970). Cryogel monolith has been developed for the separation of cells in 3D matrix. Cryogel monolith with boronate as a ligand has been demonstrated for the separation of yeast cells and fractionation of adherent and nonadherent cells (Srivastava et al. 2012). PAAm-based cryogel was synthesized at subzero temperature and then phenylboronate was chemically grafted on cryogel surface. Boronate-grafted PAAm cryogel is able to bind yeast cells at basic pH and captured cells were eluted by the use of 0.1 M fructose in phosphate buffer saline (PBS). More than 90% of cells were recovered from the column. This again verified the ability of cryogel monoliths for the separation of large molecules such as cells.

3.5   SEPARATION OF MAMMALIAN CELLS

As affinity chromatography of supermacroporous monolithic columns proved to be an efficient approach for the separation of individual microbial cell types, the concept was equally efficient and more promising for the separation of specific types of mammalian cells, namely, T and B lymphocytes and other cell types. Fractionation techniques that take advantage of cell surface antigens are more selective and efficient. The approach is based on the selective interaction of the antibodies to these cell surface antigens. Antibodies are highly suitable as ligands for cell separation owing to their great diversity and specificity. Separation methods for isolation of particular cell populations generally use antibodies against differentially expressed cell surface molecules as targets. The concept of affinity separation can be used either for negative selection or for positive selection of cells. In negative selection, unwanted cells adhere to the affinity ligand on the support matrix and the target cells pass through the matrix unretained. On the other hand, positive selection of target cells involves the binding of specific target cells to affinity adsorbent and, after washing, the bound cells are released from the matrix in a purified form.

However, this approach has been utilized here in a novel way so that affinity chromatographic matrix can be utilized for the separation of different cell types in a more generic way where the antibody to the surface antigen is available. A cryogel affinity adsorbent has been developed based on the interaction of immobilized protein A with cells bearing IgG antibodies on the surface. Protein A is a protein from *Staphylococcus aureus* that binds to the Fc portion of the IgG from a wide range of species. When covalently coupled to monolithic cryogel, it can be used as an efficient adsorbent for cells that have been coated with a specific antibody (IgG type) and can thus separate them from cells that lack the surface antigen against which the antibody is directed. Protein A was coupled to the cryogel matrix. After treating the cells with antibodies (IgG type) to the cell surface receptors, the cell mixture is loaded onto the cryogel–protein A column. The antibody-labeled cells bound to the protein A through Fc region, whereas nonlabeled cells remained unbound and passed through the column (Figure 3.3).

The generic cell separation protocol was developed using affinity cryogel matrix. CD34+ cells were separated in purified form directly from the umbilical cord blood. The mechanical elution procedure was successfully demonstrated and the result showed that the purified CD34+ cells were live and retained their phenotypic properties (Kumar and Srivastava 2010).

In another example, protein A covalently coupled to supermacroporous matrix specifically bound more than 90% of IgG-labeled B lymphocytes, while nonlabeled T lymphocytes passed through the column (Figure 3.4a). The bound lymphocytes were eluted with 60%–70% recoveries without significantly impairing the cell viability (Kumar et al. 2003). Similar results were obtained when studying CD34+ cells (Kumar et al. 2005). In another study, polyvinyl alcohol (PVA) cryogel beads and dimethylacrylamide (DMAAm) monolith were compared for their binding to human acute myeloid leukemia KG-1 cells, which is having CD34 cell surface marker. DMAAm monolith binds around 95% cells in comparison to 76% cell binding to

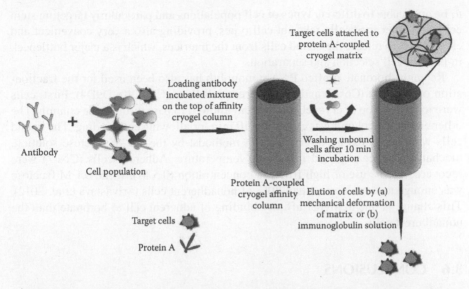

FIGURE 3.3 Schematic showing cell separation strategy used for the separation of stem cells from umbilical cord blood. Monoclonal antibody incubated with cell population to label specific cell type. The cell population containing labeled target cells was then passed through protein A-coupled epoxy-activated cryogel column. After 10 min of incubation, the affinity column was washed with buffer to remove unbound cells. (From Kumar, A. and Srivastava, A., *Nat. Protoc.* 5, 1737, 2010.)

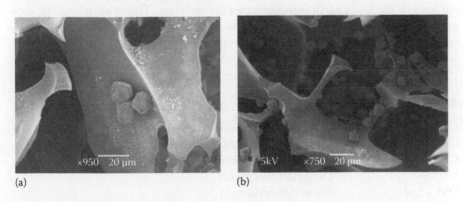

(a)   (b)

FIGURE 3.4 Scanning electron micrograph of (a) affinity-bound lymphocytes in the inner part of the supermacroporous cryogel–protein A and (b) affinity bound CD34+ human acute myeloid leukemia KG-1 cells inside the protein A-cryogel matrix. (a: From Kumar, A. et al., *J. Immunol. Methods*, 283, 185, 2003; b: From Kumar, A. et al., *J. Mol. Recognit.*, 18, 84, 2005.)

PVA beads (Figure 3.4b). Moreover, more than 90% cells were recovered from the monolith without affecting cell viability (Kumar et al. 2005). A breakthrough study showed that these cryogel matrices can be used for releasing the cells by squeezing the gels when the cells are affinity bound on such matrices (Dainiak et al. 2006b). These initial results have shown tremendous potential of such cell separation strategy to be applicable to different types of cell populations and particularly targeting stem cells and other medically relevant cell types, providing also a very convenient and elegant way to release the bound cells from the matrices, which is a major bottleneck in positive cell selection and separation.

Recently, boronate grafted PAAm monolith has also been used for the fractionation of adherent (Cos-7) and nonadherent cell lines (CC9C10, D9D4). First, cells were loaded on top of cryogel column and allowed to pass through the column. The other end of the column was closed for 10 min for allowing cell binding. The bound cells were recovered by three different methods: by the use of fructose solution, mechanical squeezing, and increasing temperature. Adherent cells (Cos-7) were recovered by the use of high fructose concentration (0.5 M), while 0.1 M fructose was enough to detach more than 80% of nonadherent cells (Srivastava et al. 2012). This change is due to higher affinity binding of adherent cell to boronate than the nonadherent cells.

3.6 CONCLUSIONS

Macroporous monolithic matrices as stationary phase represent a new generation type of column for rapid chromatographic analysis of cells and particulate matter. In contrast to conventional columns, mobile phase can flow easily through the monolith with convective flow and this flow can decrease mass transfer resistance in the column. Monoliths are mechanically stable and can withstand significant amount of pressure in chromatography. Moreover, they are compatible with different organic phases in harsh chromatographic conditions. Therefore, monoliths show great potential in type-specific separation of cells and particulate matter and can be used for purification and analysis of complex biomolecules and for integrated bioseparations.

ABBREVIATIONS

CCG	Critical concentration of gelation
CIM	Convective interaction media
CLC	Capillary liquid chromatography
DEAE	Diethylaminoethanol
DMAAm	Dimethylacrylamide
EDTA	Ethylenediamine tetraacetic acid
EGDMA	Ethyleneglycol dimethacrylate
HPLC	High pressure liquid chromatography
IDA	Iminodiacetic acid
IMAC	Immobilized metal affinity chromatography
MS	Mass spectrometry
NFLMP	Non-frozen liquid microphase
p(HEMA-co-VPBA)	Poly(hydroxyethyl methacrylate-co-vinylphenylboronate)
PAAm	Polyacrylamide
PVA	Polyvinyl alcohol

REFERENCES

Ahlqvist, J., A. Kumar, H. Sundstrom, E. Ledung, E.G. Hornsten, S.O. Enfors, and B. Mattiasson. 2006. Affinity binding of inclusion bodies on supermacroporous monolithic cryogels using labeling with specific antibodies. *J. Biotechnol.* 122: 216–225.

Annaka, M., T. Matsuura, M. Kasai, T. Nakahira, Y. Hara, and T. Okano. 2003. Preparation of comb-type *N*-isopropylacrylamide hydrogel beads and their application for size-selective separation media. *Biomacromolecules* 4: 395–403.

Aprilita, N.H., C.W. Huck, R. Bakry, I. Feuerstein, G. Stecher, S. Morandell, H.L. Huang, T. Stasyk, L.A. Huber, and G.K. Bonn. 2005. Poly(glycidyl methacrylate/divinylbenzene)-IDA-FeIII in phosphoproteomics. *J. Proteome Res.* 4: 2312–2319.

Arvidsson, P., F.M. Plieva, V.I. Lozinsky, I.Y. Galaev, and B. Mattiasson. 2003. Direct chromatographic capture of enzyme from crude homogenate using immobilized metal affinity chromatography on a continuous supermacroporous adsorbent. *J. Chromatogr. A* 986: 275–290.

Arvidsson, P., F.M. Plieva, I.N. Savina, V.I. Lozinsky, S. Fexby, L. Bulow, I.Y. Galaev, and B. Mattiasson. 2002. Chromatography of microbial cells using continuous supermacroporous affinity and ion-exchange columns. *J. Chromatogr. A* 977: 27–38.

Bhattacharyya, A. and C.M. Klapperich. 2008. Microfluidics-based extraction of viral RNA from infected mammalian cells for disposable molecular diagnostics. *Sens. Actuators B: Chem.* 129: 693–698.

Bisjak, C.P., R. Bakry, C.W. Huck, and G.K. Bonn. 2005. Amino-functionalized monolithic poly(glycidyl methacrylate-*co*-divinylbenzene) ion-exchange stationary phases for the separation of oligonucleotides. *Chromatographia* 62: s31–s36.

Branovic, K., D. Forcic, J. Ivancic, A. Strancar, M. Barut, T. Kosutic Gulija, R. Zgorelec, and R. Mazuran. 2004. Application of short monolithic columns for fast purification of plasmid DNA. *J. Chromatogr. B: Analyt. Technol. Biomed. Life Sci.* 801: 331–337.

Brne, P., A. Podgornik, K. Bencina, B. Gabor, A. Strancar, and M. Peterka. 2007. Fast and efficient separation of immunoglobulin M from immunoglobulin G using short monolithic columns. *J. Chromatogr. A* 1144: 120–125.

Cabrera, K. 2004. Applications of silica-based monolithic HPLC columns. *J. Sep. Sci.* 27: 843–852.

Chatterjee, A., P.L. Mirer, E. Zaldivar Santamaria, C. Klapperich, A. Sharon, and A.F. Sauer-Budge. 2010. RNA isolation from mammalian cells using porous polymer monoliths: An approach for high-throughput automation. *Anal. Chem.* 82: 4344–4356.

Costioli, M.D., I. Fisch, F. Garret-Flaudy, F. Hilbrig, and R. Freitag. 2003. DNA purification by triple-helix affinity precipitation. *Biotechnol. Bioeng.* 81: 535–545.

Crooke, S.T. 1998. *Antisense Research and Application.* Berlin, Germany: Springer.

Dainiak, M.B., I.Y. Galaev, A. Kumar, F.M. Plieva, and B. Mattiasson. 2007a. Chromatography of living cells using supermacroporous hydrogels, cryogels. *Adv. Biochem. Eng. Biotechnol.* 106: 101–127.

Dainiak, M.B., I.Y. Galaev, and B. Mattiasson. 2006a. Affinity cryogel monoliths for screening for optimal separation conditions and chromatographic separation of cells. *J. Chromatogr. A* 1123: 145–150.

Dainiak, M.B., I.Y. Galaev, and B. Mattiasson. 2007b. Macroporous monolithic hydrogels in a 96-minicolumn plate format for cell surface-analysis and integrated binding/quantification of cells. *Enzyme Microb. Technol.* 40: 688–695.

Dainiak, M.B., A. Kumar, I.Y. Galaev, and B. Mattiasson. 2006b. Detachment of affinity-captured bioparticles by elastic deformation of a macroporous hydrogel. *Proc. Natl. Acad. Sci. USA* 103: 849–854.

Dainiak, M.B., A. Kumar, F.M. Plieva, I.Y. Galaev, and B. Mattiasson. 2004. Integrated isolation of antibody fragments from microbial cell culture fluids using supermacroporous cryogels. *J. Chromatogr. A* 1045: 93–98.

Dainiak, M.B., F.M. Plieva, I.Y. Galaev, R. Hatti-Kaul, and B. Mattiasson. 2005. Cell chromatography: Separation of different microbial cells using IMAC supermacroporous monolithic columns. *Biotechnol. Prog.* 21: 644–649.

Dainiak, M.B., I.N. Savina, I. Musolino, A. Kumar, B. Mattiasson, and I.Y. Galaev. 2008.
Biomimetic macroporous hydrogel scaffolds in a high-throughput screening format for
cell-based assays. *Biotechnol. Prog.* 24: 1373–1383.

Forcic, D., K. Branovic-Cakanic, J. Ivancic, R. Jug, M. Barut, and A. Strancar. 2005.
Purification of genomic DNA by short monolithic columns. *J. Chromatogr. A* 1065:
115–120.

Freitag, R. and S. Vogt. 2000. Comparison of particulate and continuous-bed columns for
protein displacement chromatography. *J. Biotechnol.* 78: 69–82.

Gilar, M. and E.S.P. Bouvier. 2000. Purification of crude DNA oligonucleotides by solid-phase
extraction and reversed-phase high-performance liquid chromatography. *J. Chromatogr.
A* 890: 167–177.

Hanora, A., I. Savina, F.M. Plieva, V.A. Izumrudov, B. Mattiasson, and I.Y. Galaev. 2006.
Direct capture of plasmid DNA from non-clarified bacterial lysate using polycation-
grafted monoliths. *J. Biotechnol.* 123: 343–355.

Holdsvendova, P., J. Suchankova, M. Buncek, V. Backovska, and P. Coufal. 2007.
Hydroxymethyl methacrylate-based monolithic columns designed for separation of oli-
gonucleotides in hydrophilic-interaction capillary liquid chromatography. *J. Biochem.
Biophys. Methods* 70: 23–29.

Hoth, D.C., J.G. Rivera, and L.A. Colon. 2005. Metal oxide monolithic columns. *J. Chromatogr.
A* 1079: 392–396.

Huber, C.G. 1998. Micropellicular stationary phases for high-performance liquid chromatography of double-stranded DNA. *J. Chromatogr. A* 806: 3–30.

Ikegami, T., E. Dicks, H. Kobayashi, H. Morisaka, D. Tokuda, K. Cabrera, K. Hosoya, and
N. Tanaka. 2004. How to utilize the true performance of monolithic silica columns.
J. Sep. Sci. 27: 1292–1302.

Isobe, K. and Y. Kawakami. 2007. Preparation of convection interaction media isobutyl disc
monolithic column and its application to purification of secondary alcohol dehydrogenase and alcohol oxidase. *J. Chromatogr. A* 1144: 85–89.

Josic, D. and J. Clifton. 2007. Use of monolithic supports in proteomics technology.
J. Chromatogr. A 1144: 2–13.

Jungbauer, A. and R. Hahn. 2004. Monoliths for fast bioseparation and bioconversion and their
applications in biotechnology. *J. Sep. Sci.* 27: 767–778.

Krajacic, M., J. Ivancic-Jelecki, D. Forcic, A. Vrdoljak, and D. Skoric. 2007. Purification of
plant viral and satellite double-stranded RNAs on DEAE monoliths. *J. Chromatogr.
A* 1144: 111–119.

Kumar, A., V. Bansal, J. Andersson, P.K. Roychoudhury, and B. Mattiasson. 2006.
Supermacroporous cryogel matrix for integrated protein isolation. Immobilized metal
affinity chromatographic purification of urokinase from cell culture broth of a human
kidney cell line. *J. Chromatogr. A* 1103: 35–42.

Kumar, A., I.Y. Galaev, and B. Mattiasson. 2007. *Cell Separation: Fundamentals, Analytical
and Preparative Methods*. Berlin, Germany: Springer-Verlag.

Kumar, A., F.M. Plieva, I.Y. Galaev, and B. Mattiasson. 2003. Affinity fractionation of lymphocytes using a monolithic cryogel. *J. Immunol. Methods* 283: 185–194.

Kumar, A., A. Rodriguez-Caballero, F.M. Plieva, I.Y. Galaev, K.S. Nandakumar, M. Kamihira,
R. Holmdahl, A. Orfao, and B. Mattiasson. 2005. Affinity binding of cells to cryogel
adsorbents with immobilized specific ligands: Effect of ligand coupling and matrix
architecture. *J. Mol. Recognit.* 18: 84–93.

Kumar, A. and A. Srivastava. 2010. Cell separation using cryogel-based affinity chromatography.
Nat. Protoc. 5: 1737–1747.

Leinweber, F.C. and U. Tallarek. 2003. Chromatographic performance of monolithic and particulate stationary phases. Hydrodynamics and adsorption capacity. *J. Chromatogr. A* 1006: 207–228.

Lesignoli, E., A. Germini, R. Corradini, S. Sforza, G. Galavema, A. Dossena, and R. Marchelli. 2001. Recognition and strand displacement of DNA oligonucleotides by peptide nucleic acids (PNAs). High-performance ion-exchange chromatographic analysis. *J. Chromatogr. A* 922: 177–185.

Levkin, P.A., S. Eeltink, T.R. Stratton, R. Brennen, K. Robotti, H. Yin, K. Killeen, F. Svec, and J.M. Frechet. 2008. Monolithic porous polymer stationary phases in polyimide chips for the fast high-performance liquid chromatography separation of proteins and peptides. *J. Chromatogr. A* 1200: 55–61.

Li, Y. and M.L. Lee. 2009. Biocompatible polymeric monoliths for protein and peptide separations. *J. Sep. Sci.* 32: 3369–3378.

Li, Y.M., J.L. Liao, K. Nakazato, J. Mohammad, L. Terenius, and S. Hjerten. 1994. Continuous beds for microchromatography: Cation-exchange chromatography. *Anal. Biochem.* 223: 153–158.

Liang, C., S. Dai, and G. Guiochon. 2003. A graphitized-carbon monolithic column. *Anal. Chem.* 75: 4904–4912.

Liao, J.-L., R. Zhang, and S. Hjerten. 1991. Continuous beds for standard and micro high-performance liquid chromatography. *J. Chromatogr. A* 586: 21–26.

Lin, L., H. Chen, H. Wei, F. Wang, and J.M. Lin. 2011. On-chip sample pretreatment using a porous polymer monolithic column for solid-phase microextraction and chemiluminescence determination of catechins in green tea. *Analyst* 136: 4260–4267.

Lin, Z., H. Huang, X. Sun, Y. Lin, L. Zhang, and G. Chen. 2012. Monolithic column based on a poly(glycidyl methacrylate-co-4-vinylphenylboronic acid-co-ethylene dimethacrylate) copolymer for capillary liquid chromatography of small molecules and proteins. *J. Chromatogr. A* 1246: 90–97.

Lozinsky, V.I. 2002. Cryogels on the basis of natural and synthetic polymers: Preparation, properties and application. *Uspekhi. Khimii.* 71: 559–585.

Lozinsky, V.I., I.Y. Galaev, F.M. Plieva, I.N. Savina, H. Jungvid, and B. Mattiasson. 2003. Polymeric cryogels as promising materials of biotechnological interest. *Trends Biotechnol.* 21: 445–451.

Luo, Q., Y. Shen, K.K. Hixson, R. Zhao, F. Yang, R.J. Moore, H.M. Mottaz, and R.D. Smith. 2005. Preparation of 20-microm-i.d. silica-based monolithic columns and their performance for proteomics analyses. *Anal. Chem.* 77: 5028–5035.

Luo, Q., H. Zou, Q. Zhang, X. Xiao, and J. Ni. 2002. High-performance affinity chromatography with immobilization of protein A and L-histidine on molded monolith. *Biotechnol. Bioeng.* 80: 481–489.

Ma, J., L. Zhang, Z. Liang, W. Zhang, and Y. Zhang. 2007. Monolith-based immobilized enzyme reactors: Recent developments and applications for proteome analysis. *J. Sep. Sci.* 30: 3050–3059.

Malvy, C., A. Harel-Bellan, and L.L. Pritchard. 1999. *Triple Helix Forming Oligonucleotides.* Springer, New York.

Miller, S. 2004. Separations in a monolith. *Anal. Chem.* 76: 99–101.

Minakuchi, H., K. Nakanishi, N. Soga, N. Ishizuka, and N. Tanaka. 1996. Octadecylsilylated porous silica rods as separation media for reversed-phase liquid chromatography. *Anal. Chem.* 68: 3498–3501.

Minakuchi, H., K. Nakanishi, N. Soga, N. Ishizuka, and N. Tanaka. 1998. Effect of domain size on the performance of octadecylsilylated continuous porous silica columns in reversed-phase liquid chromatography. *J. Chromatogr. A* 797: 121–131.

Miyamoto, K., T. Hara, H. Kobayashi, H. Morisaka, D. Tokuda, K. Horie, K. Koduki et al. 2008. High-efficiency liquid chromatographic separation utilizing long monolithic silica capillary columns. *Anal. Chem.* 80: 8741–8750.

Motokawa, M., H. Kobayashi, N. Ishizuka, H. Minakuchi, K. Nakanishi, H. Jinnai, K. Hosoya, T. Ikegami, and N. Tanaka. 2002. Monolithic silica columns with various skeleton sizes and through-pore sizes for capillary liquid chromatography. *J. Chromatogr. A* 961: 53–63.

Nawrocki, J., C. Dunlap, J. Li, J. Zhao, C.V. McNeffe, A. McCormick, and P.W. Carr. 2004a. Part II. Chromatography using ultra-stable metal oxide-based stationary phases for HPLC. *J. Chromatogr. A* 1028: 31–62.

Nawrocki, J., C. Dunlap, A. McCormick, and P.W. Carr. 2004b. Part I. Chromatography using ultra-stable metal oxide-based stationary phases for HPLC. *J. Chromatogr. A* 1028: 1–30.

Nordborg, A., E.F. Hilder, and P.R. Haddad. 2011. Monolithic phases for ion chromatography. *Annu. Rev. Anal. Chem. (Palo Alto Calif.)* 4: 197–226.

Oberacher, H., A. Krajete, W. Parson, and C.G. Huber. 2000. Preparation and evaluation of packed capillary columns for the separation of nucleic acids by ion-pair reversed-phase high-performance liquid chromatography. *J. Chromatogr. A* 893: 23–35.

Ostryanina, N.D., G.P. Vlasov, and T.B. Tennikova. 2002. Multifunctional fractionation of polyclonal antibodies by immunoaffinity high-performance monolithic disk chromatography. *J. Chromatogr. A* 949: 163–171.

Padmapriya, A.A., J. Tang, and S. Agrawal. 1994. Large-scale synthesis, purification, and analysis of oligodeoxynucleotide phosphorothioates. *Antisense Res. Dev.* 4: 185–199.

Perica, M.C., I. Sola, L. Urbas, F. Smrekar, and M. Krajacic. 2009. Separation of hypoviral double-stranded RNA on monolithic chromatographic supports. *J. Chromatogr. A* 1216: 2712–2716.

Peterka, M., M. Jarc, M. Banjac, V. Frankovic, K. Bencina, M. Merhar, V. Gaberc-Porekar, V. Menart, A. Strancar, and A. Podgornik. 2006. Characterisation of metal-chelate methacrylate monoliths. *J. Chromatogr. A* 1109: 80–85.

Plieva, F., H.T. Xiao, I.Y. Galaev, B. Bergenstahl, and B. Mattiasson. 2006. Macroporous elastic polyacrylamide gels prepared at subzero temperatures: Control of porous structure. *J. Mater. Chem.* 16: 4065–4073.

Plieva, F.M., I.Y. Galaev, and B. Mattiasson. 2007. Macroporous gels prepared at subzero temperatures as novel materials for chromatography of particulate-containing fluids and cell culture applications. *J. Sep. Sci.* 30: 1657–1671.

Plieva, F.M., M. Karlsson, M.-R. Aguilar, D. Gomez, S. Mikhalovsky, and I.Y. Galaev. 2005. Pore structure in supermacroporous polyacrylamide based cryogels. *Soft Matter* 1: 303–309.

Podgornik, A. and A. Strancar. 2005. Convective Interaction Media (CIM)—Short layer monolithic chromatographic stationary phases. *Biotechnol. Annu. Rev.* 11: 281–333.

Qiu, R. and F.E. Regnier. 2005. Use of multidimensional lectin affinity chromatography in differential glycoproteomics. *Anal. Chem.* 77: 2802–2809.

Rucevic, M., J. Clifton, F. Huang, X. Li, H. Callanan, D.C. Hixson, and D. Josic. 2006. Use of short monolithic columns for isolation of low abundance membrane proteins. *J. Chromatogr. A* 1123: 199–204.

Satterfield, B.C., S. Stern, M.R. Caplan, K.W. Hukari, and J.A. West. 2007. Microfluidic purification and preconcentration of mRNA by flow-through polymeric monolith. *Anal. Chem.* 79: 6230–6235.

Smått, J.H., F.M. Sayler, A.J. Grano, and M.G. Bakker. 2012. Formation of hierarchically porous metal oxide and metal monoliths by nanocasting into silica monoliths. *Adv. Eng. Mater.* 14: 1059–1073.

Smått, J.H., C. Weidenthaler, J.B. Rosenholm, and M. Lindén. 2006. Hierarchically porous metal oxide monoliths prepared by the nanocasting route. *Chem. Mater.* 18: 1443–1450.

Srivastava, A., A.K. Shakya, and A. Kumar. 2012. Boronate affinity chromatography of cells and biomacromolecules using cryogel matrices. *Enzyme Microb. Technol.* 51: 373–381.

Strancar, A., M. Barut, A. Podgornik, P. Koselj, H. Schwinn, P. Raspor, and D. Josic. 1997. Application of compact porous tubes for preparative isolation of clotting factor VIII from human plasma. *J. Chromatogr. A* 760: 117–123.

Strancar, A., A. Podgornik, M. Barut, and R. Necina. 2002. *Advances in Biochemical Engineering/Biotechnology.* Heidelberg, Germany: Springer-Verlag.

Svec, F. and J.M.J. Fréchet. 1992. Continuous rods of macroporous polymer as high-performance liquid chromatography separation media. *Anal. Chem.* 64: 820–822.

Svec, F. and C.G. Huber. 2006. Monolithic materials: Promises, challenges, achievements. *Anal. Chem.* 78: 2101–2107.

Taguchi, A., J.-H. Smått, and M. Lindén. 2003. Carbon monoliths possessing a hierarchical, fully interconnected porosity. *Adv. Mater.* 15: 1209–1211.

Tanaka, N., H. Kobayashi, K. Nakanishi, H. Minakuchi, and N. Ishizuka. 2001. Monolithic LC columns. *Anal. Chem.* 73: 420A–429A.

Teeters, M.A., S.E. Conrardy, B.L. Thomas, T.W. Root, and E.N. Lightfoot. 2003. Adsorptive membrane chromatography for purification of plasmid DNA. *J. Chromatogr. A* 989: 165–173.

Teilum, M., M.J. Hansson, M.B. Dainiak, R. Mansson, S. Surve, E. Elmer, P. Onnerfjord, and G. Mattiasson. 2006. Binding mitochondria to cryogel monoliths allows detection of proteins specifically released following permeability transition. *Anal. Biochem.* 348: 209–221.

Toll, H., R. Wintringer, U. Schweiger-Hufnagel, and C.G. Huber. 2005. Comparing monolithic and microparticular capillary columns for the separation and analysis of peptide mixtures by liquid chromatography-mass spectrometry. *J. Sep. Sci.* 28: 1666–1674.

Tong, S., Q. Liu, Y. Li, W. Zhou, Q. Jia, and T. Duan. 2012. Preparation of porous polymer monolithic column incorporated with graphene nanosheets for solid phase microextraction and enrichment of glucocorticoids. *J. Chromatogr. A* 1253: 22–31.

Unger, K.K., R. Skudas, and M.M. Schulte. 2008. Particle packed columns and monolithic columns in high-performance liquid chromatography-comparison and critical appraisal. *J. Chromatogr. A* 1184: 393–415.

Urthaler, J., R. Schlegl, A. Podgornik, A. Strancar, A. Jungbauer, and R. Necina. 2005. Application of monoliths for plasmid DNA purification development and transfer to production. *J. Chromatogr. A* 1065: 93–106.

Vlakh, E.G. and T.B. Tennikova. 2009. Applications of polymethacrylate-based monoliths in high-performance liquid chromatography. *J. Chromatogr. A* 1216: 2637–2650.

Wang, Q.C., F. Svec, and J.M. Frechet. 1993. Macroporous polymeric stationary-phase rod as continuous separation medium for reversed-phase chromatography. *Anal. Chem.* 65: 2243–2248.

Weith, H.L., J.L. Wiebers, and P.T. Gilham. 1970. Synthesis of cellulose derivatives containing the dihydroxyboryl group and a study of their capacity to form specific complexes with sugars and nucleic acid components. *Biochemistry* 9: 4396–4401.

Williams, S.L., M.E. Eccleston, and N.K. Slater. 2005. Affinity capture of a biotinylated retrovirus on macroporous monolithic adsorbents: Towards a rapid single-step purification process. *Biotechnol. Bioeng.* 89: 783–787.

Wu, Q., J.M. Bienvenue, B.J. Hassan, Y.C. Kwok, B.C. Giordano, P.M. Norris, J.P. Landers, and J.P. Ferrance. 2006. Microchip-based macroporous silica sol-gel monolith for efficient isolation of DNA from clinical samples. *Anal. Chem.* 78: 5704–5710.

Wuhrer, M., C.A. Koeleman, C.H. Hokke, and A.M. Deelder. 2005. Protein glycosylation analyzed by normal-phase nano-liquid chromatography—Mass spectrometry of glycopeptides. *Anal. Chem.* 77: 886–894.

Xu, H., S. Wang, G. Zhang, S. Huang, D. Song, Y. Zhou, and G. Long. 2011. A novel solid-phase microextraction method based on polymer monolith frit combining with high-performance liquid chromatography for determination of aldehydes in biological samples. *Anal. Chim. Acta* 690: 86–93.

Xu, Y., Q. Cao, F. Svec, and J.M. Frechet. 2010. Porous polymer monolithic column with surface-bound gold nanoparticles for the capture and separation of cysteine-containing peptides. *Anal. Chem.* 82: 3352–3358.

Xu, Z. and R.D. Oleschuk. 2014. A fluorous porous polymer monolith photo-patterned chromatographic column for the separation of a flourous/fluorescently labeled peptide within a microchip. *Electrophoresis.* 35: 441–449.

Yamamoto, S., M. Nakamura, C. Tarmann, and A. Jungbauer. 2007. Retention studies of DNA on anion-exchange monolith chromatography binding site and elution behavior. *J. Chromatogr. A* 1144: 155–160.

Zochling, A., R. Hahn, K. Ahrer, J. Urthaler, and A. Jungbauer. 2004. Mass transfer characteristics of plasmids in monoliths. *J. Sep. Sci.* 27: 819–827.

4 Polysaccharide-Based Composite Hydrogels for Removal of Pollutants from Water

Junping Zhang and Aiqin Wang

CONTENTS

4.1 INTRODUCTION

Hydrogels are cross-linked macromolecules with a 3D network that can entrap substantial amount of aqueous fluids. Due to the excellent properties compared to traditional water-absorbing materials, hydrogels have received much attention and are widely used in many fields, such as superabsorbents, tissue engineering, and pollutant adsorption (Lutolf and Hubbell 2005).

Hydrogels can be prepared using synthetic polymers, polysaccharides, and the mixture of them via covalent or noncovalent cross-linking. The absorption capacity and rate of hydrogels for aqueous fluids are determined by many factors including the nature of polymer precursors, the kind of hydrophilic groups, cross-linking degree, and surface morphology. Most of the hydrogels based on synthetic precursors, for example, acrylic acid (AA), N-isopropylacrylamide (NIPAm), and acrylamide (AM), have excellent properties but are poor in degradability and biocompatibility. The potential of polysaccharides for the preparation of hydrogels has been widely recognized. Polysaccharides can easily form hydrogels by hydrogen bonding, ionic interactions, or chemical cross-linking. Incorporation of natural polysaccharides, such as starch and chitosan (CTS), can not only improve properties of corresponding hydrogels but also reduce our dependence on petrochemical-derived monomers. Sun et al. reported the synthesis of hydrogels from alginate and polyacrylamide (PAM) forming ionically and covalently cross-linked networks (Figure 4.1) (Sun et al. 2012). Although such gels contain 90% water, they can be stretched beyond 20 times their initial length and have fracture energies of 9000 J/m^2.

Polymer/layered silicate nanocomposites have attracted great interest because they often exhibit improved properties compared with virgin polymer or conventional micro- and macrocomposites (Ray and Okamoto 2003). These improvements include increased strength and heat resistance and decreased gas permeability and flammability. Most of the polymer/layered silicate nanocomposites are based on vinyl polymers (e.g., methyl methacrylate, acrylonitrile and styrene [St]), polycondensates (e.g., N6, poly(ε-caprolactone), and PDMS), polyolefins (e.g., polypropylene and polyethylene), and biodegradable polymers (e.g., poly(butylene succinate), unsaturated polyester, and polyhydroxy butyrate).

The commonly used clays for the preparation of nanocomposites are layered silicates. The layer thickness is around 1 nm, and the lateral dimensions of the layers vary from 30 nm to several microns, depending on the particular layered silicate. The layered silicates are characterized by moderate surface charge and specific surface area, which make them ideal materials for the reinforcement of polymers and adsorption of pollutants.

Traditional hydrogels from synthetic polymers often have some limitations besides poor biodegradability, which restrict their practical applications. Recently, biopolymer/inorganic material composites attracted much attention owing to their unique properties. Inorganic materials including clays, hydroxyapatite (HA), silica, and carbon nanotubes have been used for preparing this class of composites. The synergistic effect of biopolymer and inorganic material could improve the mechanical properties, swelling behavior, and other properties. In addition, these properties could be further tailored by altering the type and content of inorganic materials. Similarly, clay-based composite hydrogels are a type of very important hydrogels owing to the abundance of clay in nature and its low cost (Dawson et al. 2011). Clays, including montmorillonite (MMT), kaolin, and attapulgite (APT), have already been incorporated into traditional polymeric hydrogels, and encouraging results were obtained. For example, Wang et al. prepared a hydrogel by mixing water, Laponite XLG, a dendritic macromolecule, and sodium polyacrylate (Figure 4.2) (Wang et al. 2010d). This material can be molded into shape-persistent, free-standing objects owing to its exceptionally great mechanical strength and rapidly and completely self-heal.

Polysaccharide-based composite hydrogels are a new group of materials at the interface of hydrogels, polymer/clay nanocomposites, and polysaccharides. In this chapter, we reviewed the recent progress about this type of hydrogel. We discuss the preparation and characterization of the composite hydrogels, with an emphasis on

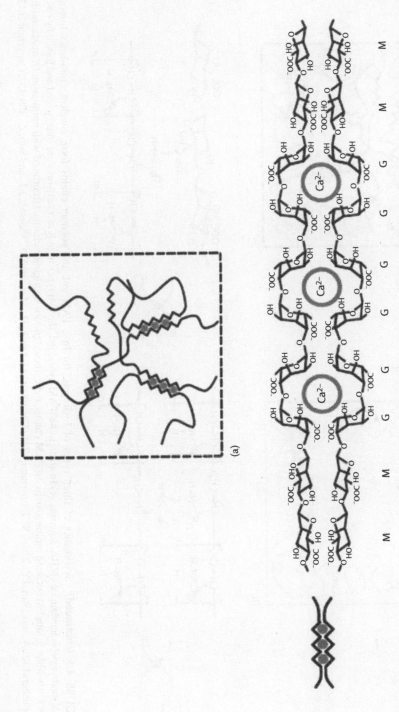

FIGURE 4.1 Schematics of three types of hydrogels. (a) In an alginate gel, the G blocks on different polymer chains form ionic cross-links through Ca²⁺ (circles).

(continued)

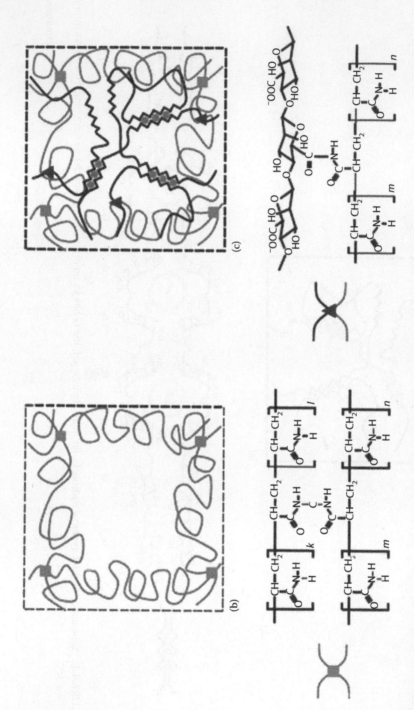

FIGURE 4.1 (continued) Schematics of three types of hydrogels. (b) In a PAM gel, the polymer chains form covalent cross-links through *N,N*-methylenebisacrylamide (MBA, squares). (c) In an alginate–PAM hybrid gel, the two types of polymer networks are intertwined and joined by covalent cross-links (triangles) between amine groups on PAM chains and carboxyl groups on alginate chains. (Reprinted with permission from Macmillan Publishers Ltd., *Nature*, Sun, J.Y., Zhao, X., Illeperuma, W.R., Chaudhuri, O., Oh, K.H., Mooney, D.J., Vlassak, J.J., and Suo, Z., 489, 133, Copyright 2012.)

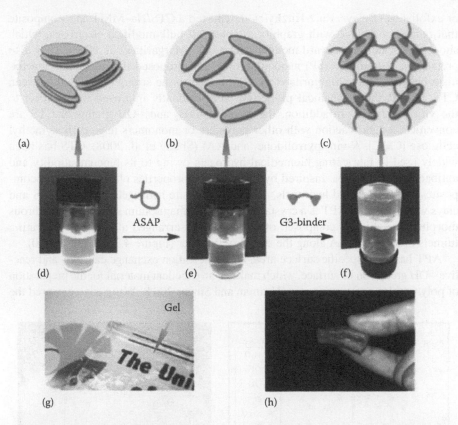

FIGURE 4.2 Noncovalent preparation of hydrogels. (a–c) Proposed mechanism for hydrogelation. CNSs, entangled with one another (a), are dispersed homogeneously by interaction of their positive-charged edge parts with anionic ASAP (b). Upon addition of Gn-binder, exfoliated CNSs are crosslinked to develop a 3D network (c). (d–f) Optical images of an aqueous suspension of CNSs (d), an aqueous dispersion of CNSs and ASAP (e), and a physical gel upon addition of G3-binder to the dispersion (f). (g, h) The gel is transparent (g) and free standing (h). (From Wang, Q.G. et al., Nature, 463, 339, 2010b. With permission.)

structure and swelling properties. Moreover, we summarized the applications of the composite hydrogels for the removal of pollutants including heavy metals, dyes, and ammonium nitrogen from water.

4.2 POLYSACCHARIDE-BASED COMPOSITE HYDROGELS

4.2.1 CHITOSAN-BASED COMPOSITE HYDROGELS

CTS is the most abundant biomass in the world. It is made by treating shrimp and other crustacean shells with sodium hydroxide. CTS is a linear polysaccharide composed of randomly distributed β-(1–4)-linked D-glucosamine and N-acetyl-D-glucosamine.

Most of the applications of CTS are based on the polyelectrolytic nature and chelating ability of the amine group. The positively charged CTS could interact with the negatively charged layers of many clay minerals, which results in the intercalation or exfoliation of clays. Ruiz-Hitzky et al. reported a CTS/Na–MMT nanocomposite that can be combined with graphite to construct bulk-modified electrodes, which show high selectivity toward monovalent anions (Margarita et al. 2003). They also prepared CTS/sepiolite (SP) nanocomposites and proposed the interaction mechanism between them (Margarita et al. 2006). Due to the strong interaction between CTS and SP, the mechanical properties were evidently improved with respect to the virgin polymer. In addition, the reactive –NH$_2$ and –OH groups of CTS are convenient for its reaction with other polymers or monomers (e.g., carboxymethyl cellulose [CMC], *N*-vinylpyrrolidone, and AA) (Shang et al. 2008). CTS has been widely used in fabricating biomedical hydrogels owing to its biocompatibility and antibacterial properties. Inspired by the excellent properties of CTS/clay nanocomposites and CTS-based hydrogels, various composite hydrogels based on CTS and clays were reported. APT is a crystalline hydrated magnesium silicate with a fibrous morphology, which is composed of talc-like units arranged alternately, generating tunnels of 3.7 Å × 6.4 Å along the *c*-axis of the fiber (Figure 4.3) (Bradley 1940).

APT has large specific surface area, moderate cation exchange capacity, and reactive –OH groups on its surface, which makes it an excellent material for the preparation of polymer/clay nanocomposites (Neaman and Singer 2004). Wang et al. prepared the

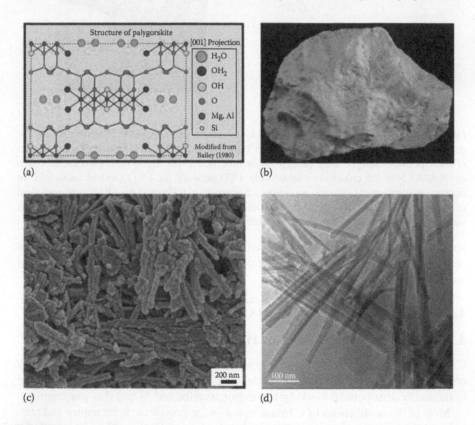

(a)

(b)

(c)

(d)

FIGURE 4.3 (a) Structure. (From http://pubs.usgs.gov/of/2001/of01-041/htmldocs/clays/ seppaly.htm.) (b) Digital. (c) SEM. (d) TEM images of APT.

CTS/APT composite microspheres via spray-drying (Wang et al. 2011c). The introduction of APT can not only enhance the isoelectric points but also achieve narrow size distribution of the microspheres. A series of ofloxacin/MMT/CTS (OFL/MMT/CTS) composite hydrogels were prepared by solution intercalation and ionic cross-linking with sodium tripolyphosphate (TPP) (Hua et al. 2010). The electrostatic interaction between CTS and MMT enhanced the stability and swelling behavior of the beads. Hua et al. designed pH-sensitive OFL/MMT/CTS nanocomposite microspheres by the solution intercalation and emulsification cross-linking techniques (Hua et al. 2008). Yang et al. developed a series of CTS/Laponite composite beads by using TPP as the ionic cross-linker (Yang et al. 2011). The exfoliated sheets of Laponite act as physical cross-linkers to facilitate the formation of network structure between CTS and Laponite.

We have prepared for the first time the CTS-g-poly(acrylic acid)/APT (CTS-g-PAA/APT) composite hydrogels (Figure 4.4) (Zhang et al. 2007b). The –OH of APT and –OH and –NH_2 of CTS participated in graft polymerization with AA. In addition, the introduced APT could enhance thermal stability and form a porous surface. By Ca^{2+} cross-linking of the CTS-g-PAA/APT composite hydrogel with sodium alginate (SA), Wang et al. obtained novel pH-sensitive composite hydrogel beads (Wang et al. 2009a). Moreover, the CTS-g-PAA/APT composite hydrogel can be easily scaled up. This composite hydrogel has been proved to be an excellent material for the removal of ammonium nitrogen (NH_4^+-N) (Zheng et al. 2009b), methylene blue (MB) (Wang et al. 2011b), and heavy metals (Wang and Wang 2010b, Wang et al. 2009g) from aqueous solution.

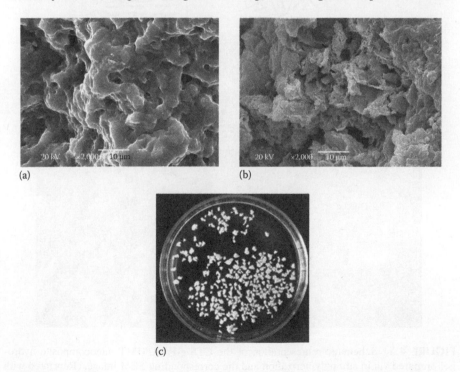

(a)

(b)

(c)

FIGURE 4.4 SEM micrographs of (a) CTS-g-PAA and (b) CTS-g-PAA/APT composite hydrogels. (c) Image of the granular composite hydrogel.

Liu et al. have studied the effect of grinding of APT on swelling properties of the CTS-*g*-PAA/APT nanocomposite hydrogels (Liu et al. 2013b). Grinding could decrease the length of single crystals and dissociate part of crystal aggregates of APT, which affect the dispersion of APT in the polymeric matrix, thus leading to change in swelling behaviors of the nanocomposites.

We also introduced MMT into the CTS-*g*-PAA hydrogel (Figure 4.5) (Zhang et al. 2007a). CTS could intercalate into the layers of MMT and form nanocomposites

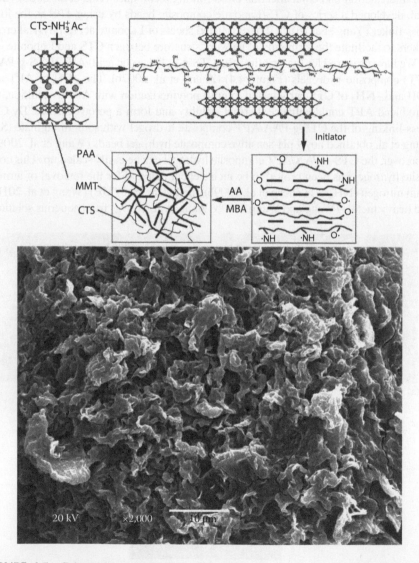

FIGURE 4.5 Schematic representation of the CTS-*g*-PAA/MMT nanocomposite hydrogel prepared via in situ polymerization and the corresponding SEM image. (Reprinted with permission from Zhang, J.P., Wang, L., and Wang, A.Q., *Ind. Eng. Chem. Res.*, 46, 2497. Copyright 2007 American Chemical Society.)

through in situ graft polymerization with AA. The surface morphology of the CTS-*g*-PAA/MMT nanocomposite is different from that of CTS-*g*-PAA. Compared to the tight surface of CTS-*g*-PAA, the surface of the CTS-*g*-PAA/MMT nanocomposite is more loose and porous, which is convenient for the penetration of water into the polymeric network and is of benefit to improve the water absorbency.

Owing to the excellent properties of CTS-based composite hydrogels, we also prepared many composite hydrogels using other clays instead of APT and MMT. Clays including rectorite (REC) (Liu and Wang 2008a,b, Zheng and Wang 2009), vermiculite (VMT) (Xie and Wang 2009a,c), SP (Xie and Wang 2009b, Xie et al. 2010), and halloysite nanotubes (HNTs) (Zheng and Wang 2010c) were used. These clays are different in crystalline structure, morphology, and properties.

REC is a kind of layered silicate with structure and characteristics like bentonite. The introduction of REC into the polymer matrix can not only improve the thermal stability of the corresponding polymer but also reduce the product cost. A series of nanocomposites based on REC, CTS, and its derivatives has been reported (Huang et al. 2012b, Li et al. 2011b, Liu et al. 2012a, 2013a, Wang et al. 2007, 2010g,h, Xu et al. 2012). CTS/organic rectorite (CTS/OREC) nanocomposite films were obtained by a casting/solvent evaporation method. The addition of OREC to pure CTS film influenced many of the properties. Liu et al. reported the preparation of the exfoliated quaternized carboxymethyl CTS/REC (QCMC/REC) nanocomposite via microwave irradiation method, which was performed in only water without any additional plasticizer (Liu et al. 2012a). Two types of interactions of hydrogen bond and electrostatic attraction exist in the QCMC/REC nanocomposite. Liu and Wang prepared the CTS-*g*-PAA/REC composite hydrogels through graft polymerization among REC, CTS, and AA in aqueous solution (Liu and Wang 2008a). They also studied the effect of modification of REC with hexadecyltrimethylammonium bromide (HDTMABr) on properties of the composite hydrogel (Liu and Wang 2008b). CTS-*g*-PAA/OREC nanocomposites show an exfoliated nanostructure. The water absorbency of CTS-*g*-PAA/OREC increased with increasing organification degree of OREC owing to the improved compatibility of polymer matrix and OREC.

VMT is a layered aluminum silicate with exchangeable cations and reactive – OH groups on the surface (Zheng et al. 2007). The CTS/VMT nanocomposites were prepared by the solution mixing process of CTS with three different modified VMTs (HVMT, NVMT, and organo-VMT [OVMT]), which were treated by HCl, NaCl, and HDTMABr, respectively (Zhang et al. 2009). The modification and the nanoscale dispersion of the modified VMTs were confirmed by x-ray diffraction (XRD) and TEM. The thermal stability of CTS/HVMT, CTS/NVMT, and CTS/OVMT nanocomposites were significantly improved compared to that of neat CTS. Introducing HVMT into CTS matrix can enhance the thermal stability due to the well dispersion of HVMT and better interaction between HVMT and CTS. VMT also can improve water absorbency of the composites (Zheng et al. 2007).

The CTS-*g*-PAA/unexpanded VMT (CTS-*g*-PAA/UVMT) composite hydrogels were prepared by graft polymerization (Xie and Wang 2009c). The reactions between AA and –OH groups of UVMT could improve the polymeric network and then enhance the water absorbency. They also studied the effects of acid-activated, ion-exchanged, and organic-modified VMT on properties of the CTS-*g*-PAA/VMT composite hydrogels (Xie and Wang 2009c). The modification of VMT could improve water absorbency and swelling rate. By dispersing microparticles of the CTS-g-PAA/VMT composite hydrogel into SA aqueous solution and then cross-linked with Ca²⁺, Wang et al. prepared pH-sensitive composite hydrogel beads (Wang et al. 2010b).

SP is a hydrated magnesium silicate clay mineral with microfibrous morphology. A significant number of silanol groups are present at the surface of these minerals, and these groups are directly accessible to various reagents (Tekin et al. 2006). Huang et al. prepared the CTS/PVA nanocomposite films reinforced with SP (Huang et al. 2012a). The mechanical properties were improved with an increase in SP loading, but the moisture uptake decreased.

A CTS-g-PAA/SP composite hydrogel was prepared by graft polymerization among CTS, AA, and SP. CTS and SP participated in graft polymerization with AA. The introduced SP enhanced swelling rate and water absorbency (Xie and Wang 2009b). In addition, the modification of SP (acid-activated SP and cation-exchanged SP) also has influenced the water absorbency and swelling behavior (Xie et al. 2010). Water absorbency and swelling behavior depend strongly on HCl concentration and the kinds of cation exchanged. The introduction of suitable amount of acid-activated and cation-exchanged SP could not only enhance water absorbency but also improve the swelling rate. Shi et al. found that SP could improve the adsorption rate, gel strength, and acid and salt resistance of the CTS-g-PAA/SP composite hydrogels (Shi et al. 2011a).

HNTs, a kind of aluminosilicate mineral, are chemically similar to kaolin, but morphologically have a hollow tubular structure. Due to special hollow tubular structure and excellent properties, HNTs have potential applications in many fields and have attracted considerable attention (Shchukin et al. 2008). Various CTS/HNT nanocomposites have been prepared. Deen et al. reported electrophoretic deposition of composite CTS–HNT–HA films (Figure 4.6) (Deen et al. 2012). The use of CTS as a dispersing and charging agent for both HNTs and HA allowed the formation of CTS–HNT–HA monolayers, which showed corrosion protection of the stainless steel substrates.

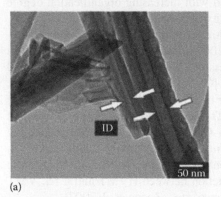 

(a) (b)

FIGURE 4.6 (a) TEM and (b) SEM of the CTS–HNT–HA film. (Reprinted from *Colloids Surf. A*, 410, Deen, I., Pang, X., and Zhitomirsky, I., Electrophoretic deposition of composite chitosan–halloysite nanotube–hydroxyapatite films, 38–44, Copyright 2012, with permission from Elsevier.)

Zheng et al. grafted PAA onto CTS to form granular hydrogel composites with HNTs (Zheng and Wang 2010c). HNTs are in the form of partially hydrated state and no additional interaction is present among the reactants. The sample surface gets coarse and many micropores are visible by introducing HNTs into the hydrogel. Wang et al. prepared the CTS-g-PAA/HNT/SA composite hydrogel by the ionic gelation method (Wang et al. 2010c). The HNT content obviously influences the swelling ratio and cumulative release of diclofenac sodium (DS).

Muscovite (MVT, mica) is a tetrahedral–octahedral–tetrahedral-structured phyllosilicate (Klien et al. 2008). There are reactive –OH groups on the surface of MVT and these groups are accessible to prepare organic–inorganic nanocomposites (Xie and Wang 2010). Huang et al. reported an anti-UV CTS/mica copolymer (Huang et al. 2010). The modified mica distributes homogeneously on the surface of the copolymer films. Xie et al. introduced MVT during the free radical graft polymerization of CTS and AA and got granular composite hydrogels (Xie and Wang 2010). AA was grafted onto CTS and –OH groups of MVT participated in the reaction. Ion-exchanged MVT could improve water absorbency and swelling rate compared to the pristine MVT.

Sodium humate (SH) is composed of multifunctional aliphatic components and aromatic constituents. SH contains a large number of functional groups, for example, carboxylates and phenolic hydroxyls (Hayes et al. 1989). A series of superabsorbents based on SH was prepared considering that SH can regulate plant growth, accelerate root development, enhance photosynthesis, improve soil cluster structures, and enhance the absorption of nutrient elements (Zhang et al. 2005).

On the basis of SH-based superabsorbents, we introduced SH into the CTS-g-PAA system instead of clay minerals. Liu et al. reported a novel CTS-g-PAA/SH composite hydrogel via graft polymerization (Liu et al. 2007). CTS and SH participated in graft polymerization reaction with AA. The introduced SH could enhance water absorbency. APT and SH were also incorporated together into the CTS-g-PAA system (Zhang et al. 2011, Zhang and Wang 2010). The synergistic effect of APT and SH on swelling capacity was observed.

In spite of unique properties, CTS is only soluble in acidic solution, which hindered its applications. Chemical modification is a frequently used method to make it soluble in water. Various studies were conducted to make derivatives of CTS, such as PEG grafting, sulfonation, N- and O-hydroxylation, and carboxymethyl CTS (Wang et al. 2006). The derivatives of CTS were also used for preparing composite hydrogels. A novel N-succinylCTS-g-polyacrylamide/APT (NSC-g-PAM/APT) composite hydrogel was prepared by inverse suspension polymerization (Li et al. 2007b). –OH of APT and –OH and –NHCO of NSC participated in graft polymerization with AM. The introduced APT could enhance the thermal stability of the hydrogel. The composite hydrogel has a microporous surface and shows higher swelling rate compared to that of without APT. Ma et al. prepared pH- and temperature-responsive semi-interpenetrating polymer network (semi-IPN) hydrogels of carboxymethyl CTS with poly(N-isopropylacrylamide) (PNIPAm) crosslinked by Laponite XLG (Ma et al. 2007). The hydrogels exhibited a volume phase transition temperature (VPTT) around 33°C with no significant deviation from the conventional PNIPAm hydrogels.

4.2.2 STARCH-BASED COMPOSITE HYDROGELS

Starch is among the most abundant and inexpensive biopolymers. Most starches are composed of two structurally distinct molecules: amylose, a linear or lightly branched (1 → 4)-linked α-glucopyranose, and amylopectin, a highly branched molecule of (1 → 4)-linked α-glucopyranose with α-(1 → 6) branch linkages.

Starch is one of the ideal biopolymers for the preparation of biopolymer/clay nanocomposites. Various clays (e.g., Laponite, kaolin, and MMT) and starches (e.g., cassava starch, cationic starch, and cornstarch) have been used for the preparation of starch-based nanocomposites. Chivrac et al. reported a new approach to elaborate exfoliated starch-based nanobiocomposites (Chivrac et al. 2008). They used cationic starch as a new clay organomodifier to better match the polarity of the matrix and thus to facilitate the clay exfoliation process.

In 2000, Wu et al. reported synthesis and properties of starch-g-PAM/clay composite hydrogels (Wu et al. 2000). Clays including bentonite and sercite are used for preparing the composite hydrogels. The kind of clay is an important factor affecting absorbent properties since the composite was formed by a reaction between –OH groups of clay and organic groups of starch and AM. In addition, the hydration of clay is also an important factor influencing the swelling capacity.

Li et al. prepared the starch-g-PAM/APT composite hydrogels via graft polymerization reaction (Li et al. 2005). The graft polymerization between –OH groups on APT and monomers took place during the reaction. The introduction of 5% APT could increase the water absorbency from 545 g/g of the neat starch-g-PAM to 1317 g/g (Li et al. 2004, Lin et al. 2001). Zhang et al. prepared the starch phosphate-g-PAM/APT composite hydrogels (Zhang et al. 2006). The composite hydrogel acquired the highest water absorbency of 1268 g/g when 10% APT was incorporated. The introduced starch phosphate and APT endowed the composite with a higher thermal stability, greatly improved water absorbency, and salt-resistant properties.

AA is also a frequently used monomer besides AM to prepare starch- or clay-based composite hydrogels. Li et al. synthesized the starch-g-PAA/APT composite hydrogels with a water absorbency of 1077 in distilled water (Li et al. 2007a). This composite hydrogel could be useful in agricultural and horticultural applications. Luo et al. prepared starch-g-P(AM-co-AA)/MMT nanosuperabsorbent via γ-ray irradiation (Luo et al. 2005). The layers of MMT are exfoliated and uniformly dispersed in the polymer matrix. Hua et al. prepared the starch-g-PAA/SH composite hydrogels (Hua and Wang 2008). SH can improve the swelling rate and reswelling capability.

4.2.3 CELLULOSE-BASED COMPOSITE HYDROGELS

Cellulose, the most abundant natural material, is a frequently used natural polymer for the preparation of nanocomposites. Various clays including MMT and Laponite have been used. Liu et al. reported a clay nanopaper with tough nanofibrillated cellulose matrix for fire retardancy and gas barrier functions, which is of interest in self-extinguishing composites and in oxygen barrier layers (Liu et al. 2011d). Perotti et al. reported bacterial cellulose/Laponite clay nanocomposites with greatly enhanced Young modulus and tensile strength (Perotti et al. 2011).

To prepare hydrogels, derivatives of cellulose soluble in aqueous phase are necessary, which is convenient for the graft polymerization of monomers onto the backbone. CMC is a representative cellulose derivative. It can be easily produced by the alkali-catalyzed reaction of cellulose with chloroacetic acid. The polar carboxyl groups render cellulose soluble in aqueous solution, chemically reactive, and hydrophilic (Wang and Wang 2010a). Li et al. prepared the CMC/MMT composite hydrogels for the controlled release of acetochlor (Li et al. 2008). The performance of inorganic clays on slowing the release of acetochlor is related to their sorption capacities. Wang et al. reported a nanocomposite hydrogel of CMC and APT by free radical polymerization of CMC, AA, and APT (Wang et al. 2010a). APT nanofibers were retained in nanocomposite and uniformly dispersed in the CMC-g-PAA matrix. The thermal stability and water absorption of the nanocomposites were improved by APT.

REC, medical stone (MS), and SH were also introduced into CMC-g-PAA instead of APT and promising results were obtained (Chen et al. 2010, Wang and Wang 2011, Wang et al. 2011f). REC was exfoliated and uniformly dispersed in the CMC-g-PAA matrix (Figure 4.7). The thermal stability and swelling capability were improved by introducing REC. The nanocomposite showed excellent responsive properties and reversible on–off switching characteristics in various saline, pH, and organic solvent/water solutions (Wang and Wang 2011). The incorporation of 20% MS evidently enhanced the water absorbency by 100% (Wang et al. 2011f). Chen et al. introduced APT and SH simultaneously into the CMC-g-PAA matrix (Chen et al. 2010). The composite hydrogel acquired the highest water absorbency of 582 g/g when 5% HA and 30% APT were introduced. Wang et al. prepared the CMC-g-PAA/APT/SA composite hydrogel beads by ionic cross-linking (Wang et al. 2012a). Introducing 20% APT into CMC-g-PAA

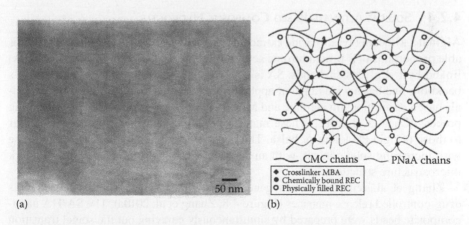

(a)   (b)

FIGURE 4.7 (a) TEM image of CMC-g-PAA/REC (5 wt% REC) composite hydrogel and (b) the schematic network structure. (Wang, W.B. and Wang, A.Q., Preparation, swelling, and stimuli-responsive characteristics of superabsorbent nanocomposites based on carboxymethyl cellulose and rectorite. *Polym. Adv. Technol.* 2011. 22. 1602–1611. Copyright Wiley-VCH Verlag GmbH & Co. KGaA. Reproduced with permission.)

hydrogel could change the surface structure of the composite hydrogel beads, decrease the swelling ability, and relieve the burst release of DS.

Ma et al. prepared a semi-IPN CMC/PNIPAm/clay nanocomposite hydrogel using Laponite XLG (Ma et al. 2008). The clay was substantially exfoliated to form nanodimension platelets dispersed homogeneously in the hydrogels and acted as a multifunctional cross-linker. Hydroxyethyl cellulose (HEC) is another representative derivative of cellulose with excellent water solubility and biocompatibility. Because of the existence of abundant reactive –OH groups, HEC is liable to be modified by grafting polymerization with vinyl monomers to derive new materials. APT, VMT, and MS have been used for preparing composite hydrogels together with HEC and AA (Wang et al. 2010a, 2011a,g). The introduction of 5% APT into HEC-g-PAA polymeric network could improve both water absorbency and water absorption rate (Wang et al. 2010a). AA could graft onto HEC chains and the –OH groups of VMT participate in polymerization reaction. Modification of VMT by proper acidification (acidified vermiculite [AVMT]) or organification (OVMT) can improve both water absorbency and initial water absorption rate. HEC-g-PAA/OVMT exhibited the highest swelling capability and initial swelling rate in contrast to HEC-g-PAA/VMT and HEC-g-PAA/AVMT. Introducing 10% MS greatly enhanced the swelling capacity from 162 to 810 g/g (Wang et al. 2011e). Methylcellulose (MC) is a hydrophilic methyl ether of cellulose and has been used in pharmaceutical and food industries. Wang et al. prepared the MC-g-PAA/APT composite hydrogels via free radical polymerization (Wang et al. 2012b). The composite has the highest swelling ability with a weight ratio of AA to MC of 5:1 and 20% APT. Ni et al. reported an environmentally friendly slow-release nitrogen fertilizer based on APT, ethyl cellulose film, and CMC/HEC hydrogel (Ni et al. 2011). The product can reduce nutrient loss, improve use efficiency of water, and prolong irrigation cycles in drought-prone environments.

4.2.4 Sodium Alginate-Based Composite Hydrogels

Alginate, a linear polysaccharide extracted from brown seaweed, is composed of variable proportions of β-D-mannuronic acid (M block) and α-L-guluronic acid (G block) linked by 1–4 glycosidic bonds. SA is a polyelectrolyte with negative charges on its backbone and has a wide range of applications. Low flammability, foamlike materials based on ammonium alginate and MMT were fabricated through a freeze-drying process (Chen et al. 2012). These materials exhibit mechanical properties similar to those of rigid PU foams or balsa. The compressive modulus and density increase with increasing solid content, with an associated change from a layered to network microstructure structure.

Zhang et al. reported in situ generation of SA/HA nanocomposite beads as drug-controlled release matrices (Figure 4.8, Zhang et al. 2010a). The SA/HA nanocomposite beads were prepared by simultaneously carrying out the solgel transition of SA and in situ generation of HA microparticles. This is a new and simple way to decrease the release rate of drugs and overcoming the frequently observed burst release problem of drugs from hydrogel matrices. The HA microparticles were successfully generated in the beads and had a clear influence on the micromorphology. The interfacial interaction between SA and HA was greatly improved owing to the

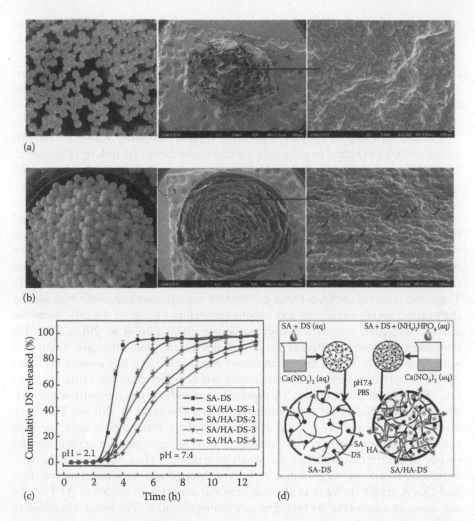

FIGURE 4.8 Digital photographs and SEM images of (a) SA and (b) SA/HA nanocomposite beads. (c) Effect of HA content on in vitro cumulative release profiles of DS from the nanocomposite beads in pH 2.1 and pH 7.4 PBS. (d) Schematic illustration of the role of HA in the nanocomposite beads. (Reprinted from *Acta Biomater.*, 6, Zhang, J.P., Wang, Q., and Wang, A.Q., In situ generation of sodium alginate/hydroxyapatite nanocomposite beads as drug controlled release matrices, 445–454, Copyright 2010, with permission from Elsevier.)

in situ generation of HA. The uniformly dispersed HA microparticles acted as inorganic cross-linkers in the nanocomposite beads and could restrict the movability of the SA polymer chains and then slow down their swelling and dissolution rates, which is also the main reason for the improved drug loading and controlled release behavior. The growth conditions of HA influence the entrapment efficiency and release rate of DS.

Clays including MMT, APT, and hydrotalcite are the most frequently used inorganic materials for the reinforcement of polymeric materials (Ray and Okamoto 2003).

Most of these clays have exchangeable cations between the anionic silicate sheets except for hydrotalcite with exchangeable anions between layers. This unique property has made it receive considerable attention recently and has been used as catalysts and organic–inorganic nanocomposites. However, the hydrotalcite-like materials, layered double hydroxides (LDHs), were frequently used instead of hydrotalcite because LDHs have the advantages of easy to be prepared and tailored, biocompatibility, low cytotoxicity, and low cost.

LDHs were also used to hybrid with SA to prepare composite hydrogels. The pH-sensitive SA/LDH hybrid beads were prepared via surface cross-linking (Zhang et al. 2010b). The positively charged LDHs are adsorbed on the negatively charged SA chains through electrostatic interaction and act as inorganic cross-linkers. Hua et al. also prepared the SA/LDH composite hydrogels by reconstruction of calcined hydrotalcite (Hua et al. 2012). Yang et al. reported a pH-sensitive nanocomposite hydrogel based on SA and APT (Yang et al. 2012). The water absorbency is as high as 694 g/g, which increased by 35% in contrast to the APT-free sample. Shi et al. prepared an SA-based composite hydrogel, SA-g-P(AA-co-St)/APT, by grafted copolymerization of AA and St onto the SA backbones in the presence of APT (Shi et al. 2012). They also reported an SA-g-P(AA-co-St)/APT superporous composite hydrogel by the grafting copolymerization and micelle templating formed by the self-assembled anionic surfactant sodium n-dodecyl sulfonate (SDS) (Shi et al. 2013). The SDS concentration strongly affected the morphologies and pore structure. Compared with SA-g-PAA hydrogel, the simultaneous introduction of a tiny amount of St and APT not only enhanced the swelling ratio but also increased the initial swelling rate. Wang et al. prepared the SA-g-poly(AA-co-sodium p-styrenesulfonate)/APT, SA-g-P(AA-co-NaSS)/APT, composite hydrogels (Wang et al. 2013b). The introduction of NaSS and APT could also improve the surface morphology, swelling capacity, and swelling rate. APT prefers to exist as aggregates or crystal bundles owing to electrostatic and van der Waals interactions. Thus, the unique nanometer characteristics of APT cannot be fully developed and its application was limited (Xu et al. 2011). Wang et al. disaggregated the crystal bundles of APT by the assistance of ethanol during high-pressure homogenization. The better dispersion of APT in SA-g-P(AA-co-St) improved the gel strength and swelling properties of the nanocomposite hydrogels (Figure 4.9) (Wang et al. 2013a).

MS is a special igneous rock composed of silicic acid, alumina oxide, and other elements. MS has excellent multicomponent characteristic, biological activity, and safety and has been widely applied in food science, medicine, environmental sanitation and mineral water, etc. (Juan et al. 2008). Gao et al. prepared a pH-sensitive composite hydrogel based on SA and MS (SA-g-PAA/MS) (Gao et al. 2011). The surface morphology and thermal stability and swelling capacity and rate of the hydrogel were clearly improved by introducing MS.

Li et al. prepared pH- and temperature-responsive semi-IPN magnetic nanocomposite hydrogels by using SA, PNIPAm, and Fe_3O_4 nanoparticles with Laponite XLG as a cross-linker (Li et al. 2012b). The hydrogels exhibited a VPTT at 32°C with no significant deviation from the conventional chemically cross-linked PNIPAm hydrogels. The swelling ratio was much higher than that of PNIPAm hydrogel. Moreover, the swelling ratio of nanocomposite hydrogels gradually decreased with

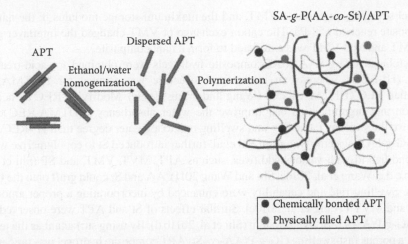

FIGURE 4.9 Schematic illustration of the formation of SA-*g*-P(AA-*co*-St)/APT nanocomposite hydrogels. (From Wang, Y.Z. et al., *Polym. Bull.*, 70, 1181, 2013b. With permission.)

increasing the content of clay and increased with increasing the content of SA. The nanocomposite hydrogels had a much better mechanical property than the PNIPAm hydrogels. The incorporation of clay did not affect the saturation magnetization of the hydrogels. A nanocomposite hydrogel with semi-interpenetrating polymer network was synthesized by in situ polymerization of NIPAm in an aqueous Laponite suspension containing SA (Wang et al. 2011d). This nanocomposite hydrogel reserved thermoresponsibility and high mechanical performance of PNIPAm–Laponite.

4.2.5 GUAR GUM-BASED COMPOSITE HYDROGELS

Guar gum (GG), a representative natural vegetable gum, is a branched polymer with β-D-mannopyranosyl units linked (1–4) with single-membered α-D-galactopyranosyl units occurring as side branches (Wang and Wang 2009b). Recently, various inorganic materials including carbon nanotubes, silica, and clays were used to hybrid with GG for preparing functional materials (Li et al. 2012a, Singh et al. 2009). Zhang et al. prepared the MMT/carrageen/GG gel beads (Zhang et al. 2007c). The thermal stability was better than those of carrageen and GG because of the addition of MMT. MMT also reduced their solidification temperature and swelling ratio.

Wang et al. prepared a series of composite hydrogels by introducing GG and various additives (Wang et al. 2008b, Wang and Wang 2009a,b,c,d,e,f, Wang et al. 2009b, 2011g). APT, SH, MMT, VMT, REC, and MS have been incorporated into the system. The weight ratio of AA to GG and the content of additives have great influences on the swelling ability of the composite hydrogels. Wang et al. also studied the effects of modification of VMT, including organo-VMT and cation-modified VMT, on properties of the composite hydrogels (Wang et al. 2009b, 2011g). HDTMABr intercalated into the gallery layers of VMT. The intercalated VMT has been exfoliated during polymerization and uniformly dispersed in the GG-*g*-PAA matrix. HDTMA–VMT improved the swelling properties more remarkably than VMT. In addition, organo-VMT improved the gel strength compared to VMT, and the maximum storage modulus of the nano-composite reached 658 Pa. The cation exchange of VMT changed the interlayer gap of VMT and M^{n+} VMT was exfoliated to form a nanocomposite.

Exfoliated GG-g-PAA/REC composite hydrogels were obtained for acid-treated REC (H^+-REC), whereas an intercalated structure was formed for HDTMABr-modified REC (HDTMA-REC) (Wang and Wang 2009e). Modifying REC by acidi-fication and organification can improve the water absorbency. HDTMA-REC can improve the swelling capability and swelling rate to a greater degree than H^+-REC. In the GG-g-PAA system, Wang and Shi et al. further introduced St to copolymerize with AA and hybrid with various additives, such as APT, MVT, VMT, and SP (Shi et al. 2011b,c,d,e, Wang et al. 2010e, Shi and Wang 2011). AA and St could graft onto the GG chain. Swelling rate and capability were enhanced by incorporating a proper amount of St and MVT (Wang et al. 2010e). Similar effects of St and APT were observed in the GG-g-P(AA-co-St)/APT system (Shi et al. 2011b,d). By using surfactant as the tem-plate, a porous fast-swelling GG-g-P(AA-co-St)/APT composite hydrogel was prepared (Shi et al. 2011b). Proper amount of SDS and nonionic surfactant p-octyl poly(ethylene glycol)phenyl ether can simultaneously enhance the swelling capacity and swelling rate. Yang et al. developed the GG-g-PAA/APT/SA pH-sensitive composite hydrogel beads for controlled drug delivery via ionic gelation (Yang et al. 2013).

4.2.6 OTHER POLYSACCHARIDE-BASED COMPOSITE HYDROGELS

Psyllium (PSY) is the common name used for several members of the plant genus Plantago. Psyllium is a gel-forming mucilage composed of a highly branched ara-binoxylan. The backbone consists of xylose units, while arabinose and xylose form the side chains. The hydrogels derived from PSY have also been developed as colon-specific drug delivery agent. Singh et al. developed PSY-based hydrogels through graft copolymerization of PSY, PVA, and AA (Singh and Sharma 2010). The use of a very small amount of AA has developed the low-energy, cost-effective, bio-degradable, and biocompatible material for potential biomedical applications. An et al. prepared the PSY-g-PAA hydrogels by graft copolymerization (An et al. 2010). On the basis of the PSY-g-PAA hydrogels, they also reported a series of PSY-based composite hydrogels by incorporating various clays such as APT, biotite, and MMT (An et al. 2011a,b, 2012). AA grafted onto PSY and the –OH groups of clays partici-pated in graft polymerization. The nanocomposite containing 10% APT gives the highest water absorbency (An et al. 2012).

Kappa-carrageenan (κC), a sulfated polysaccharide, belongs to the large family of carrageenans, which are extracted from red seaweeds (An et al. 2010). Due to their excellent physical properties and biological activities, κC is extensively utilized in the food, cosmetics, textile, and pharmaceutical industries. Nanocomposite hydrogels were synthesized by copolymerization of AM and AA in the presence of carra-geenan and MMT (Mahdavinia et al. 2010). The optimum water absorbency was obtained at 10% clay, 10% carrageenan, and 1:1 of monomers. They also prepared the carrageenan-g-PAM/Laponite RD composite hydrogels (Reza et al. 2012a).

Collagen is the most abundant protein in mammals and is composed of glycine–proline–(hydroxy)proline repeats. Collagen is frequently used for preparing hydrogels because of its unique properties and biocompatibility. Collagen is also used to prepare nanocomposites with clays. The polypeptide/MMT nanocomposites were prepared using fish-based collagen peptide (Teramoto et al. 2007). Marandi reported the collagen-*g*-P(AA-*co*-AM)/MMT nanocomposite hydrogels (Marandi et al. 2011). MMT can improve water retention of the hydrogels under heating.

4.3   APPLICATIONS IN REMOVAL OF POLLUTANTS

Nowadays, pollution becomes a serious problem accompanying the rapid development of economy. Water pollution causes approximately 14,000 deaths per day, mostly due to the contamination of drinking water by untreated sewage in developing countries including China, India, and Zambia. Pollution in China is one aspect of the broader topic of environmental issues in China. The specific contaminants leading to water pollution include a wide spectrum of chemicals, pathogens, and physical or sensory changes. The chemical contaminants may include organic (e.g., detergents, insecticides, and dyes) and inorganic (e.g., heavy metals and ammonia) substances.

Effective pollution control is necessary to minimize the adverse effects of water pollution. Several methods including adsorption, coagulation, and precipitation have been developed to remove pollutants from wastewaters (Demirbas 2008). The adsorption method is the most frequently used because of its flexibility in design and operation as well as easy regeneration of the adsorbent. It is gradually recognized that using low-cost and environmentally friendly adsorbents to remove pollutants is an effective and economical method of water decontamination. Many materials have been used as adsorbents including CTS, clays, zeolite, and others. However, low adsorption capacity and rate are the frequently met problems for these low-cost adsorbents, which seriously restrict their applications. Approximately 3 h are needed to reach adsorption equilibrium using acid-activated APT for the removal of Cu(II) from aqueous solution (Chen et al. 2007).

Recently, the application of hydrogels as adsorbents for the removal of heavy metal ions, ammonium nitrogen, and dyes has been paid much attention (Wang et al. 2009c, Zheng and Wang 2009). Materials including CTS (Gao et al. 2013), hemicelluloses (Peng et al. 2012), polydopamine (Gao et al. 2013), and poly(dimeth yldiallylammoniumchloride)/PAM (Zheng and Wang 2010a) have been used to prepare hydrogels for the removal of pollutants in water. The flexible polymer chains and functional groups of a hydrogel make it swell partly once immersed in aqueous solution. This is convenient for the penetration of pollutants with water into the 3D network of hydrogels and bind with the functional groups. On the basis of our previous work about preparation of polysaccharide-based composite hydrogels, we studied the application of these hydrogels for the removal of pollutant from aqueous solution.

4.3.1   REMOVAL OF HEAVY METALS

The main threats to human health from heavy metals are associated with exposure to lead, cadmium, mercury, and arsenic. Although several adverse health effects of heavy metals have been well known, exposure to heavy metals continues and is even increasing in some parts of the world. Meanwhile, the problem of coping with the presence of heavy metals has become a top priority in water treatment. Various materials such as mineral apatite, mesoporous silicas, and CTS have been used for the removal of heavy metals from water (Chen et al. 1997). The transformed quantum dots showed characteristic color development, with Hg^{2+} being exceptionally identifiable due to the visible bright yellow color formation. However, there are some other problems for these adsorbent besides the aforementioned low adsorption capacity and rate. For example, CTS is an excellent adsorbent for Hg(II); however, its solubility in acid restricts its practical applications.

Hydrogels are promising materials for the removal of heavy metals. Zheng et al. reported a granular hydrogel that is an excellent adsorbent for the removal of Cu(II) and Ni(II) (Zheng and Wang 2012). Liu et al. studied the adsorption of Cu(II) using cross-linked CTS hydrogels impregnated with Congo red (CR) by ion-imprint technology (Liu et al. 2012b). Wang et al. reported the adsorption of Hg(II) ions from aqueous solution using the CTS-g-PAA/APT composite hydrogel (Wang and Wang 2010b). The adsorption of Hg(II) on the composites is chemical adsorption. All the –COOH, –NH_2, and –OH groups in the composites participate in the adsorption process. The composite can also adsorb a large amount of Hg(II) even at the very low pH_0 (pH_0 2.00). The adsorption of the composites for Hg(II) were all better fitted for the pseudo-second-order model and the Langmuir model, respectively. The values of ΔG of the composites for Hg(II) are all negative, which indicates the spontaneous nature of the adsorption process. The positive ΔH confirms that the adsorption of the composites for Hg(II) are all endothermic processes. The introduction of APT into CTS-g-PAA not only improves the adsorption ability and rate but also reduced the cost.

The CTS-g-PAA/APT composite hydrogels are also quite effective adsorbents for the removal of Cu(II) (Wang et al. 2009g). The introduction of APT into CTS-g-PAA could generate a loose and porous surface, which improved the adsorption ability to some extent. Compared with other adsorbents listed in Table 4.1, the equilibrium adsorption capacity of the composites are quite high. The monolayer coverage of Cu(II) on the surface of the composites was in the ascendant. The high adsorption capacity and average desorption efficiency during the consecutive five-time adsorption–desorption processes of CTS-g-PAA/APT composites implied that the composites possess the potential of regeneration and reuse.

Wang et al. also studied the performance of the CTS-g-PAA/APT composite hydrogel for the removal of Cd(II) (Wang and Wang 2010c). When $Cd(CH_3COO)_2$ was used as the solute, the equilibrium adsorption capacity was evidently larger than that of the other three cadmium salts ($Cd(NO_3)_2$, $CdCl_2$, $CdSO_4$) (Table 4.2). Results from kinetic experiments showed that Cd(II) adsorption rate on the composite was quite fast and more than 90% of Cd(II) adsorption occurred within the initial 3 min, and the adsorption equilibrium may be reached within 10 min.

The adsorption capacity of the CTS-g-PAA/VMT composites for Cu(II) decreases with the increase of the VMT content, but the adsorption capacity of sample with 30% VMT is still as high as 220 mg/g (Wang et al. 2010f). Wang et al. also compared the abilities of the CTS-g-PAA/VMT composite hydrogels to remove Pb(II) and Cd(II) ions (Wang and Wang 2012). The adsorption is fast. Ninety percent of the total adsorption was reached in 3 min for both Pb(II) and Cd(II) at 303 K, and the equilibrium was reached in 15 min. The maximum Pb(II) adsorption capacity

TABLE 4.1
Adsorption Capacity of the Adsorbents for Cu(II)

Adsorbents	Adsorption Capacity (mg/g)	Adsorbents	Adsorption Capacity (mg/g)
CTS	17.79	CTS-g-PAA/10% APT	243.76
N,O-carboxymethyl CTS	162.5	CTS-g-PAA/20% APT	226.95
Alumina/CTS membranes	200	CTS-g-PAA/50% APT	170.65
Nylon membrane	10.80	PAM/APT	104.22–115.03
Clinoptilolite	25.42	St-g-PAA	177.94
AC	33.12	St-g-PAA/5% SH	179.85
Modified pine bark	44.49	CTS-g-PAA/30% VMT	220
Thiourea-modified CTS	66.09	MC-g-PAA/5% APT	253.55
Cu(II)-imprinted composite	71.18	MC-g-PAA/10% APT	242.06
Acid-activated APT	32.24	MC-g-PAA/20% APT	231.55
CTS-g-PAA	262.25	MC-g-PAA/30% PT	166.45

Sources: Data from Zheng, Y.A. and Wang, A.Q., *J. Hazard. Mater.*, 171, 671, 2009; Zheng Y.A. et al., *Desalination*, 263, 170, 2010.

TABLE 4.2
Effect of Species of Cadmium Salts on Adsorption Capacity for Cd(II)

Sample	Maximum Adsorption Capacity for Cd(II) (mg/g)			
	$Cd(NO_3)_2$	$CdCl_2$	$CdSO_4$	$Cd(CH_3COO)_2$
CTS-g-PAA/30% APT	133.7	126.8	130.2	323.2

Source: Data from Wang, X.H. et al., *China Mining Mag.*, 19, 101, 2010f.

(3.08 mmol/g) of the composite is only a little more than that of Cd(II) (2.98 mmol/g). However, the desorption efficiency of Pb(II) (63.27%) is much lower than that of Cd(II) (86.26%) using 0.1 mol/L HNO_3. The adsorption of Pb(II) and Cd(II) ions by the composite seemed to involve ion exchange, chelation, electrostatic attraction, or adsorption. $-NH_2$, $-COOH$, and $-OH$ groups are involved in the adsorption process. Moreover, only a few of the $-NH_2$ groups participate in the reaction with Pb(II) ions.

The CTS-g-PAA/SP adsorbent has a coarse, porous, and accidented surface, which is convenient for the adsorption of pollutants (Zheng et al. 2009a). The adsorption capacity of the adsorbent for Pb(II) is 638.9 mg/g, about three times of SP. After five cycles of adsorption–desorption, the adsorption capacity decreased to 489.2 mg/g, 76.6% of its original adsorption capacity. However, no adsorption can be observed after three cycles of adsorption–desorption when SP was used as the adsorbent. Shi et al. prepared the CTS-g-PAA/SP composite hydrogel with a specific surface area of 82.17 m²/g, which increases the adsorption capacity (Shi et al. 2011a). The saturated adsorption capacities of the composites for Pb(II) and Hg(II) are 286.95 and 371.89 mg/g.

The CTS-g-PAA/APT/SH composite hydrogels are excellent adsorbents for the removal of Pb(II) (Zhang et al. 2011, Zhang and Wang 2010). The adsorbed Pb(II) could be desorbed from the composite hydrogel efficiently using 0.05 mol/L HCl solution. In addition, the APT and SH are helpful for the recycling of the composite hydrogels, and the adsorption capacity is kept higher than 590 mg/g after five times of consecutive adsorption–desorption. The results indicate that the composite hydrogels are of high-performance, low-cost, and recyclable green adsorbents.

Szymon et al. studied the adsorption of polyvalent metal cation on the graft copolymers of potato starch with AM and AA or N-vinylformamide in the presence of 1%–16% MMT (Szymon et al. 2013). The adsorption of the composite was higher for Cd(II) than for Fe(III). Especially distinct differences were found for starch-based copolymers of poly(N-vinylformamide). Practically no influence of MMT content in the copolymers on the cation sorption was observed.

The St-g-PAA/SH composite hydrogels are effective for the removal of Cu(II) (Zheng et al. 2010). The composite hydrogel has higher adsorption capacity than the other adsorbents reported (Table 4.1). The addition of 5% SH into St-g-PAA not only improves the adsorption rate and capacity for Cu(II) but also makes the regeneration easier. The St-g-PAA/SH hydrogels exhibited high adsorption capacity in a wide pH range. The presence of Na^+ has no obvious effects on Cu(II) adsorption, whereas the presence of Pb(II) in the solution greatly affects the adsorption capacity. This is because the composite hydrogel adsorbs cations via complexation.

Liu et al. studied Pb(II) adsorption using the CMC-g-PAA/APT composite hydrogel (Liu et al. 2010a). The adsorption of Pb(II) is quite fast. Adsorption equilibrium could be reached within 60 min, and more than 90% of the equilibrium adsorption was achieved in 10 min. The adsorption process can be well described by the pseudo-second-order kinetic model, and the equilibrium adsorption isotherm was closely fitted with the Langmuir model. Complexation is the main adsorption mechanism of the composite hydrogel for Pb(II). The MC-g-PAA/APT composite hydrogels are ideal materials for the removal of cations (Wang et al. 2012a). Figure 4.10 shows the equilibrium absorption capacity of active carbon (AC) and the MC-g-PAA/APT nanocomposites with 5%, 10%, 20%, and 30% APT for Ni(II), Cu(II), and Zn(II). The adsorption capacities of the nanocomposite with 5% APT for Ni(II), Cu(II), and Zn(II) were 9.86, 7.66, and 21.86 times greater than AC, respectively. This confirmed that the chemical complexation adsorption of the nanocomposite is advantageous over the physical adsorption interaction of AC. The adsorption capacities of nanocomposite are still 6.7 (for Ni(II)), 5.01 (for Cu(II)), and 16.4 (for Zn(II)) times of AC even if the amount of APT reached 30%, which is beneficial to reduce the production cost.

Gao et al. studied the removal of heavy metal ions using the SA-g-PAA/MS composite hydrogels (Gao et al. 2011). Compared to AC, the adsorption capacity of the composite hydrogels for Ni(II), Cu(II), Zn(II), and Cd(II) ions increased by 10.4, 8.0, 23.0, and 14.3 fold, respectively. An et al. prepared the PSY-g-PAA/MMT composite hydrogels by free radical graft copolymerization, which are effective in adsorbing Cu(II) ions (An et al. 2011b).

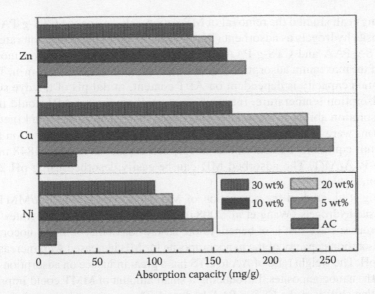

FIGURE 4.10 Absorption capacities of AC and the MC-*g*-PAA/APT nanocomposites with 5%, 10%, 20%, and 30% APT for Ni(II), Cu(II), and Zn(II) ions, respectively. (From Wang, W.B. et al., *J. Macromol. Sci. Part A*, 49, 306, 2012b. With permission.)

4.3.2 Removal of Dyes

Toxic chemical dyes are frequently used in the textile, printing, and paper industries. In China, about 1.6 billion tons of dye-containing wastewater is produced every year, but only a small proportion is recycled. Being highly colored, dyes are readily apparent in wastewater even in very low concentration. In addition, the colored wastewater can reduce the light penetration thus limiting the depuration by natural oxidation. Consequently, the presence of toxic dyes in aqueous streams poses a direct threat to the environment and humans and has gained worldwide attention. Many techniques such as ion exchange, chemical precipitation, coagulation, ozonation, and adsorption (Yan et al. 2009, Zhang et al. 2009) have been developed to remove dyes from aqueous solution. Adsorption is considered as a preferred and effective technique among these techniques as it can deal with various concentrations of dyes and it does not induce the formation of hazardous materials. The most widely employed adsorbent is AC, but the cost and difficulty in separation after adsorption hinder its large-scale application. Thus, many researchers focused on more effective adsorbents to treat dye-containing wastewater.

Despite that several low-cost adsorbents such as VMT, hazelnut shell, and rice husk ash have been investigated for decontamination purpose, the adsorption capacities of some adsorbents were restricted. So, more effective adsorbents should be attempted. Recently, because of their excellent characteristics such as high-molecular and oxygen permeability and low interfacial tension, hydrogels have attracted extensive attention for dye removal.

Wang et al. studied the removal of MB cationic dye using the CTS-*g*-PAA/APT composite hydrogels as adsorbent (Wang et al. 2011b). The adsorption rate of MB on CTS-*g*-PAA and CTS-*g*-PAA/APT with 30% APT was fast, and more than 90% of the maximum adsorption capacities were achieved within 15 min. The MB adsorption capacity is dependent on APT content, initial pH of the dye solution, and adsorption temperature. Introducing a small amount of APT could improve the adsorption ability of the CTS-*g*-PAA composite. The adsorption kinetics and isotherms were in good agreement with a pseudo-second-order equation and the Langmuir equation. The maximum adsorption capacities reached 1848 mg/g for CTS-*g*-PAA/APT. The adsorbed MB can be easily desorbed using pH 2.0 HCl solution.

Wang et al. studied the adsorption of MB using the CTS-*g*-PAA/MMT nanocomposite hydrogels (Wang et al. 2008a). The introduced MMT generates a loose and porous surface that is of benefit to the adsorption ability of the nanocomposite. The adsorption capacity of the nanocomposite for MB increased with increasing the initial pH. The weight ratio of AA to CTS has a great influence on adsorption capacities of the nanocomposites. Introducing a small amount of MMT could improve the adsorption ability of the CTS-*g*-PAA hydrogel. The maximum adsorption capacity was 1859 mg/g for CTS-*g*-PAA/MMT with 30% MMT. In desorption studies, comparatively high desorption of dyes was obtained using pH 2.0 acidic solution. So, the CTS-*g*-PAA/MMT nanocomposite hydrogel is a very effective adsorbent for the removal of MB from aqueous solution.

Liu et al. studied the adsorption of MB by the CTS-*g*-PAA/VMT composite hydrogels (Liu et al. 2010b). By introducing 10% VMT into the CTS-*g*-PAA polymeric network, the composite hydrogel showed the highest adsorption capacities for MB and could be used as a potential adsorbent for the removal of cationic dye in wastewater. They also studied the adsorption of MB by the CTS-*g*-PAA/biotite composite hydrogels (Liu et al. 2011c). The adsorption capacity calculated from the Langmuir isotherm was 2125.70 mg/g for CTS-*g*-PAA/10% biotite at 30°C. The adsorption capacity was much higher compared with other hydrogels with the same content of other clays including MMT, VMT, and APT (Table 4.3) (Liu et al. 2010b, Wang et al. 2008a, 2011b).

A granular hydrogel based on CTS and AA was prepared using Fenton reagent as the redox initiator under the ambient temperature in an air atmosphere (Figure 4.11) (Zheng et al. 2012a). The resulting networks serve as the micro- or nanoreactors for producing highly stable silver nanoparticles via in situ reduction of $AgNO_3$ with $NaBH_4$ as reducing agent and Al^{3+} as surface cross-linking agent. MB and CR with an initial concentration of 20 mg/L can be reduced completely within 30 min, and the catalytic activity is independent on initial pH and ion strength. In addition, the Ag-entrapped hydrogel shows excellent reusability for 10 successive cycles, with no appreciable decrease in the catalytic effects.

The NSC-*g*-PAM/APT composite hydrogels were quite effective adsorbents for the removal of MB (Li et al. 2011a). All the adsorption processes were better fitted by pseudo-second-order equation and the Langmuir equation. The adsorption capacity of the composite hydrogel was higher than those of CTS, APT, and many other adsorbents.

TABLE 4.3
Adsorption Capacity of Several Hydrogels for MB

Adsorbents	Adsorption Capacity (mg/g)	Adsorbents	Adsorption Capacity (mg/g)
Raw date pits	281	CTS-g-PAA/10% MMT	1875
H-mag	173	CTS-g-PAA/10% VMT	1612.32
Na-mag	331	CTS-g-PAA/10% APT	1870
HA-AM-PAA-B	242.4	CTS-g-PAA/10% BT	2125.70
Poly(N,N-dimethylacrylamide-co-SA)	600	CTS-g-PAA/10% HNTs	1034
Polymer modified biomass	869.6	CMC-g-PAA/5% APT	2094.80
MWS	450.0 ± 14.4	CMC-g-PAA/20% APT	1979.48
CMC-g-PAA	2153.50		

Source: Data from Liu, A.D. et al., *Biomacromolecules*, 12, 633, 2011a.

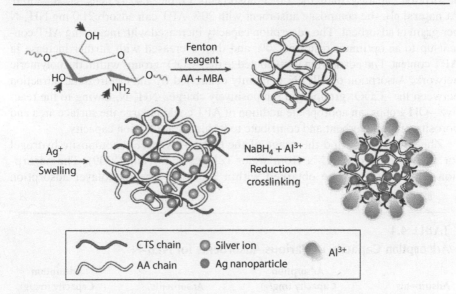

FIGURE 4.11 Schematic representation for the Ag nanoparticles formation within the hydrogel network. (Reprinted from *Chem. Eng. J.*, 179, Zheng, Y.A., Xie, Y.T., and Wang, A.Q., Rapid and wide pH-independent ammonium-nitrogen removal using a composite hydrogel with three-dimensional networks, 90–98, Copyright 2012, with permission from Elsevier.)

The CMC-g-PAA/APT composite hydrogel was used for the adsorption of MB (Table 4.3) (Liu et al. 2011b). The adsorption of the hydrogel for MB was quite fast, and the adsorption equilibrium could be reached within 30 min. The composite hydrogel has very high adsorption capacity for MB. The adsorption capacity is still as high as 1979.5 mg/g at 30°C even when 20% APT was introduced. The adsorbents exhibited excellent affinity for MB and can be applied to treat wastewater containing basic dyes.

4.3.3 REMOVAL OF AMMONIUM NITROGEN AND OTHERS

(NH_4^+-N) is an important member of the nitrogen-containing compounds that act as nutrients for aquatic plants and algae. However, if the (NH_4^+-N) level in water is too high, they can be toxic to some aquatic organisms, and if enough nutrients are present, eutrophication may occur, which is a serious problem throughout the world and poses a direct threat to public health. Therefore, the total removal or at least a significant reduction of (NH_4^+-N) in water is thus obligatory prior to disposal into streams, lakes, and seas. As the convenient and effective approach, natural or modified zeolite and molecular sieve have been used as the adsorbents to remove (NH_4^+-N) in water. However, the adsorption capacity is too low to satisfy the effective demand.

Hydrogels were also promising materials for the removal of (NH_4^+-N). Yuan and Kusuda used a poly(NIPAm-co-chlorophyllin) hydrogel to adsorb (NH_4^+-N). The hydrogel had an adsorption capacity of 0.18 mg/g for (NH_4^+-N) (Yuan and Kusuda 2005). Zheng et al. reported that the adsorption of CTS-g-PAA/APT composite hydrogel for (NH_4^+-N) is pH dependent and has faster adsorption kinetics and higher adsorption capacity than the other adsorbents (Table 4.4) (Zheng et al. 2009). At natural pH, the composite adsorbent with 20% APT can adsorb 21.0 mg NH_4^+-N per gram of adsorbent. The adsorption capacity increased with increasing APT content up to an optimum result of 20% and then decreased with further increase in APT content. The adsorption is attributed to the $-COO^-$ groups within the polymeric network. Adsorption of NH_4^+-N is mainly controlled by the electrostatic attraction between the $-COO^-$ groups and the positively charged NH_4^+-N. Owing to the reactive $-OH$ groups, an appropriate addition of APT could enlarge the surface area and porosity of the adsorbent and contribute to its higher adsorption capacity.

Zheng et al. reported the usage of the CTS-g-PAA/REC composite hydrogel for the removal of NH_4^+-N from water (Zheng and Wang 2009). The adsorption equilibrium can be obtained within 3–5 min. The monolayer adsorption

TABLE 4.4
Adsorption Capacity of Various Adsorbents for NH_4^+-N

Adsorbents	Adsorption Capacity (mg/g)	Adsorbents	Adsorption Capacity (mg/g)
CTS-g-PAA/20% APT	21.0	Zeolite 13X	8.61
MMT	7.8	Activated sludge	0.4–0.5
Kaolin	5.5	PNIPAm	0.08
APT	8.5	Poly(NIPAm-co-chlorophyllin)	0.18
REC	6.4	CTS-g-PAA	109.2
PAC	5.7	CTS-g-PAA/10% REC	123.8
Natural zeolite	1.0–1.5	CTS-g-PAA/30% REC	61.95
Clinoptilolite	8.96	CTS-g-PAA/HNTs	27.7

Source: Data from Zheng, Y.A. et al., *Chem. Eng. J.*, 155, 215, 2009b.

is 109.2, 123.8, and 61.95 mg/g for CTS-g-PAA, CTS-g-PAA/10% REC, and CTS-g-PAA/30%REC, respectively, meaning high adsorption capacity for NH_4^+ removal. This hydrogel can be extensively used in a wide pH range from 4.0 to 9.0, and external temperature has little effects on the adsorption capacity, suggesting that the adsorbent is applicable to a large class of adsorption systems. Cations, especially multivalent cations, which coexisted with NH_4^+ in the solution, can have some negative effects on the adsorption capacity of CTS-g-PAA/REC. However, the adsorption capacity for NH_4^+-N is still higher than 30 mg/g even with the existence of high-concentration saline solutions (0.5 mmol/L). The adsorbent can be regenerated by desorption of NH_4^+-N. The regeneration condition is mild and the recovered adsorbent can be used again. In addition, more adsorption sites would be created during the regeneration.

The CTS-g-PAA/HNT composite hydrogels also can remove NH_4^+-N from synthetic wastewater (Zheng and Wang 2010c). By introducing HNTs into the polymeric networks, the composite shows comparable adsorption capacity to that of pure polymer hydrogel. The adsorption equilibrium can be achieved within 5 min with the equilibrium adsorption capacity of 27.7 mg/g. The adsorption process is pH independent within 4.0–7.0, and the adsorption capacity of the as-prepared adsorbent is not affected over the five cycles of adsorption.

Zheng et al. found that the CTS-g-PAA/VMT composite hydrogels are rapid and wide pH-independent NH_4^+-N adsorbents (Figure 4.12) (Zheng et al. 2012). The adsorption equilibrium can be achieved in a few minutes and the maximum, constant adsorption capacity can be observed in a wide pH range of 4.0–8.0. The adsorbent can exclusively adsorb NH_4^+-N in the mixing solution containing NH_4^+-N and PO_4^{3-} P, suggesting that the electrostatic attraction dominates the whole adsorption process.

The ion cross-linked CTS-g-PAA/VMT composite hydrogels were applied for the removal of phosphate ions from aqueous solution (Zheng and Wang 2010b). The trivalent ion cross-linked hydrogel exhibited a potential for the removal of phosphate ions. A lower pH is convenient for the adsorption of phosphate ions onto the adsorbent. The maximum adsorption capacity of 22.64 mg/g was comparable with that reported for other adsorbents (Table 4.5).

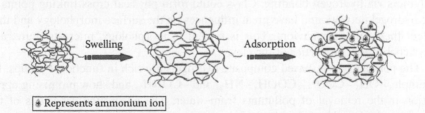

FIGURE 4.12 Model scheme for the adsorption of NH_4^+-N onto CTS-g-PAA/VMT. (Reprinted from *Chem. Eng. J.*, 179, Zheng, Y.A., Xie, Y.T., and Wang, A.Q., Rapid and wide pH-independent ammonium-nitrogen removal using a composite hydrogel with three-dimensional networks, 90–98, Copyright 2012, with permission from Elsevier.)

TABLE 4.5
Adsorption Capacity of Various Adsorbents for Phosphate

Adsorbents	Adsorption Capacity (mg/g)	Reference
Al-CPV	22.64	Zheng and Wang (2010a)
Za-ZFA	35.31	Wu et al. (2006)
Iron oxide tailing	8.21	Zeng et al. (2004)
Aluminum oxide	35.03	Borggaard et al. (2005)
Steel furnace slag	1.43	Sakadevan and Bavor (1998)
Natural zeolite	0.46	Drizo (1998)
DMAEMA-g-PP	11.4[a]	Taleb et al. (2008)
Ammonium-functionalized MCM-48	35.4[b]	Saad et al. (2007)
Mesoporous ZrO_2	29.71	Liu et al. (2008)
Iron oxide/fly ash composite	27.39	Yao et al. (2009)

Source: Zheng, Y.A. and Wang, A.Q., *Adsorpt. Sci. Technol.*, 28, 89, 2010b.
[a] Initial phosphate concentration, 100 mg/L.
[b] Initial phosphate concentration, 300 mg/L.

4.4 CONCLUDING REMARKS AND PERSPECTIVES

Polysaccharide-based composite hydrogels with novel properties develop very quickly recently. Polysaccharides including CTS, cellulose, starch, SA, and GG have been used for preparing the composite hydrogels. Various clays with different structure and properties such as APT, MMT, VMT, and Laponite were introduced into the polymeric 3D network.

Owing to the synergistic effect among polysaccharide, vinyl monomer, and clay mineral, many of the physicochemical properties such as swelling ratio and rate, thermostability, and gel strength of the composite hydrogels are superior to their counterparts. The –OH and –NH₂ groups of many polysaccharides take part in graft polymerization with the vinyl monomers. The –OH groups of clay minerals may also react with the monomers and interact with polysaccharides and grafted hydrophilic polymers via hydrogen bonding. Clays could form physical cross-linking points in the hydrogel network and have great influences on the surface morphology and then affect the swelling behaviors. This is related to morphology, microstructure, and physicochemical properties of the clay.

The polysaccharide-based composite hydrogels are rich in functional groups, for example, –OH, –COO⁻, –COOH, –NH₂, and –CONH₂, and show promising application in the removal of pollutants from water. The adsorption capacities of the composite hydrogels are very high for heavy metal ions, dyes, and NH_4^+-N compared to the traditional adsorbents such as AC and clays. Swelling of the composite hydrogels is helpful for the penetration of pollutant molecules into the hydrogel network and their interaction between the functional groups. This makes the adsorption very fast and the adsorption capacity very high.

Most of the studies introduce clays directly into the polymeric network. Thus, the compatibility of polymer and clays remains to be improved. Modification of clay with proper reagents is helpful to improve their compatibility.

Before using for practical applications, some problems need to be solved for the polysaccharide-based composite hydrogels. Effort in this area is still needed in order to further optimize the composite hydrogel's performance in the future. (1) A proper swelling degree of the composite is helpful to enhance the adsorption capacity and rate. However, too high swelling ratio will result in handling difficulty after adsorption. Thus, further cross-linking points are necessary to maintain proper swelling ratio while keeping their excellent adsorption capacity. (2) Most of the reports studied the adsorption capacity with very high concentration of the pollutants, which is far from practical applications. The concentration of metal ions in industrial wastewater is not very high, for example, below 10 ppm for Ni (II). Thus, further studies should focus on the application of the composite hydrogels for the removal of pollutants at low concentration. (3) The reusability still needs to be improved, although many composite hydrogels show good adsorption capacity after five adsorption–desorption recycles. Stability for even more cycles should be investigated. In addition, the repeating adsorption–desorption will result in decrease of their gel strength and partly loss of the composite hydrogels, which should be addressed in further studies. The dissolved composite hydrogels may result in secondary pollution and must be avoided. (4) The selectivity of the composite hydrogels for pollutants is also very important for their practical applications. The coexistence of many other metal ions and compounds in various forms will make the adsorption process very complicated. Thus, composite hydrogels with excellent selectivity for pollutants are in high demand and should become a focus in further studies.

ABBREVIATIONS

AA	Acrylic acid
AC	Active carbon
AM	Acrylamide
APT	Attapulgite
CMC	Carboxymethyl cellulose
CR	Congo red
CTS	Chitosan
CV	Crystal violet
$CVNH_4^+$-N	Crystal violet ammonium nitrogen
DS	Diclofenac sodium
GDL	Gluconolactone
GG	Guar gum
HA	Hydroxyapatite
HDTMABr	Hexadecyltrimethylammonium bromide
HEC	Hydroxyethyl cellulose
HEMA	2-Hydroxyethyl methacrylate

HES	Hydroxyethyl starch
HESDS	Hydroxyethyl starch diclofenac sodium
HNTs	Halloysite nanotubes
κC	Kappa-carrageenan
LDHs	Layered double hydroxides
MB	Methylene blue
MBA	N,N'-methylenebisacrylamide
MC	Methylcellulose
MMT	Montmorillonite
MS	Medical stone
NH_4^+-N	Ammonium nitrogen
NIPAm	N-Isopropylacrylamide
OFL	Ofloxacin
PAM	Polyacrylamide
PSY	Psyllium
REC	Rectorite
SA	Sodium alginate
SDS	Sodium n-dodecyl sulfonate
SH	Sodium humate
SP	Sepiolite
St	Styrene
TPP	Sodium tripolyphosphate
TPS	Thermoplastic starch
VMT	Vermiculite
VPTT	Phase transition temperature

REFERENCES

An, J.K., W.B. Wang, and A.Q. Wang. 2010. Preparation and swelling properties of a pH-sensitive superabsorbent hydrogel based on psyllium gum. *Starch/Stärke* 62: 501–507.

An, J.K., W.B. Wang, and A.Q. Wang. 2011a. Preparation and swelling properties of ph responsive psyllium-graft-poly(acrylic acid)/biotite superabsorbent composites. *Polym. Mater. Sci. Eng.* 27: 31–34.

An, J.K., W.B. Wang, and A.Q. Wang. 2011b. Preparation of psyllium gum-*g*-poly(acrylic acid)/Na-montmorillonite supersorbent composites and its adsorption performance for Cu^{2+}. *J. Funct. Polym.* 24. 186–190.

An, J.K., W.B. Wang, and A.Q. Wang. 2012. Preparation and swelling behavior of a pH-responsive psyllium-*g*-poly(acrylic acid)/attapulgite superabsorbent nanocomposite. *Int. J. Polym. Mater.* 61: 906–918.

Borggaard, O.K., B. Raben-Lange, A.L. Gimsing, and B.W. Strobel. 2005. Influence of humic substances on phosphate adsorption by aluminium and iron oxides. *Geoderma* 127: 270–279.

Bradley, W.F. 1940. The structural scheme of attapulgite. *Am. Miner.* 25: 405–410.

Chen, H., W.B. Wang, and A.Q. Wang. 2010. Preparation and swelling behaviors of CMC-*g*-PAA/APT/HA superabsorbent composites. *Humic Acid* 5: 5–10.

Chen, H., Y.G. Zhao, and A.Q. Wang. 2007. Removal of Cu(II) from aqueous solution by adsorption onto acid-activated palygorskite. *J. Hazard. Mater.* 149: 346–354.

Chen, H.B., Y.T. Wang, M. Sánchez-Sotoc, and D.A. Schiraldi. 2012. Low flammability, foam-like materials based on ammonium alginate and sodium montmorillonite clay. *Polymer* 53: 5825–5831.

Chen, X.B., J.V. Wright, J.L. Conca, and L.M. Peurrung. 1997. Effects of pH on heavy metal sorption on mineral apatite. *Environ. Sci. Technol.* 31: 624–631.

Chivrac, F., E. Pollet, M. Schmutz, and L. Avérous. 2008. New approach to elaborate exfoli-ated starch-based nanobiocomposites. *Biomacromolecules* 9: 896–900.

Dawson, J.I., J.M. Kanczler, X.B.B. Yang, G.S. Attard, and R.O.C. Oreffo. 2011. Skeletal regeneration: Application of nanotopography and biomaterials for skeletal stem cell based bone repair. *Adv. Mater.* 23: 3304–3308.

Deen, I., X. Pang, and I. Zhitomirsky. 2012. Electrophoretic deposition of composite chitosan–halloysite nanotube–hydroxyapatite films. *Colloids Surf. A* 410: 38–44.

Demirbas, A. 2008. Heavy metal adsorption onto agro-based waste materials: A review. *J. Hazard. Mater.* 157: 220–229.

Drizo, A., C.A. Frost, K.A. Smith, and J. Grace. 1998. Phosphate and ammonium removal by constructed wetlands with horizontal subsurface flow, using shale as a substrate. *Water Res.* 35: 95–102.

Gao, H.C., Y.M. Sun, J.J. Zhou, R. Xu, and H.W. Duan. 2013. Mussel-inspired synthesis of polydopamine-functionalized graphene hydrogel as reusable adsorbents for water puri-fication. *ACS Appl. Mater. Interfaces* 5: 425–432.

Gao, T.P., W.B. Wang, and A.Q. Wang. 2011. A pH-sensitive composite hydrogel based on sodium alginate and medical stone: Synthesis, swelling and heavy metal ions adsorption properties. *Macromol. Res.* 19: 739–748.

Hayes, M.H.B., P. MacCarthy, R.L. Malcolm, and R.S. Swift (eds.). *Humic Substances II: In Search of Structure*. Chichester, U.K.: Wiley; 1989.

Hua, S.B. and A.Q. Wang. 2008. Preparation and properties of superabsorbent containing starch and sodium humate. *Polym. Adv. Technol.* 19: 1009–1014.

Hua, S.B., H.X. Yang, and A.Q. Wang. 2008. A pH-sensitive nanocomposite microsphere based on chitosan and montmorillonite with in vitro reduction of the burst release effect. *Drug Dev. Ind. Pharm.* 36: 1106–1114.

Hua, S.B., H.X. Yang, W.B. Wang, and A.Q. Wang. 2010. Controlled release of ofloxacin from chitosan–montmorillonite hydrogel. *Appl. Clay Sci.* 50: 112–117.

Hua, S.B., H.X. Yang, J.P. Zhang, and A.Q. Wang. 2012. Preparation and swelling properties of pH-sensitive sodium alginate/layered double hydroxides hybrid beads for controlled release of diclofenac sodium. *Drug Dev. Ind. Pharm.* 38: 728–734.

Huang, D.J., B. Mu, and A.Q. Wang. 2012a. Preparation and properties of chitosan/poly (vinyl alcohol) nanocomposite films reinforced with rod-like sepiolite. *Mater. Lett.* 86: 69–72.

Huang, W., H.J.L. Xu, Y. Xue, R. Huang, H.B. Deng, and S.Y. Pan. 2012b. Layer-by-layer immobilization of lysozome-chitosan-organic rectorite composites on electrospun nano-fibrous mats for pork preservation. *Food Res. Int.* 48: 784–791.

Huang, Y.S., S.H. Yu, Y.R. Sheu, and K.S. Huang. 2010. Preparation and thermal and anti-UV properties of chitosan/mica copolymer. *J. Nanomater.* 2010: 65–71.

Juan, L., P.Y. Zhang, Y. Gao, X.G. Song, and J.H. Dong. 2008. Overview of Maifanshi: Its physi-chemical properties and nutritious function in drinking water. *Environ. Sci. Technol.* 31: 63–67.

Klien, C., B. Dutrow, and J.D. Dana. *Manual of Mineral Science*, 23rd edn. Hoboken, NJ: John Wiley & Sons, Inc.,; 2008.

Li, A., R.F. Liu, and A.Q. Wang. 2005. Preparation of starch–graft–poly (acrylamide)/attapulg-ite superabsorbent composite. *J. Appl. Polym. Sci.* 98: 1351.

Li, A., A.Q. Wang, and J.M. Chen. 2004. Studies on poly(acrylic acid)/attapulgite superabsor-bent composite. I. Synthesis and characterization. *J. Appl. Polym. Sci.* 92: 1596–1603.

Li, A., J.P. Zhang, and A.Q. Wang. 2007a. Utilization of starch and clay for the preparation of superabsorbent composite. *Bioresour. Technol.* 98: 327–332.

Li, J.F., Y.M. Li, and H.P. Dong. 2008. Controlled release of herbicide acetochlor from clay/carboxylmethilcellulose gel formulations. *J. Agric. Food Chem.* 56: 1336–1342.

Li, P., J.P. Zhang, and A.Q. Wang. 2007b. A novel succinyl-chitosan-g-polyacrylamide/attapulgite composite hydrogel prepared through inverse suspension polymerization. *Macromol. Mater. Eng.* 292: 962–969.

Li, Q., Y.H. Zhao, L. Wang, and A.Q. Wang. 2011a. Adsorption characteristics of methylene blue onto the N-succinyl-chitosan-g-polyacrylamide/attapulgite composite. *Korean J. Chem. Eng.* 28: 1658–1664.

Li, X.X., X.Y. Li, B.L. Ke, X.W. Shi, and Y.M. Du. 2011b. Cooperative performance of chitin whisker and rectorite fillers on chitosan films. *Carbohydr. Polym.* 85: 747–752.

Li, Y., P.R. Chang, P.W. Zheng, and X.F. Ma. 2012a. Characterization of magnetic guar gum-grafted carbon nanotubes and the adsorption of the dyes. *Carbohydr. Polym.* 87: 1919–1924.

Li, Z.Q., J.F. Shen, H.W. Ma, X. Lu, M. Shi, N. Li, and M.X. Ye. 2012b. Preparation and characterization of sodium alginate/poly(N-isopropylacrylamide)/clay semi-IPN magnetic hydrogels. *Polym. Bull.* 68: 1153–1169.

Lin, J.M., J.H. Wu, Z.F. Yang, and M.L. Pu. 2001. Synthesis and properties of poly(acrylic acid)/mica superabsorbent nanocomposite. *Macromol. Rapid Commun.* 22: 422–424.

Liu, A.D., A. Walther, O. Ikkala, L. Belova, and A. Lars Berglund. 2011a. Clay nanopaper with tough cellulose nanofiber matrix for fire retardancy and gas barrier functions. *Biomacromolecules* 12: 633–641.

Liu, B., X.Y. Wang, X.Y. Li, X.J. Zeng, R.C. Sun, and J.F. Kennedy. 2012a. Rapid exfoliation of rectorite in quaternized carboxymethyl chitosan. *Carbohydr. Polym.* 90: 1826–1830.

Liu, B., X.Y. Wang, C.S. Pang, J.W. Luo, Y.Q. Luo, and R.C. Sun. 2013a. Preparation and antimicrobial property of chitosan oligosaccharide derivative/rectorite nanocomposite. *Carbohydr. Polym.* 92: 1078–1085.

Liu, H., X. Sun, C. Yin, and C. Hu. 2008. Removal of phosphate by mesoporous ZrO$_2$. *J. Hazard. Mater.* 151: 616–622.

Liu, J.H. and A.Q. Wang. 2008a. Synthesis & water retention of chitosan-g-poly(acrylic acid)/rectorite superabsorbent composites. *Non-Metallic Mines* 31: 37–40.

Liu, J.H. and A.Q. Wang. 2008b. Synthesis, characterization and swelling behaviors of chitosan-g-poly(acrylic acid)/organo-rectorite nanocomposite superabsorbents. *J. Appl. Polym. Sci.* 110: 678–686.

Liu, J.H., Q. Wang, and A.Q. Wang. 2007. Synthesis and characterization of chitosan-g-poly(acrylic acid)/sodium humate superabsorbent. *Carbohydr. Polym.* 70: 166–173.

Liu, Y., H. Chen, J.P. Zhang, and A.Q. Wang. 2013b. Effect of number of grindings of attapulgite on enhanced swelling properties of the superabsorbent nanocomposites. *J. Compos. Mater.* 47: 969–978.

Liu, Y., Y.R. Kang, D.J. Huang, and A.Q. Wang. 2012b. Cu^{2+} removal from aqueous solution by modified chitosan hydrogels. *J. Chem. Technol. Biotechnol.* 87: 1010–1016.

Liu, Y., W.B. Wang, Y.L. Jin, and A.Q. Wang. 2011b. Adsorption behavior of methylene blue from aqueous solution on the hydrogel composites based on carboxymethyl cellulose and attapulgite. *Sep. Sci. Technol.* 46: 858–868.

Liu, Y., W.B. Wang, and A.Q. Wang. 2010a. Adsorption of lead ions from aqueous solution by using carboxymethyl cellulose-g-poly (acrylic acid)/attapulgite hydrogel composites. *Desalination* 259: 258–264.

Liu, Y., Y.A. Zheng, and A.Q. Wang. 2010b. Enhanced adsorption of methylene blue from aqueous solution by chitosan-g-poly (acrylic acid)/vermiculite hydrogel composites. *J. Environ. Sci.* 22: 486–493.

Liu, Y., Y.A. Zheng, and A.Q. Wang. 2011c. Effect of biotite of hydrogels on enhanced removal of cationic dye from aqueous solution. *Ionics* 17: 535–543.

Liu, Y., Y.A. Zheng, and A.Q. Wang. 2011d. Response surface methodology for optimizing adsorption process parameters for methylene blue removal by hydrogel composite. *Adsorpt. Sci. Technol.* 28: 913–922.

Luo, W., W.A. Zhang, and P. Chen. 2005. *Co*-(acrylic acid)/montmorillonite nanosuperabsorbent via gamma-ray irradiation technique. *J. Appl. Polym. Sci.* 96: 1341–1346.

Lutolf, M.P. and J.A. Hubbell. 2005. Synthetic biomaterials as instructive extracellular microenvironments for morphogenesis in tissue engineering. *Nat. Biotechnol.* 23: 47–55.

Ma, J.H., Y.J. Xu, Q.S. Zhang, L.S. Zha, and B. Liang. 2007. Preparation and characterization of pH- and temperature-responsive semi-IPN hydrogels of carboxymethyl chitosan with poly(*N*-isopropylacrylamide) crosslinked by clay. *Colloid Polym. Sci.* 285: 479–484.

Ma, J.H., L. Zhang, B. Fan, Y.J. Xu, and B.R. Liang. 2008. A novel sodium carboxymethylcellulose/poly(*N*-isopropylacrylamide)/clay semi-IPN nanocomposite hydrogel with improved response rate and mechanical properties. *J. Polym. Sci. Part B: Polym. Phys.* 46: 1546–1555.

Marandi, G.B., G.R. Mahdavinia, and S. Ghafary. 2011. Collagen-*g*-poly(sodium acrylate-*co*-acrylamide)/sodium montmorillonite superabsorbent nanocomposites: Synthesis and swelling behavior. *J. Polym. Res.* 18: 1487–1499.

Margarita, D., L.B. Mar, A. Pilar, J.A. Antonio, B. Julio, and R.H. Eduardo. 2006. Microfibrous chitosan-sepiolite nanocomposites. *Chem. Mater.* 18: 1602–1610.

Margarita, D., C. Montserrat, and R.H. Eduardo. 2003. Biopolymer-clay nanocomposites based on chitosan intercalated in montmorillonite. *Chem. Mater.* 15: 3774–3780.

Neaman, A. and A. Singer. 2004. Possible use of the Sacalum (Yucatan) palygorskite as drilling muds. *Appl. Clay Sci.* 25: 121–124.

Ni, B.L., M.Z. Liu, S.Y. Lü, L.H. Xie, and Y.F. Wang. 2011. Environmentally friendly slow-release nitrogen fertilizer. *J. Agric. Food Chem.* 59: 10169–10175.

Peng, X.W., L.X. Zhong, J.L. Ren, and R.C. Sun. 2012. Synthesis and characterization of amphoteric xylan-type hemicelluloses by microwave irradiation. *J. Agric. Food Chem.* 60: 3909–3916.

Perotti, G.F., H.S. Barud, Y. Messaddeq, S.J.L. Ribeiro, and R.L. Constantino Vera. 2011. Bacterial cellulose–laponite clay nanocomposites. *Polymer* 52: 157–163.

Ray, S.S. and M. Okamoto. 2003. Polymer/layered silicate nanocomposites: Structure formation, interactions and deformation mechanisms. *Prog. Polym. Sci.* 28: 1539–1641.

Reza, M.G., M. Abdolhossein, B. Ali, and M. Bakhshali. 2012a. Novel carrageenan-based hydrogel nanocomposites containing laponite RD and their application to remove cationic dye. *Iran. Polym. J.* 21: 609–619.

Saad, R., K. Belkacemi, and S. Hamoudi. 2007. Adsorption of phosphate and nitrate anions on ammonium-functionalised MCM-48: Effects of experimental conditions. *J. Colloid Interface Sci.* 311: 375–381.

Sakadevan, K. and H. Bavor. 1998. Phosphate adsorption characteristics of soils, slags and zeolite to be used as substrates in constructed wetland systems. *Water Res.* 32: 393–398.

Shang, J., Z.Z. Shao, and X. Chen. 2008. Electrical behavior of a natural polyelectrolyte hydrogel: Chitosan/carboxymethylcellulose hydrogel. *Biomacromolecules* 9: 1208–1213.

Shchukin, D.G., S.V. Lamaka, K.A. Yasakau, M.L. Zheludkevich, M.G.S. Ferreira, and H. Möhwald. 2008. Active anticorrosion coatings with halloysite nanocontainers. *J. Phys. Chem. C* 112: 958–964.

Shi, J.W. 2011. Synthesis and adsorption performance of CTS-*g*-poly(acrylic acid)/sepiolite composites. Master's thesis, Chengdu University of Technology, Chengdu, China, p. 6.

Shi, X.N. and A.Q. Wang. 2011. Synthesis and swelling properties of guar gum-*g*-poly(sodium acrylate-*co*-styrene)/sepiolite superabsorbent composite. *J. Chem. Ind. Eng.* 62: 864–869.

Shi, X.N, W.B. Wang, Y.R. Kang, and A.Q. Wang. 2012. Enhanced swelling properties of a novel sodium alginate-based superabsorbent composites: NaAlg-*g*-poly(NaA-*co*-St)/APT. *J. Appl. Polym. Sci.* 125: 1822–1832.

Shi, X.N., W.B. Wang, and A.Q. Wang. 2011a. Effect of surfactant on porosity and swelling behaviors of guar gum-*g*-poly(sodium acrylate-*co*-styrene)/attapulgite superabsorbent hydrogels. *Colloids Surf. B* 88: 279–286.

Shi, X.N., W.B. Wang, and A.Q. Wang. 2011b. Swelling behavior of guar gum-*g*-poly(sodium acrylate-*co*-styrene)/attapulgite superabsorbent composites. *J. Macromol. Sci. B* 50: 1847–1863.

Shi, X.N., W.B. Wang, and A.Q. Wang. 2011c. Synthesis and enhanced swelling properties of a guar gum-based superabsorbent composite by the simultaneous introduction of styrene and attapulgite. *J. Polym. Res.* 18: 1705–1713.

Shi, X.N, W.B. Wang, and A.Q. Wang. 2011d. Synthesis, characterization and swelling behaviors of guar gum-*g*-poly(sodium acrylate-*co*-styrene)/vermiculite superabsorbent composites. *J. Compos. Mater.* 45: 2189–2198.

Shi, X.N., W.B. Wang, and A.Q. Wang. 2013. pH-responsive sodium alginate-based superporous hydrogel generated by an anionic surfactant micelle templating. *Carbohydr. Polym.* 94: 449–455.

Singh, B. and V. Sharma. 2010. Design of psyllium-PVA-acrylic acid based novel hydrogels for use in antibiotic drug delivery. *Int. J. Pharm.* 389: 94–106.

Singh, V., S. Pandey, S.K. Singh, and R. Sanghi. 2009. Removal of cadmium from aqueous solutions by adsorption using poly(acrylamide) modified guar gum–silica nanocomposites. *Sep. Purif. Technol.* 67: 251–261.

Sun, J.Y., X. Zhao, W.R. Illeperuma, O. Chaudhuri, K.H. Oh, D.J. Mooney, J.J. Vlassak, and Z. Suo. 2012. Highly stretchable and tough hydrogels. *Nature* 489: 133–136.

Taleb, M.F.A., G.A. Mahmoud, S.M. Elsigeny, and E.A. Hegazy. 2008. Adsorption and desorption of phosphorus and nitrate ions using quaternary (polypropylene-*g*-*N,N*-dimethylaminoethyl methacrylate) graft copolymer. *J. Hazard. Mater.* 159: 372–379.

Tekin, N., A. Dinçer, Ő. Demirbaş, and M. Alkan. 2006. Adsorption of cation polyacrylamide onto sepiolite. *J. Hazard. Mater.* B134: 211–219.

Teramoto, N., D. Uchiumi, A. Niikura, Y. Someya, and M. Shibata. 2007. Polypeptide/layered silicate nanocomposites using fish-based collagen peptide: Effect of crosslinking and chain extension of the collagen peptide. *J. Appl. Polym. Sci.* 106: 4024–4030.

Wang, A.Q. and W.B. Wang. 2009a. Superabsorbent materials. *Kirk-Othmer Encyclopedia of Chemical Technology.* Wiley, New York, pp. 1–34.

Wang, J.L., W.B. Wang, and A.Q. Wang. 2010a. Synthesis, characterization and swelling behaviors of hydroxyethyl cellulose-*g*-poly(acrylic acid)/attapulgite superabsorbent composite. *Polym. Eng. Sci.* 50: 1019–1027.

Wang, J.L., W.B. Wang, Y.A. Zheng, and A.Q. Wang. 2011a. Effects of modified vermiculite on the synthesis and swelling behaviors of hydroxyethyl cellulose-*g*-poly(acrylic acid)/vermiculite superabsorbent nanocomposites. *J. Polym. Res.* 18: 401–408.

Wang, L. and A.Q. Wang. 2008. Adsorption behaviors of Congo red on the N, *O*-carboxymethyl-chitosan/montmorillonite nanocomposite. *Chem. Eng. J.* 143: 43–50.

Wang, L., J.P. Zhang, and A.Q. Wang. 2008a. Removal of methylene blue from aqueous solution using chitosan-*g*-poly(acrylic acid)/montmorillonite superadsorbent nanocomposite. *Colloids Surf. A.* 322: 47–53.

Wang, L., J.P. Zhang, and A.Q. Wang. 2011b. Fast removal of methylene blue from aqueous solution by adsorption onto chitosan-*g*-poly (acrylic acid)/attapulgite composite. *Desalination* 266: 33–39.

Wang, Q., W.B. Wang, J. Wu, and A.Q. Wang. 2012a. Effect of attapulgite contents on release behaviors of a pH sensitive carboxymethyl cellulose-*g*-poly (acrylic acid)/attapulgite/ sodium alginate composite hydrogel bead containing diclofenac. *J. Appl. Polym. Sci.* 124: 4424–4432.

Wang, Q., J. Wu, W.B. Wang, and A.Q. Wang. 2011c. Preparation, characterization and drug-release behaviors of crosslinked chitosan/attapulgite hybrid microspheres by a facile spray-drying technique. *J. Biol. Nanobiotechnol.* 2: 250–257.

Wang, Q., X.L. Xie, X.W. Zhang, J.P. Zhang, and A.Q. Wang. 2010b. Preparation and swelling properties of pH sensitive composite hydrogel beads based on chitosan-*g*-poly (acrylic acid)/vermiculite and sodium alginate for diclofenac controlled release. *Int. J. Biol. Macromol.* 46: 356–362.

Wang, Q., J.P. Zhang, and A.Q. Wang. 2009a. Preparation and characterization of a novel pH-sensitive chitosan-*g*-poly(acrylic acid)/attapulgite/sodium alginate composite hydrogel bead for controlled release of diclofenac sodium. *Carbohydr. Polym.* 78: 731–737.

Wang, Q., J.P. Zhang, and A.Q. Wang. 2010c. Preparation and properties of drug loaded hydrogel beads based on halloysite. *Chem. Res. Appl.* 22: 858–863.

Wang, Q.G., J.L. Mynar, M. Yoshida, E. Lee, M. Lee, K. Okuro, K. Kinbara, and T. Aida. 2010d. High-water-content mouldable hydrogels by mixing clay and a dendritic molecular binder. *Nature* 463: 339–343.

Wang, T., D. Liu, C.X. Lian, S.D. Zheng, X.X. Liu, C.Y. Wang, and Z. Tong. 2011d. Rapid cell sheet detachment from alginate semi-interpenetrating nanocomposite hydrogels of PNIPAm and hectorite clay. *React. Funct. Polym.* 71: 447–454.

Wang, W.B., Y.R. Kang, and A.Q. Wang. 2010e. Synthesis, characterization and swelling properties of the guar gum-*g*-poly(sodium acrylate-*co*-styrene)/muscovite superabsorbent composites. *Sci. Technol. Adv. Mater.* 11: 025006.

Wang, W.B. and A.Q. Wang. 2009b. Effects of crosslinking degree on the properties of guar gum-*g*-poly(acrylic acid)/sodium humate superabsorbents. *Polym. Mater. Sci. Eng.* 25: 41–44.

Wang, W.B. and A.Q. Wang. 2009c. Preparation and slow released fertilizer properties of GG-*g*-PAA/SH superabsorbents. *Humic Acid* 1: 19–23.

Wang, W.B. and A.Q. Wang. 2009d. Preparation, characterization and properties of superabsorbent nanocomposites based on natural guar gum and modified rectorite. *Carbohydr. Polym.* 77: 891–897.

Wang, W.B. and A.Q. Wang. 2009e. Synthesis and swelling properties of guar gum-*g*-poly(sodium acrylate)/Na-montmorillonite superabsorbent nanocomposite. *J. Compos. Mater.* 43: 2805–2819.

Wang, W.B. and A.Q. Wang. 2009f. Synthesis, swelling behaviors and slow-release characteristics of guar gum-*g*-poly(sodium acrylate)/sodium humate superabsorbent. *J. Appl. Polym. Sci.* 112: 2102–2111.

Wang, W.B. and A.Q. Wang. 2010a. Adsorption and rheological studies of sodium carboxymethyl cellulose onto kaolin: Effect of degree of substitution. *Carbohydr. Polym.* 82: 83–91.

Wang, W.B. and A.Q. Wang. 2011. Preparation, swelling, and stimuli-responsive characteristics of superabsorbent nanocomposites based on carboxymethyl cellulose and rectorite. *Polym. Adv. Technol.* 22: 1602–1611.

Wang W.B., J. Wang, Y.R. Kang, and A.Q. Wang. 2011e. Synthesis, swelling and responsive properties of a new composite hydrogel based on hydroxyethyl cellulose and medicinal stone. *Composites Part B Eng.* 42: 809–818.

Wang, W.B., J. Wang, and A.Q. Wang. 2012b. pH-responsive nanocomposites from methylcellulose and attapulgite nanorods: Synthesis, swelling and absorption performance on heavy metal ions. *J. Macromol. Sci. Part A* 49: 306–315.

Wang, W.B., J.X. Xu, and A.Q. Wang. 2011f. A pH-, salt- and solvent-responsive carboxymethylcellulose-*g*-poly(sodium acrylate)/medical stone superabsorbent composite with enhanced swelling and responsive properties. *Express Polym. Lett.* 5: 385–400.

Wang, W.B., J.P. Zhang, and A.Q. Wang. 2009b. Preparation and swelling properties of super-absorbent nanocomposites based on natural guar gum and organo-vermiculite. *Appl. Clay Sci.* 46: 21–26.

Wang, W.B., N.H. Zhai, and A.Q. Wang. 2011g. Preparation and swelling characteristics of a superabsorbent nanocomposite based on natural guar gum and cation-modified vermiculite. *J. Appl. Polym. Sci.* 119: 3675–3686.

Wang, W.B., Y.A. Zheng, and A.Q. Wang. 2008b. Syntheses and properties of the superabsorbent composites based on natural guar gum and attapulgite. *Polym. Adv. Technol.* 19: 1852–1859.

Wang, X.H. and A.Q. Wang. 2010b. Adsorption characteristics of chitosan-*g*-poly(acrylic acid)/attapulgite hydrogel composite for Hg(II) ions from aqueous solution. *Sep. Sci. Technol.* 45: 2086–2094.

Wang, X.H. and A.Q. Wang. 2010c. Removal of Cd(II) from aqueous solution by the composite hydrogel based on attapulgite. *Environ. Technol.* 31: 745–753.

Wang, X.H. and A.Q. Wang. 2012. Equilibrium isotherm and mechanism studies of Pb(II) and Cd(II) ions onto hydrogel composite based on vermiculite. *Des. Water Treat.* 48: 38–49.

Wang, X.H., Y.T. Xie, and A.Q. Wang. 2010f. Adsorption behaviors of copper ion(II) on chitosan-*g*-poly(acrylic acid)/vermiculite composites. *China Mining Mag.* 19: 101–104.

Wang, X.H., Y.A. Zheng, and A.Q. Wang. 2009c. Fast removal of copper ions from aqueous solution by chitosan-*g*-poly(acrylic acid)/attapulgite composites. *J. Hazard. Mater.* 168: 970–977.

Wang, X.Y., Y.M. Du, J.W. Luo, B.F. Lin, and J. F. Kennedy. 2007. Chitosan/organic rectorite nanocomposite films: Structure, characteristic and drug delivery behavior. *Carbohydr. Polym.* 69: 41–49.

Wang, X.Y., B. Liu, J.L. Ren, C.F. Liu, X.H. Wang, J. Wu, and R.C. Sun. 2010g. Preparation and characterization of new quaternized carboxymethyl chitosan/rectorite nanocomposite. *Composited Sci. Technol.* 70: 1161–1167.

Wang, X.Y., S.P. Strand, Y.M. Du, and M. Kjell. 2010h. Chitosan–DNA–rectorite nanocomposites: Effect of chitosan chain length and glycosylation. *Carbohydr. Polym.* 79: 590–596.

Wang, Y.Z., X.N. Shi, W.B. Wang, and A.Q. Wang. 2013a. Ethanol-assisted dispersion of attapulgite and its effect on improving properties of alginate-based superabsorbent nanocomposite. *J. Appl. Polym. Sci.* 129: 1080–1088.

Wang, Y.Z., W.B. Wang, X.N. Shi, and A.Q. Wang. 2013b. Enhanced swelling and responsive properties of an alginate-based superabsorbent hydrogel by sodium *p*-styrenesulfonate and attapulgite nanorods. *Polym. Bull.* 70: 1181–1193.

Wu, D., B. Zhang, C. Li, Z. Zhang, and H. Kong. 2006. Simultaneous removal of ammonium and phosphate by zeolite synthesized from fly ash as influenced by salt treatment. *J. Colloid Interface Sci.* 304: 300–306.

Wu, J.H., J.M. Lin, M. Zhou, and C.R. Wei. 2000. Synthesis and properties of starch-graft-polyacrylamide/clay superabsorbent composite. *Macromol. Rapid Commun.* 21: 1032–1034.

Xie, Y.T. and A.Q. Wang. 2009a. Effects of modified vermiculite on water absorbency and swelling behavior of chitosan-g-poly(acrylic acid)/vermiculite superabsorbent composite. *J. Compos. Mater.* 43: 2401–2417.

Xie, Y.T. and A.Q. Wang. 2009b. Preparation and properties of chitosan-g-poly(acrylic acid)/sepiolite superabsorbent composites. *Polym. Mater. Sci. Eng.* 25: 129–132.

Xie, Y.T. and A.Q. Wang. 2009c. Synthesis, characterization and performance of chitosan-g-poly(acrylic acid)/vermiculite superabsorbent composites. *J. Polym. Res.* 16: 143–150.

Xie, Y.T. and A.Q. Wang. 2010. Preparation and swelling behaviour of cts-*g*-poly(acrylic acid)/muscovite superabsorbent composites. *Iran. Polym. J.* 19: 131–141.

Xie, Y.T., A.Q. Wang, and G. Liu. 2010. Effects of modified sepiolite on water absorbency and swelling behavior of chitosan-*g*-poly (acrylic acid)/sepiolite superabsorbent composite. *Polym. Composit.* 31: 89–96.

Xu, J.X., J.P. Zhang, Q. Wang, and A.Q. Wang. 2011. Disaggregation of palygorskite crystal bundles via high-pressure homogenization. *Appl. Clay Sci.* 54: 118–123.

Xu, R.F., S.J. Xin, X. Zhou, W. Li, F. Cao, X.Y. Feng, and H.B. Deng. 2012. Quaternized chitosan–organic rectorite intercalated composites based nanoparticles for protein controlled release. *Int. J. Pharm.* 438: 258–265.

Yan, L., Q. Shuai, X. Gong, Q. Gu, and H. Yu. 2009. Metallic iron filters for universal access to safe drinking water. *Clean-Soil Air Water* 37: 392–397.

Yang, H.X., S.B. Hua, W.B. Wang, and A.Q. Wang. 2011. Composite hydrogel beads based on chitosan and laponite: Preparation, swelling, and drug release behavior. *Iran. Polym. J.* 20: 479–490.

Yang, H.X., W.B. Wang, and A.Q. Wang. 2012. A pH-sensitive biopolymer-based superabsorbent nanocomposite from sodium alginate and attapulgite: Synthesis, characterization and swelling behaviors. *J. Dispersion Sci. Technol.* 33: 1154–1162.

Yang, H.X., W.B. Wang, J.P. Zhang, and A.Q. Wang. 2013. Preparation, characterization and drug-release behaviors of a pH-sensitive composite hydrogel bead based on guar gum, attapulgite and sodium alginate. *Int. J. Polym. Mater.* 62: 369–376.

Yao, S., J. Li, and Z. Shi. 2009. Phosphate ion removal from aqueous solution using an iron oxide-coated fly ash adsorbent. *Adsorpt. Sci. Technol.* 27: 603–608.

Yuan, L. and T. Kusuda. 2005. Adsorption of ammonium and nitrate ions by poly(N-isopropylacrylamide) gel and poly(N-isopropylacrylamide-co-chlorophyllin) gel in different states. *J. Appl. Polym. Sci.* 96: 2367–2372.

Zeng, L., X. Li, and J. Liu. 2004. Adsorptive removal of phosphate from aqueous solutions using iron oxide tailings. *Water Res.* 38: 1318–1326.

Zhang, J., K.H. Lee, L. Cui, and T. Jeong. 2009. Formation of organic nanoparticles by freeze-drying and their controlled release. *J. Ind. Eng. Chem.* 15: 185–189.

Zhang, J., W.Q. Wang, Y.P. Wang, J.Y. Zeng, S.T. Zhang, Z.Q. Lei, and X.T. Zhao. 2007c. Preparation and characterization of montmorillonite/carrageen/guar gum gel spherical beads. *Polym. Polym. Compos.* 15: 131–136.

Zhang, J.P., Y.L. Jin, and A.Q. Wang. 2011. Rapid removal of Pb(II) from aqueous solution by chitosan-g-poly(acrylic acid)/attapulgite/sodium humate composite hydrogels. *Environ Technol.* 32: 523–531.

Zhang, J.P., A. Li, and A.Q. Wang. 2005. Swelling behaviors and application of poly(acrylic acid-co-acrylamide)/sodium humate/attapulgite superabsorbent composite. *Polym. Adv. Technol.* 16: 813–820.

Zhang, J.P., A. Li, and A.Q. Wang. 2006. Study on superabsorbent composite. VI. Preparation, characterization and swelling behaviors of starch phosphate-graft-acrylamide/attapulgite superabsorbent composite. *Carbohydr. Polym.* 65: 150–158.

Zhang, J.P. and A.Q. Wang. 2010. Adsorption of Pb(II) from aqueous solution by chitosan-g-poly(acrylic acid)/attapulgite/sodium humate composite hydrogels. *J. Chem. Eng. Data* 55: 2379–2384.

Zhang, J.P., L. Wang, and A.Q. Wang. 2007a. Preparation and properties of chitosan-g-poly(acrylic acid)/montmorillonite superabsorbent nanocomposite via in situ intercalative polymerization. *Ind. Eng. Chem. Res.* 46: 2497–2502.

Zhang, J.P., Q. Wang, and A.Q. Wang. 2007b. Synthesis and characterization of chitosan-g-poly(acrylic acid)/attapulgite superabsorbent composites. *Carbohydr. Polym.* 68: 367–374.

Zhang, J.P., Q. Wang, and A.Q. Wang. 2010a. In situ generation of sodium alginate/hydroxyapatite nanocomposite beads as drug controlled release matrices. *Acta Biomater.* 6: 445–454.

Zhang, J.P., Q. Wang, X.L. Xie, X. Li, and A.Q. Wang. 2010b. Preparation and swelling properties of pH-sensitive sodium alginate/layered double hydroxides hybrid beads for controlled release of diclofenac sodium. *J. Biomed. Mater. Res. B* 92B: 205–214.

Zheng, Y.A., S.B. Hua, and A.Q. Wang. 2010. Adsorption behavior of Cu^{2+} from aqueous solutions onto starch-g-poly(acrylic acid)/sodium humate hydrogels. *Desalination* 263: 170–175.

Zheng, Y.A., P. Li, and J.P. Zhang. 2007. Synthesis, characterization and swelling behaviors of poly(sodium acrylate)/vermiculite superabsorbent composites. *Eur. Polym. J.* 43: 1691–1698.

Zheng, Y.A. and A.Q. Wang. 2009. Evaluation of ammonium removal using a chitosan-g-poly (acrylic acid)/rectorite hydrogel composite. *J. Hazard. Mater.* 171: 671–677.

Zheng, Y.A. and A.Q. Wang. 2010a. Nitrate adsorption using poly(dimethyl diallyl ammonium chloride)/polyacrylamide hydrogel. *J. Chem. Eng. Data.* 55: 3494–3500.

Zheng, Y.A. and A.Q. Wang. 2010b. Potential of phosphate removal using Al^{3+}-crosslinking chitosan-g-poly(acrylic acid)/vermiculite ionic hybrid. *Adsorpt. Sci. Technol.* 28: 89–99.

Zheng, Y.A. and A.Q. Wang. 2010c. Enhanced adsorption of ammonium using hydrogel composites based on chitosan and halloysite. *J. Macromol. Sci. Part A Pure Appl. Chem.* 47: 33–38.

Zheng, Y.A. and A.Q. Wang. 2012. Granular hydrogel initiated by Fenton reagent and their performance on Cu(II) and Ni(II) removal. *Chem. Eng. J.* 200: 601–610.

Zheng, Y.A., Y.T. Xie, and A.Q. Wang. 2009a. Adsorption of Pb^{2+} onto chitosan-grafted-poly (acrylic acid)/sepiolite composite. *Environ. Sci.* 30: 2575–2579.

Zheng, Y.A., Y.T. Xie, and A.Q. Wang. 2012. Rapid and wide pH-independent ammonium-nitrogen removal using a composite hydrogel with three-dimensional networks. *Chem. Eng. J.* 179: 90–98.

Zheng, Y.A., J.P. Zhang, and A.Q. Wang. 2009b. Fast removal of ammonium-nitrogen from aqueous solution using chitosan-g-poly (acrylic acid)/attapulgite composite. *Chem. Eng. J.* 155: 215–222.

5 Iron Oxide Magnetic Composite Adsorbents for Heavy Metal Pollutant Removal

Doina Hritcu, Gianina Dodi,
and Marcel Ionel Popa

CONTENTS

5.1 INTRODUCTION

The term "heavy metal" is generally applied to the chemical elements having atomic weight between 63.5 and 200.6 and an atomic density greater than 6 g/cm^3. While a large number of atomic species fall into this category, the ones that are relevant in environmental context due to their highly toxic effect are cadmium, lead, mercury, arsenic, chromium, copper, nickel, and zinc, with the first three being considered harmful even at trace levels (O'Connell et al. 2008, Barakat 2011). Their presence in groundwater and soil places both humans and ecosystems in danger because heavy metals are not biodegradable and tend to accumulate in soft tissues producing long-term damage, as described in detail in numerous toxicology studies (Iarc 1990, Nordberg et al. 2007, Thompson and Bannigan 2008).

The main source of contamination with heavy metals in groundwater, aquatic system and soil is the industrial activity. Electrochemical processes such as electroplating, electrolysis, anodizing–cleaning, battery manufacturing, and mineral processing result in important quantities of wastewater containing heavy metals or their ions. Moreover, the wood and paper industry, tanneries, petroleum refineries, inorganic pigments and paints producers, and printed circuit board manufacturers are also sources of metal contamination (Manahan 2001).

Conventional methods for heavy metal ion removal from aqueous effluents include the following (Fu and Wang 2011):

- Chemical precipitation (using hydroxide, carbonate, sulfide, or chelating agents)
- Membrane filtration (ultrafiltration, nanofiltration, reverse osmosis, electrodialysis)
- Coagulation and flocculation (using aluminum or iron salts and, respectively, polymers)
- Flotation (dissolved air, ion, or precipitation flotation)
- Electrochemical methods (electrocoagulation, electroflotation, electrodeposition)
- Adsorption (including biosorption, a relatively new method)

Adsorption techniques are recognized as very efficient, especially when addressing relatively low metal ion concentrations that are otherwise hard to decrease. Adsorption is a fundamental thermodynamic property of interfaces caused by a discontinuity in intermolecular or interatomic forces that result in the distribution of the dissolved chemical species between the liquid and the solid phase. The forces that drive the adsorption phenomenon, either physical or chemical in nature, are mainly controlled by the properties of the adsorbate and adsorbent and also by the process parameters such as temperature, pH, contact time, particle size, agitation rate, and concentrations of pollutants. Experimental equilibrium data regarding the amount of adsorbed substance versus its solution concentration fitted onto theoretical models such as Langmuir, Freundlich, Redlich–Peterson, Dubinin–Kaganer–Radushkevich (DKR), Temkin, and Sips adsorption isotherms are used to evaluate the thermodynamic and kinetic parameters in order to optimize large-scale applications. Low-cost adsorbents proven for their efficiency in heavy metal removal include chitosan, zeolites, agricultural waste materials (rice husk, fruit peel, coconut shell, and seafood processing waste), fly ash, peat moss, clays, and several industrial by-products such as sawdust, lignin, red mud, and ferric hydroxide (Babel and Kurniawan 2003). A new generation of adsorbents with special properties (high surface area, fast separation) is currently evolving due to the progress in nanomaterial technology (Ali 2012). This review presents recent developments in composite adsorbents containing magnetic iron oxide nanoparticles.

5.2 MAGNETIC IRON OXIDE PARTICLES AS ADSORBENTS

5.2.1 PROPERTIES

Nanosized metal oxides, such as manganese, cerium, magnesium, aluminum, titanium, and iron oxides, are a novel class of highly efficient adsorbents for heavy metal ions. Among them, iron oxides are especially interesting due to the fact that iron is nontoxic; its oxides are readily available in natural minerals or are relatively easy to prepare, resulting in environmentally friendly and low-cost sorbents (Hu et al. 2012). Moreover, the iron oxides with magnetic properties, namely, magnetite and maghemite, offer the advantage of facile separation from the liquid phase.

Magnetite (Fe_3O_4) is a naturally occurring iron oxide that contains both ferrous and ferric ions on the octahedral sites of the spinel lattice. Synthetic magnetic nanoparticles may reach a theoretical saturation magnetization of about 97 emug^{-1} that is reduced with decreasing crystallite size. Maghemite (γ-Fe_2O_3) is the low-temperature oxidation product of magnetite. It has the same structure as magnetite, with randomly distributed Fe^{2+} vacancies and a theoretical saturation magnetization of about 87 emug^{-1} (Murbe et al. 2008). Both magnetite and maghemite particles become superparamagnetic when their size is in the nanometer range (approximately below 25 nm). Unlike their ferromagnetic counterparts that possess permanent magnetic dipoles, superparamagnetic particles show no magnetization in the absence of a magnetic field, while reaching large saturation values in its presence. Due to this interesting property, superparamagnetic colloids in dilute suspension do not interact with each other to produce aggregates in the absence of the magnetic field and, therefore, a large surface area is available for adsorption. After binding the target ions, the particles may readily be separated from dispersion using magnetic gradients. Their motion, called magnetophoresis, is significantly faster under high local gradients, due to the formation of chain-like particle aggregates. This process is known as high-gradient magnetic separation (HGMS) (Faraudo et al. 2013). For recycling the adsorbent particles, it is important that the aggregation is reversible and the colloid redisperses well once the magnetic field is removed. A HGMS device acts like a filter containing a mesh composed of magnetically susceptible wires placed inside an electromagnet. The magnetic field applied across the mesh causes strong gradients that attract and retain the particles from the liquid dispersion flowing through the separator. The magnetic force that moves the particles toward the wires and, subsequently, traps them has to be large enough to overturn the fluid drag, gravitation, and both inertial and respectively diffusional forces (Ambashta and Sillanpää 2010).

Due to the presence of surface oxygen atoms that may be polarized under various pH values to yield either positive or negative charges, both magnetite and maghemite nanoparticles present physical adsorption behavior toward heavy metal ions. The driving forces for binding are electrostatic in nature and are therefore nonspecific and variable in real-life applications due to environmental conditions such as ionic strength, pH, and the presence of humic substances (Tang and Lo 2013). Moreover, the bare iron oxide particles are colloidally stabilized only by their surface charges. During the desorption and regeneration stage, when the pH is reversed compared to the adsorption phase, they are prone to irreversible aggregation that results in diminished performance due to the decreased surface area. Surface modification or incorporation in nanocomposite materials overcomes these disadvantages by contributing to colloidal stabilization and by adding specific functional groups.

5.2.2 Preparation Methods

Both magnetite and maghemite may be found in natural iron ores. Coarse magnetite particles collected from black sands or obtained by milling may be used as adsorbents after careful cleaning and sieving. Magnetic iron oxide nanoparticles are relatively easy to prepare by various synthetic routes, such as alkaline coprecipitation from a solution containing ferrous and ferric salts, thermal decomposition of organometallic compounds (iron carbonyls) in organic solvents containing surfactants, oxidation of ferrous hydroxide, oxidation of ferrous salts in alkaline solution by potassium nitrate, micelle method (water in oil droplets in emulsion used as nanoreactors), spray and laser pyrolysis, and reduction reaction between ferric salts and polyols (Petrova et al. 2011, Li et al. 2013).

Nanocomposite magnetic adsorbents are prepared using silica, carbon, and natural or synthetic polymers as matrices for the iron oxide particles. They combine the advantages of the magnetic core with the colloidal stability and the functional groups brought by the shell to yield materials with superior performance. The nanostructured iron oxides either are synthesized ex situ and subsequently embedded in the matrix or are obtained in situ in the presence of the support. A large amount of research is currently focused on developing new methods to obtain low-cost, efficient magnetic nanocomposites for environmental applications, as outlined in recent reviews (Zhao et al. 2011, Giakisikli and Anthemidis 2013, Zhu et al. 2013).

5.2.3 APPLICATIONS

Some examples of bare or surface-modified iron oxide particulate adsorbents applied in heavy metal ion separations are presented in Table 5.1. Natural magnetite particles obtained by milling and sieving, although larger in size and thus having a lower surface area, proved more effective in some separations than synthetic nanoparticles, when used bare. For large-scale applications, micron-sized particles might also involve lower operation costs by using magnetically stabilized beds (Petrova et al. 2011).

Nanocomposite adsorbents containing iron oxide nanoparticles, obtained by coprecipitation or high-temperature treatment, embedded in carbon or zeolite matrices are also described in recent studies, as summarized in Table 5.2. Magnetic activated charcoal materials present special interest due to their low cost.

5.3 MAGNETIC NANOCOMPOSITE ADSORBENTS WITH POLYMERIC SUPPORTS

Polymeric matrices that are suitable for embedding iron oxide nanoparticles are extremely diverse. They protect the magnetic core from oxidation and contribute to colloidal stabilization against aggregation. Moreover, the functional groups add specificity to the resulting hybrid adsorbent. Such an approach bridges the current gap between two well-established technologies, namely, the production of magnetic nanoparticles and the synthesis of polymeric functional sorbents. The application of polymer-supported iron oxide particles in water decontamination processes represents a fast and cost-effective separation method for the target compound from the mixture without the need of filtration or centrifugation. Once separated from the liquid mixture using magnetic field, the sorbent is ready for regeneration and reuse.

The main classes of polymers used as matrices are described in the following sections.

TABLE 5.1

Iron Oxide Adsorbents

Adsorbent	Ion Removed	Method to Produce	Reference
Natural and synthetic magnetite (coarse and nanoparticles)	Radioactive contaminants, Cs^+, Ce^{4+}, Co^{2+}, UO_2^{2+}, Eu^{2+}, Cr^{6+}, Se^{4+}, AsO_2^-, $HAsO_4^{2-}$	Various methods (review article)	Petrova et al. (2011)
Hematite-coated Fe_3O_4 particles produced by ball milling of magnetite powder	As^{3+} and As^{5+}	Natural magnetite with or without surfactant added during milling	Simeonidis et al. (2011)
Cetyltrimethylammonium bromide (CTAB)-modified Fe_3O_4 nanoparticles	$HAsO_4^{2-}$	Coprecipitation	Jin et al. (2012)
Ascorbic-acid-coated Fe_3O_4 nanoparticles	As^{3+} and As^{5+}	Hydrothermal method	Feng et al. (2012)
2-Mercaptobenzthiazole-modified Fe_3O_4 nanoparticles	Hg^{2+}	Coprecipitation, followed by modification with complexing agent for mercury	Parham et al. (2012)
Natural magnetite	$Ag(S_2O_3)_2^{3-}$	Sieved natural magnetite sand, subsequently surface cleaned	Petrova et al. (2012)
Nanoscale ferrites: $MeFe_2O_4$ (Me = Mn, Co, Cu, Mg, Zn, Ni)	Cr^{6+}	Coprecipitation	Hu et al. (2007)
Magnetite–maghemite mixed nanoparticles	As^{3+} and As^{5+}	Electric wire explosion (EWE) in air	Song et al. (2013)

5.3.1 NATURAL POLYMERS

Natural polymers are renewable materials, generally nontoxic and biodegradable, with excellent functional properties and sustainable development features. Polysaccharides are biopolymers composed of repeating units of mono- or disaccharides linked by glycosidic bonds, forming either linear or branched structures. The versatility of polysaccharides as starting materials for adsorbents stems from the variety of the available building blocks and also from the abundance of hydroxyl, acetamido, and amino functional groups that generate their hydrophilicity and metal chelation capacity (Crini 2005).

The polysaccharides most commonly used for coating iron oxide magnetic particles include agarose, alginate, carrageenan, chitosan, dextran, gum arabic, pectin, pullulan, starch, and cyclodextrin. They can be obtained from numerous sources, each bringing distinct properties (Dias et al. 2011). The magnetic iron oxide nanoparticles may be attached to the support by one of the following mechanisms: electrostatic

TABLE 5.2
Magnetic Nanocomposite Adsorbents with Inorganic Supports

Adsorbent	Ion Removed	Method to Produce	Reference
Magnetic graphene oxide nanocomposite	Cd^{2+}	Coprecipitation of iron oxide nanoparticles on the surface of graphene oxide	Deng et al. (2013)
Magnetic nanoparticle–graphitic carbon nanostructures	Ag^+, Au^{3+}	Solid-state pyrolysis of ion exchange resins complexed with iron salt	Wang et al. (2012)
Zeolite–magnetite nanocomposite	U^{6+}	Iron oxide nanoparticles mixed with zeolite in suspension	Fungaro et al. (2012)
Magnetic activated carbon	Hg^{2+}, As^{5+}, Au^{3+}	Activated carbon modified with magnetite, maghemite, or ferrite nanoparticles obtained by coprecipitation or by high-temperature treatment	Safarik et al. (2012)

interaction, ligand exchange (surface complexation), hydrophobic interaction, entropic effect, hydrogen bonding, cation bridging, or covalent bond, as summarized in Table 5.3.

Agarose is a naturally occurring polysaccharide consisting of repeating agarobiose monomeric units, with many useful features such as hydrophilicity, chemical resistance in the pH range of 0–14, biocompatibility, and biodegradability. As the

TABLE 5.3
Strategies for Preparing Magnetic Polysaccharide Composite Particles

Biopolymer	Details of the Method	References
Agarose	Encapsulation or pore accumulation, chemical coprecipitation	Safdarian et al. (2013)
Alginate	Chemical coprecipitation	Ma et al. (2007)
Carrageenan	Chemical coprecipitation	Al-Assaf et al. (2009)
Chitosan	Ionotropic gelation, chemical coprecipitation, hydrothermal, thermo-decomposition, copolymerization, reverse-phase suspension cross-linking method, in situ microemulsion method	Hu et al. (2009), Denkbas et al. (2002), Li et al. (2006), Li et al. (2008), Hritcu et al. (2011)
Cyclodextrin	Electrostatic interaction	Wang and Brusseau (1995)
Dextran	Chemical coprecipitation, covalently bound	Mornet et al. (2005)
Gums	Chemical coprecipitation, covalently bound, electrostatic interaction	Shashwat and Chen (2007)
Pectin	Electrostatic interaction	Gong et al. (2012)
Pullulan	Covalently bound	Gao et al. (2010)
Starch	Chemical coprecipitation	Dung et al. (2009)

hydroxyl groups on agarose can act as chelation sites, agarose is already proven to be an excellent sorbent for metal ions in aqueous solutions. Safdarian et al. (2013) proposed a one-step method for the synthesis of nanomagnetic agarose microflake particles used in the separation, preconcentration, and analysis of Pd^{2+} ions in aqueous solutions. Agarose-functionalized magnetic microspheres with a core–shell structure were prepared for the removal of radionuclides from aqueous solutions (Li et al. 2011). The high sorption affinity for U^{6+} (1.151 mmol/g) and Eu^{3+} ions (1.276 mmol/g) was achieved through the strong complexation of radionuclides with the hydroxyl groups located on the macromolecular backbone of agarose shell.

A novel development of a hybrid system of polymer–magnetite cryobead for the recovery of hexavalent uranium from the aqueous subsurfaces was presented by Tripathi et al. (2013). The alginate–agarose–magnetite composite beads with large surface area and high interconnected porosity were synthesized by cryotropic gelation at sub-zero temperature. The maximum uranium adsorption (97% ± 2%), the passive endothermic behavior, and the recovery yield (69% ± 3%) prove potential applications in the recovery of uranium from contaminated aqueous subsurfaces.

Alginate is a linear polymer composed of β-D-mannuronate and α-L-guluronate units linked by β-1,4 and α-1,4 glycosidic bonds. This natural polysaccharide extracted from brown seaweeds is one of the most extensively investigated biopolymers for the removal of pollutants from aqueous solutions, since it is inexpensive, nontoxic, and efficient. Carboxylate groups are negatively charged in neutral and alkaline media and hence show high affinity for cations. An innovative technology that has lately gained attention is the use of magnetic alginate composite materials for their good adsorption capacity, selectivity, and magnetic properties. Bée et al. (2011) developed a clean and safe magnetic adsorbent for water pollution remediation. This "ecoconception" approach yielded magnetic nanoparticles embedded in a calcium-alginate matrix. Their adsorption capacity has been investigated by measuring the removal of Pb^{2+} ions. The sorbent was found to be efficient in a pH range of 2.3–6. A series of batch experiments have been described in a second study in which the ability of magnetic alginate beads to remove Ni^{2+}, Co^{2+}, As^{3+}, and As^{5+} ions from aqueous solutions was investigated (Ngomsik et al. 2009). The obtained results provided encouraging premises for the application of these magnetic beads to the recovery of heavy metal ions from aqueous solutions in large-scale operations.

Chitosan is a polymer obtained by alkaline deacetylation of chitin, the polysaccharide found in the exoskeleton of shellfish and crustaceans, the second most abundant biopolymer next to cellulose (Wan Ngah et al. 2011). Due to its chemical structure, consisting of β(1–4) 2-amino-deoxy-D-glucopyranose and N-acetyl-D-glucosamine residues (Mourya and Inamdar 2008), the chitosan macromolecule has high nitrogen content and it is therefore suitable for metal ion sorption (Varma et al. 2004) by several mechanisms. The protonation of amine groups in acidic solutions gives the polymer a cationic behavior and, consequently, the potential for attracting metal anions. Chelation, electrostatic attraction, or ion exchange may occur at the amino group, depending on the metal ion and the solution pH (Guibal 2004), but hydroxyl groups (especially those located in the C-3 position) may also contribute to the adsorption process. Wan Ngah et al. (2011) recently reviewed novel chitosan composite formulations with various materials, including magnetic iron oxides, for adsorbents intended for wastewater treatment. Chitosan magnetic particles combine the low cost and biodegradability brought by the natural polymeric support with a high surface area and easy magnetic separation, thus enhancing the reusability of the composite in many applications (Hu et al. 2009). Several methods for obtaining magnetic chitosan adsorbents have been developed, as summarized in Table 5.3.

Magnetic chitosan particles were used to remove various toxic pollutants such as Cd^{2+}, Cu^{2+}, Hg^{2+}, Co^{2+} ions and dyes by adsorption followed by magnetic field separation. Noble metal ions, namely, gold and silver, may be recovered from aqueous solutions by the same method.

A large number of recent studies describe the preparation of chitosan/magnetite composites for the removal of heavy metals from wastewater. Podzus et al. (2009) reported the preparation of microspheres composed of superparamagnetic iron oxide nanoparticles and chitosan cross-linked with glutaraldehyde, demonstrating their ability to adsorb Cu^{2+} from aqueous solutions. Thiourea-modified magnetic chitosan microspheres were prepared and investigated for the adsorption of Hg^{2+}, Cu^{2+}, and Ni^{2+} ions. The chemical modification improved the selectivity and adsorption capacity (Zhou et al. 2009). Also, Monier et al. (2010) and Chen and Wang (2012) studied the adsorption of Co^{2+} and Ni^{2+} ions onto modified magnetic chitosan chelating resins and xanthate-modified magnetic chitosan, respectively. Tran et al. (2010) prepared a hydrogel consisting of 2-acrylamido-2-methyl-1-propanesulfonic acid cross-linked with chitosan-embedding magnetic particles that was effective in the removal of Pb^{2+} and Ni^{2+}. Chitosan magnetic nanoparticles were also investigated by Liu et al. (2009) for the efficient removal of Pb^{2+}, Cu^{2+}, and Cd^{2+} from aqueous solutions. A simple method for preparing magnetite/chitosan composite microparticles using an in situ method was developed and optimized in our group (Hritcu et al. 2011). The magnetic nanoparticles were produced by the oxidation of ferrous ions initially dispersed within the polysaccharide solution. The advantages of the method reside in the following features:

- The iron oxide nanoparticles are already encapsulated upon formation, thus avoiding agglomeration and mutual magnetization cancellation.
- The composite particles are produced in a single step.
- The ratio between matrix and magnetic material may be varied according to the desired properties of the composite.
- The material may be tailored for optimum separation time depending on the magnetic field strength in real-life applications.

The novel adsorbents were able to remove uranyl and thorium ions (Hritcu et al. 2012a), also Co^{2+} and Ni^{2+} ions (Hritcu et al. 2012b) from simulated wastewater solutions. The addition of a synthetic (glycidyl methacrylate-*co*-ethylene glycol dimethacrylate-*co*-poly(ethylene glycol) methyl methacrylate) copolymer shell functionalized with ethylenediamine allowed control of surface functional groups and subsequently improved their accessibility, demonstrated by the increased adsorption capacity for Cu^{2+} ions (Dodi et al. 2012). The properties of both types of materials synthesized in our laboratory are summarized in Table 5.4.

TABLE 5.4
Magnetic Chitosan Adsorbents Prepared and Tested in Our Laboratory

Adsorbent Type	Magnetization (emu/g)	Ion Removed	Adsorption Capacity (mg/g)	Reference
Chitosan-encapsulated magnetic particles produced by in situ oxidation of ferrous ions	24	Th^{4+}	312.5	Hritcu et al. (2012a,b)
		UO_2^{2+}	666.7	
		Co^{2+}	558.3	
		Ni^{2+}	833.4	
		Cu^{2+}	234	
Core–shell magnetic chitosan particles functionalized by grafting	13	Cu^{2+}	574	Dodi et al. (2012)

β-Cyclodextrin is a cyclic oligosaccharide that consists of seven glucopyranoside units connected through α-(1,4) linkages, with primary and secondary hydroxyl groups located outside and inside the toroidal-shaped cavity, respectively. Recent reports on environmental applications of cyclodextrins for heavy metal removal evidenced that metal ions are complexed by cyclodextrins through hydroxyl groups. Substituting functional groups located outside the cyclodextrin ring may alter metal complexation potential. For example, carboxymethyl-β-cyclodextrin has the ability to complex heavy metals such as cadmium, nickel, strontium, and mercury through the interactions between the metal ions and the –COOH functional groups (Wang and Brusseau 1995). Recently, Hu et al. prepared a novel magnetic composite material by grafting β-cyclodextrin onto multiwalled carbon nanotubes/iron oxides that was demonstrated to be a promising adsorbent for Pb^{2+} ion removal (Hu et al. 2010). Badruddoza et al. successfully synthesized a magnetic nanoadsorbent comprising Fe_3O_4 nanoparticles modified with carboxymethyl-β-cyclodextrin via the carbodiimide method (Badruddoza et al. 2011). The adsorption of Cu^{2+} onto these magnetic nanoadsorbents was found to be pH and temperature dependent, with the equilibrium data fitted onto the Langmuir isotherm model. The results of this work suggest that the magnetic composites are promising adsorbents for the removal of heavy metal ions from wastewater using the technology of magnetic separation.

Pectin is a structural polysaccharide found in plant cell walls, and which contains partially esterified polygalacturonic acid residues. Its potential as a biosorbent for removing heavy metal ions from aqueous solutions has been demonstrated (Panchev et al. 2010). The preparation of hybrid nanomaterials using pectin and iron oxide magnetic nanoparticles has been reported in the literature. Gong et al. synthesized an effective pectin-coated iron oxide magnetic adsorbent for the removal of Cu^{2+} from aqueous systems (Gong et al. 2012). The interaction between copper ions and pectin surface during the adsorption process was attributed to the ion exchange and electrostatic attraction mechanisms. Due to the high adsorption capacity and separation convenience, magnetic-pectin composite materials are considered a promising alternative for Cu^{2+} removal from wastewater.

Natural gums are polysaccharides produced by plants, consisting of multiple sugar units with heterogeneous composition. Gums represent one of the most abundant industrial raw materials and have been the subject of intensive research due to their excellent properties, such as low cost, sustainability, biodegradability, and biocompatibility. Gums are a promising class of strong adsorbents that fulfill their role via interactions such as hydrogen bonding and hydrophobic forces. The differences in the affinities of polysaccharides toward hydroxyl species of different metals lead to selective adsorption that may be combined with flotation method, as documented by Bicack et al. (2007).

Gum kondagogu is a novel partially acetylated biopolymer, a rhamnogalacturonan-type polysaccharide, rich in acidic sugar residues such as galacturonic and glucuronic acids; neutral sugars such as glucose, galactose, rhamnose, and arabinose; but also tannins, proteins, and minerals. Based on the literature reports, this gum contains carboxyl, hydroxyl, ether, acetyl, aliphatic, and carbonyl functional groups and can be used as a potential biosorbent for the removal of toxic metal contaminants (Vinod and Sashidhar 2009). The adsorption capacity and the separation rate could be improved by functionalizing the gum network with the iron oxide nanoparticles. Saravanan et al. (2012) synthesized, characterized, and successfully exploited gum-kondagogu-modified magnetic iron oxide nanoparticles for the removal of hazardous metal cations and found the following order in efficiency: $Cd^{2+} > Cu^{2+} > Pb^{2+} > Ni^{2+} > Zn^{2+} > Hg^{2+}$.

Gum arabic is a slightly acidic polysaccharide consisting in three main fractions, namely, a highly branched polysaccharide of a β-(1–3) galactose backbone with linked branches of arabinose and rhamnose terminating in glucuronic acid and peptide moieties (Al-Assaf et al. 2009). Gum arabic is a natural environmentally friendly polymer with excellent adsorption properties that contains active functionalities such as carboxylate and amine groups. A novel magnetic nanoadsorbent for the separation of metal ions was developed by surface modification of Fe_3O_4 nanoparticles with gum arabic via the interaction between the carboxylic groups belonging to gum molecules and the surface hydroxyl groups located on the Fe_3O_4 particles (Shashwat and Chen 2007). The adsorption of copper ions on the resulting composite material was attributed to the surface complexation with the gum arabic amine groups, favored by high specific surface area and the absence of internal diffusion resistance. The adsorption behavior and the investigated mechanisms confirmed that an efficient magnetic nanoadsorbent for the fast removal of copper ions from aqueous solutions has been obtained.

A composition of gellan gum and Fe_3O_4 nanoparticles was employed as a novel adsorbent for heavy metal removal in aqueous solutions (Wang et al. 2009). The material couples magnetic separation with ionic exchange for heavy metal removal in an efficiency order of $Pb^{2+} > Cr^{3+} > Mn^{2+}$, under the studied experimental parameters.

5.3.2 MAGNETIC COMPOSITES CONTAINING SYNTHETIC POLYMERS

The use of synthetic matrices for magnetic iron oxide nanoparticles offers the advantage of tailoring specific adsorbents for certain applications. Moreover, a porous morphology with increased surface area is also relatively easy to produce. Composites may be prepared either by coating a magnetic core with the polymeric matrix or by filling the iron oxide in the pores of a preformed resin, produced by well-established dispersion, suspension, or emulsion polymerization techniques.

Magnetic particles functionalized with synthetic polymers, such as poly(glycidyl methacrylate) (Horak et al. 2007), poly(methyl methacrylate) (Denizli et al. 2000), polyethylenimine, poly(vinyl alcohol) (Lee et al. 1996), poly(vinylpyrrolidone), and poly(methacrylic acid) are reported. A series of core–shell nanomagnetic polymer adsorbents coupled with different multiamine compounds, that is, ethylenediamine, diethylenetriamine, triethylenetetramine, and tetraethylenepentamine, have been prepared and used for the investigation of the adsorption mechanism of Cu^{2+} and Cr^{6+} coexisting in aqueous solutions. The results showed that the adsorption mechanism of Cu^{2+} and Cr^{6+} ions is related to coordination interactions, electrostatic attraction, and ion exchange processes (Shena et al. 2012).

Sun et al. prepared polyethylenimine-functionalized magnetic porous microspheres using glycidyl methacrylate dispersion polymerization with ring opening and ethylenediamine chemical modification, followed by in situ coprecipitation of Fe_3O_4 into the pores and subsequent Michael addition of amino groups with methyl methacrylate and ester group amidation with polyethyleneimine (Sun et al. 2013). The resulting microspheres were used for the efficient removal of Cr^{6+} ions from aqueous solution. Zhao et al. also demonstrated the removal of Cr^{6+} using ethylenediamine-functionalized magnetic polymers (Zhao et al. 2010).

5.4 MAGNETIC COMPOSITES CONTAINING AGRICULTURAL WASTE

Agricultural low-cost materials, such as rice bran/husk, wheat bran/husk, sawdust of various plants, bark of the trees, groundnut shells, coconut shells, black gram husk, hazelnut shells, walnut shells, cottonseed hulls, waste tea leaves, maize corn cob, jatropha deoiled cakes, sugarcane, apple, banana, orange peels, soybean hulls, grape stalks, water hyacinth, sugar-beet pulp, sunflower stalks, coffee beans, arjun nuts, and cotton stalks, are already known as good adsorbents for Cd^{2+}, Cu^{2+}, Pb^{2+}, Zn^{2+}, Ni^{2+}, Cr^{2+}, and other heavy metal ions (Demirbas 2008). The functional groups of these economic and ecofriendly biosorbents facilitate metal complexation through chemisorption, adsorption on surface, or diffusion through pores (Farooq et al. 2010). The addition of a magnetic component is advantageous due to adsorbent easy removal and reuse.

Orange peel is a low-cost, nontoxic biosorbent containing active hydroxyl and carboxyl functional groups that are present in its cellulose, hemicellulose, and pectin components. A novel magnetic nanoadsorbent for the removal of toxic Cd^{2+} metal ions was developed by covalently binding the surface hydroxyl groups belonging to Fe_3O_4 nanoparticles with the carboxyl groups located in orange peel powder (Gupta and Nayak 2012). Comparative batch studies demonstrated that at optimal process parameters, a maximum metal sorption capacity was reached through a complexation mechanism. The newly developed magnetic material has evidenced not only high adsorption efficiency and faster kinetics, but also additional benefits like facile synthesis, easy recovery, the absence of secondary pollutants, cost-effectiveness, and environmental friendliness.

An inexpensive and effective adsorbent obtained from an agricultural material that requires little processing is tea waste, mainly containing cellulose and hemicelluloses, lignin, condensed tannins, and structural proteins. The responsible functional groups form relatively strong chemical/physical linkages with metal ions from solutions and wastewaters. The efficiency of magnetic nanoparticles impregnated onto tea waste in removing Ni^{2+} ions from aqueous solutions has been investigated by Panneerselvam and coauthors (Panneerselvam et al. 2011). The adsorption data revealed that such a magnetic material might be an attractive option for Ni^{2+} removal from industrial effluents.

5.5 CONCLUSIONS AND FUTURE OUTLOOK

Iron oxide nanocomposite adsorbents add a magnetic component to the separation of heavy metal ions by adsorption and, therefore, have the potential to lower operation costs for wastewater treatment. Although a large number of materials have already been developed and proved extremely efficient in laboratory batch tests, the reported large-scale tests are still few. In order for these adsorbents to become economically interesting, they have to prove that shorter operation time resulting from facile separation, regeneration, and reuse compensates the additional cost of a high-gradient magnetic separator in real applications. The scale-up work needs interdisciplinary teams consisting in both chemists and physicists in order to design the composite adsorbents with optimum performance in terms of low cost, colloidal stability, adsorption/desorption rate, and dispersion flow properties in magnetic field. In terms of the available matrices, natural polysaccharides are probably the promising option, with chitosan presenting the most advantageous features.

Since the technology is still relatively new, it is expected that the necessary steps will be taken to make the newly synthesized materials useful for industrial applications.

ACKNOWLEDGMENT

This chapter was supported by a grant of the Ministry of National Education, CNCS-UEFISCDI, project number PN-II-ID-PCE-2012-4-0433.

REFERENCES

Al-Assaf, S., M. Sakata, C. McKenna, E.H. Aoki, and G.O. Phillips. 2009. Molecular associations in acacia gums. *Struct. Chem.* 20: 325–336.
Ali, I. 2012. New generation adsorbents for water treatment. *Chem. Rev.* 112: 5073–5091.
Ambashta, R.D. and M. Sillanpää. 2010. Water purification using magnetic assistance: A review. *J. Hazard. Mater.* 180: 38–49.
Babel, S. and T.A. Kurniawan. 2003. Low-cost adsorbents for heavy metals uptake from contaminated water: A review. *J. Hazard. Mater.* B97: 219–240.
Badruddoza, A.Z.M., A.S.H. Tay, P.Y. Tan, K. Hidajat, and M.S. Uddin. 2011. Carboxymethyl-β-cyclodextrin conjugated magnetic nanoparticles as nano-adsorbents for removal of copper ions: Synthesis and adsorption studies. *J. Hazard. Mater.* 185: 1177–1186.
Banerjee, S.S. and D.-H. Chen. 2007. Fast removal of copper ions by gum arabic modified magnetic nano-adsorbent. *J. Hazard. Mater.* 147: 792–799.

Barakat, M.A. 2011. New trends in removing heavy metals from industrial wastewater: A review. *Arab. J. Chem.* 4: 361–377.

Bée, A., D. Talbot, S. Abramson, and V. Dupuis. 2011. Magnetic alginate beads for Pb(II) ions removal from wastewater. *J. Colloid Interface Sci.* 362: 486–492.

Bicak, O., Z. Ekmekci, D.J. Bradshaw, and P.J. Harris. 2007. Adsorption of guar gum and CMC on pyrite. *Miner. Eng.* 20: 996–1002.

Chen, Y. and J. Wang. 2012. The characteristics and mechanism of Co(II) removal from aqueous solution by a novel xanthate-modified magnetic chitosan. *Nucl. Eng. Des.* 242: 452–457.

Crini, G. 2005. Recent developments in polysaccharide-based materials used as adsorbents in wastewater treatment. *Prog. Polym. Sci.* 30: 38–70.

Demirbas, A. 2008. Heavy metal adsorption onto agro-based waste materials: A review. *J. Hazard. Mater.* 157: 220–229.

Deng, J.-H., X.-R. Zhang, G.-M. Zeng, J.-L. Gong, Q.-Y. Niu, and J. Liang. 2013. Simultaneous removal of Cd(II) and ionic dyes from aqueous solution using magnetic graphene oxide nanocomposite as an adsorbent. *Chem. Eng. J.* 226: 189–200.

Denizli, A., G. Ozkan, and M.Y. Arica. 2000. Preparation and characterization of magnetic polymethylmethacrylate microbeads carrying ethylenediamine for removal of Cu(II), Cd(II), Pb(II), and Hg(II) from aqueous solutions. *J. Appl. Polym. Sci.* 78: 81–89.

Denkbas, E.B., E. Kilicay, C. Birlikseven, and E. Oztürk. 2002. Magnetic chitosan microspheres: Preparation and characterization. *React. Funct. Polym.* 50: 225–232.

Dias, A.M.G.C., A. Hussain, A.S. Marcos, and A.C.A. Roque. 2011. A biotechnological perspective on the application of iron oxide magnetic colloids modified with polysaccharides. *Biotechnol. Adv.* 29: 142–155.

Dodi, G., D. Hritcu, G. Lisa, and M.I. Popa. 2012. Core–shell magnetic chitosan particles functionalized by grafting: Synthesis and characterization. *Chem. Eng. J.* 203: 130–141.

Dung, T., T. Danh, L. Ho, D. Chien, and N. Due. 2009. Structural and magnetic properties of starch coated magnetic nanoparticles. *J. Exp. Nanosci.* 4: 259–267.

Faraudo, J., J.S. Andreu, and J. Camacho. 2013. Understanding diluted dispersions of superparamagnetic particles under strong magnetic fields: A review of concepts, theory and simulations. *Soft Matter* 9: 6654–6664. doi: 10.1039/c3sm00132f.

Farooq, U., J.A. Kozinski, M.A. Khan, and M. Athar. 2010. Biosorption of heavy metal ions using wheat based biosorbents—A review of the recent literature. *Bioresour. Technol.* 101: 5043–5053.

Feng, L., M. Cao, X. Ma, Y. Zhu, and C. Hu. 2012. Superparamagnetic high-surface-area Fe_3O_4 nanoparticles as adsorbents for arsenic removal. *J. Hazard. Mater.* 217–218: 439–446.

Fu, F. and Q. Wang. 2011. Removal of heavy metal ions from wastewaters: A review. *J. Environ. Manag.* 92: 407–418.

Fungaro, D.A., M. Yamaura, and G.R. Craesmeyer. 2012. Uranium removal from aqueous solution by zeolite from fly ash-iron oxide magnetic nanocomposite. *Int. Rev. Chem. Eng. (I.RE.CH.E.)* 4: 353–358.

Gao, F., Y. Cai, J. Zhou, X. Xie, W. Ouyang, and Y. Zhang. 2010. Pullulan acetate coated magnetite nanoparticles for hyper-thermia: Preparation, characterization and in vitro experiments. *Nano Res.* 3: 23–31.

Giakisikli, G. and A.N. Anthemidis. 2013. Magnetic materials as sorbents for metal/metalloid preconcentration and/or separation. A review. *Anal. Chim. Acta* 789: 1–16.

Gong, J.L., X.Y. Wang, G.M. Zeng, L. Chen, J.H. Deng, X.R. Zhang, and Q.Y. Niu. 2012. Copper (II) removal by pectin–iron oxide magnetic nanocomposite adsorbent. *Chem. Eng. J.* 185–186: 100–107.

Guibal, E. 2004. Interactions of metal ions with chitosan-based sorbents: A review. *Sep. Purif. Technol.* 38: 43–74.

Gupta, V.K. and A. Nayak. 2012. Cadmium removal and recovery from aqueous solutions by novel adsorbents prepared from orange peel and Fe_2O_3 nanoparticles. *Chem. Eng. J.* 180: 81–90.

Horak, D., E. Petrovsky, and A. Kapicka. 2007. Synthesis and characterization of magnetic poly(glycidyl methacrylate) microspheres. *J. Magn. Magn. Mater.* 311: 500–506.

Hritcu, D., G. Dodi, D. Humelnicu, and M.I. Popa. 2012a. Magnetic chitosan composite particles: Evaluation of thorium and uranyl ion adsorption from aqueous solutions. *Carbohydr. Polym.* 87: 1185–1191.

Hritcu, D., G. Dodi, and M.I. Popa. 2012b. Heavy metal ion adsorption on chitosan-magnetite microspheres. *Int. Rev. Chem. Eng.* 4: 364–368.

Hritcu, D., G. Dodi, M. Silion, N. Popa, and M.I. Popa. 2011. Composite magnetite–chitosan microspheres: In-situ preparation and characterization. *Polym. Bull.* 67: 177–186.

Hu, J., I.M.C. Lo, and G. Chen. 2007. Comparative study of various magnetic nanoparticles for Cr(VI) removal. *Sep. Purif. Technol.* 56: 249–256.

Hu, J., D. Shao, C. Chen, G. Sheng, J. Li, X. Wang, and M. Nagatsu. 2010. Plasma-induced grafting of cyclodextrin onto multiwall carbon nanotube/iron oxides for adsorbent application. *J. Phys. Chem. B* 114: 6779–6785.

Hu, M., S. Zhang, B. Pan, W. Zhang, L. Lv, and Q. Zhang. 2012. Heavy metal removal from water/wastewater by nanosized metal oxides: A review. *J. Hazard. Mater.* 211–212: 317–331.

Hu, Y., Y. Bo, L. Yaobo, and C. Rong Shi. 2009. Preparation of magnetic chitosan microspheres and its applications in wastewater treatment. *Sci. China Ser. B: Chem.* 52: 249–256.

IARC. 1990. Monograph on the evaluation of carcinogenic risk to humans. Chromium, nickel and welding. *Int. Agency Res. Cancer.* 49: 187–208.

Jin, Y., F. Liu, M. Tong, Y. Hou, and Y. Jin. 2012. Removal of arsenate by cetyltrimethylammonium bromide modified magnetic nanoparticles. *J. Hazard. Mater.* 227–228: 461–468.

Lee, J., T. Isobe, and M. Senna. 1996. Preparation of ultrafine Fe_3O_4 particles by precipitation in the presence of PVA at high pH. *J. Colloid Interface Sci.* 177: 490–494.

Li, B., D. Jia, Y. Zhou, Q. Hu, and W. Cai. 2006. In situ hybridization to chitosan/magnetite nanocomposite induced by the magnetic field. *J. Magn. Magn. Mater.* 306: 223–227.

Li, G., Y. Jiang, K. Huang, P. Ding, and J. Chen. 2008. Preparation and properties of magnetic Fe_3O_4–chitosan nanoparticles. *J. Alloys Compd.* 466: 451–456.

Li, J., Z. Guo, S. Zhang, and X. Wang. 2011. Enrich and seal radionuclides in magnetic agarose microspheres. *Chem. Eng. J.* 172: 892–897.

Li, X.S., G.T. Zhu, Y.B. Luo, B.F. Yuan, and Y.Q. Feng. 2013. Synthesis and applications of functionalized magnetic materials in sample preparation. *Trends Anal. Chem.* 45: 233–247.

Liu, X.W., Q.Y. Hu, Z. Fang, X.J. Zhang, and B.B. Zhang. 2009. Magnetic chitosan nanocomposites: A useful recyclable took for heavy metal ion removal. *Langmuir* 25: 3–8.

Ma, H., X. Qi, Y. Maitani, and T. Nagai. 2007. Preparation and characterization of superparamagnetic iron oxide nanoparticles stabilized by alginate. *Int. J. Pharm.* 333: 177–186.

Manahan, S.E. 2001. *Fundamentals of Environmental Chemistry*. Boca Raton, FL: CRC Press LLC.

Monier, M., D.M. Ayad, Y. Wei, and A.A. Sarhan. 2010. Adsorption of Cu(II), Co(II) and Ni(II) ions by modified chitosan chelating resin. *J. Hazard. Mater.* 177: 962–970.

Mornet, S., J. Portier, and E. Duguet. 2005. A method for synthesis and functionalization of ultra small superparamagnetic covalent carriers based on maghemite and dextran. *J. Magn. Magn. Mater.* 293: 127–134.

Mourya, V.K. and N.N. Inamdar. 2008. Chitosan-modifications and applications: Opportunities galore. *React. Funct. Polym.* 68: 1013–1051.

Murbe, J., A. Rechtenbach, and J. Topfer. 2008. Synthesis and physical characterization of magnetite nanoparticles for biomedical applications. *Mater. Chem. Phys.* 110: 426–433.

Ngomsik, A.F., A. Bee, J.M. Siaugue, D. Talbot, V. Cabuila, and G. Cote. 2009. Co(II) removal by magnetic alginate beads containing Cyanex 272®. *J. Hazard. Mater.* 166: 1043–1049.

Nordberg, G., B. Fowler, M. Nordberg, and L.F. Friberg. 2007. *Handbook on the Toxicity of Metals*, 3rd edn., pp. 743–758. Amsterdam, the Netherlands: Elsevier.

O'Connell, D.W., C. Birkinshaw, and T.F. O'Dwyer. 2008. Heavy metal adsorbents prepared from the modification of cellulose: A review. *Bioresour. Technol.* 99: 6709–6724.

Panchev, I.N., A. Slavov, Kr. Nikolova, and D. Kovacheva. 2010. On the water-sorption properties of pectin. *Food Hydrocolloids* 24: 763–769.

Panneerselvam, P., N. Morad, and K.A. Tan. 2011. Magnetic nanoparticle (Fe$_3$O$_4$) impregnated onto tea waste for the removal of nickel(II) from aqueous solution. *J. Hazard. Mater.* 186: 160–168.

Parham, H., B. Zargar, and R. Shiralipour. 2012. Fast and efficient removal of mercury from water samples using magnetic iron oxide nanoparticles modified with 2-mercaptobenzothiazole. *J. Hazard. Mater.* 205–206: 94–100.

Petrova, T.M., L. Fachikov, and J. Hristov. 2011. The magnetite as adsorbent for some hazardous species from aqueous solutions: A review. *Int. Rev. Chem. Eng.* 3: 134–152.

Petrova, T.M., V.A. Karadjova, L. Fachikov, and J. Hristov. 2012. Silver recovery from spent photographic solutions by natural magnetite: Attempts to estimate the process mechanism and optimal process conditions. *Int. Rev. Chem. Eng.* 4: 373–378.

Podzus, P.E., M.E. Daraio, and S.E. Jacobo. 2009. Chitosan magnetic microspheres for technological applications: Preparation and characterization. *Physica B.* 404: 2710–2712.

Safarik I., K. Horska, K. Pospiskova, and M. Safarikova. 2012. Magnetically responsive activated carbons for bio- and environmental applications. *Int. Rev. Chem. Eng.* 4: 346–352.

Safdarian, M., P. Hashemi, and M. Adeli. 2013. One-step synthesis of agarose coated magnetic nanoparticles and their application in the solid phase extraction of Pd(II) using a new magnetic field agitation device. *Anal. Chim. Acta* 774: 44–50.

Saravanan, P., V.T.P. Vinod, B. Sreedhar, and R.B. Sashidhar. 2012. Gum kondagogu modified magnetic nano-adsorbent: An efficient protocol for removal of various toxic metal ions. *Mater. Sci. Eng. C* 32: 581–586.

Shena, H., S. Pana, Y. Zhanga, X. Huang, and H. Gong. 2012. A new insight on the adsorption mechanism of amino-functionalized nano-Fe$_3$O$_4$ magnetic polymers in Cu(II), Cr(VI) co-existing water system. *Chem. Eng. J.* 183: 180–191.

Simeonidis, K., Th. Gkinis, S. Tresintsi, C. Martinez-Boubeta, G. Vourlias, I. Tsiaoussis, G. Stavropoulos, M. Mitrakas, and M. Angelakeris. 2011. Magnetic separation of hematite-coated Fe$_3$O$_4$ particles used as arsenic adsorbents. *Chem. Eng. J.* 168: 1008–1015.

Song, K., W. Kim, C.Y. Suh, D. Shin, K.S. Ko, and K. Ha. 2013. Magnetic iron oxide nanoparticles prepared by electrical wire explosion for arsenic removal. *Powder Technol.* 246: 572–574.

Sun, X., L. Yang, H. Xing, J. Zhao, X. Li, Y. Huang, and H. Liu. 2013. Synthesis of polyethylenimine-functionalized poly(glycidyl methacrylate) magnetic microspheres and their excellent Cr(VI) ion removal properties. *Chem. Eng. J.* 234: 338–345.

Tang, S.C.N. and I.M.C. Lo. 2013. Magnetic nanoparticles: Essential factors for sustainable environmental applications. *Water Res.* 47: 2613–2632.

Thompson, J. and J. Bannigan. 2008. Cadmium: Toxic effects on the reproductive system and the embryo. *Reprod. Toxicol.* 25(3): 304–315.

Tran, H.V., L.D. Tran, and T.N. Nguyen. 2010. Preparation of chitosan/magnetite composite beads and their application for removal of Pb(II) and Ni(II) from aqueous solution. *Mater. Sci. Eng. C* 30: 304–310.

Tripathi, A., J.S. Melo, and S.F. D'Souza. 2013. Uranium (VI) recovery from aqueous medium using novel floating macroporous alginate-agarose-magnetite cryobeads. *J. Hazard. Mater.* 246–247: 87–95.

Varma, A.J., S.V. Deshpande, and J.F. Kennedy. 2004. Metal complexation by chitosan and its derivatives: A review. *Carbohydr. Polym.* 55: 77–93.

Vinod, V.T.P. and R.B. Sashidhar. 2009. Solution and conformational properties of gum kondagogu (*Cochlospermum gossypium*)—A natural product with immense potential as a food additive. *Food Chem.* 116: 686–692.

Wang, L., C. Tian, G. Mu, L. Sun, H. Zhang, and H. Fu. 2012. Magnetic nanoparticles/ graphitic carbon nanostructures composites: Excellent magnetic separable adsorbents for precious metals from aqueous solutions. *Mater. Res. Bull.* 47: 646–654.

Wang, X. and M.L. Brusseau. 1995. Simultaneous complexation of organic compounds and heavy metals by a modified cyclodextrin. *Environ. Sci. Technol.* 29: 2632–2635.

Wang, X., C. Zhao, P. Zhao, P. Dou, Y. Ding, and P. Xu. 2009. Gellan gel beads containing magnetic nanoparticles: An effective biosorbent for the removal of heavy metals from aqueous system. *Bioresour. Technol.* 100: 2301–2304.

Wan Ngah, W.S., L.C. Teong, and M.A.K.M. Hanafiah. 2011. Adsorption of dyes and heavy metal ions by chitosan composites: A review. *Carbohydr. Polym.* 83: 1446–1456.

Zhao, X., L. Lv, B. Pana, W. Zhang, S. Zhang, and Q. Zhang. 2011. Polymer-supported nano-composites for environmental application: A review. *Chem. Eng. J.* 170: 381–394.

Zhao, Y.G., H.Y. Shen, S.D. Pan, and M.Q. Hu. 2010. Synthesis, characterization and proper-ties of ethylenediamine-functionalized Fe_3O_4 magnetic polymers for removal of Cr(VI) in wastewater. *J. Hazard. Mater.* 182: 295–302.

Zhou, L., Y. Wang, Z. Liu, and Q. Huang. 2009. Characteristics of equilibrium, kinetics studies for adsorption of Hg(II), Cu(II), and Ni(II) ions by thiourea-modified magnetic chitosan microspheres. *J. Hazard. Mater.* 161: 995–1002.

Zhu J., S. Wei, M. Chen, H. Gu, S.B. Rapole, S. Pallavkar, T.C. Ho, J. Hopper, and Z. Guo. 2013. Magnetic nanocomposites for environmental remediation. *Adv. Powder Technol.* 24: 459–467.

6 Biopolymer–Zeolite Composites as Biosorbents for Separation Processes

Maria Valentina Dinu and Ecaterina Stela Dragan

CONTENTS

6.1 INTRODUCTION

Biopolymer–zeolite composites are a combination of a biopolymer matrix and micro-/ nano-sized zeolites, in the form of beads, membranes, and fibers. Biopolymer– zeolite composites have received considerable attention over the last decade owing to their potential to dramatically enhance the properties relative to the raw zeolite or the pristine biopolymer matrix (Bhat and Aminabhavi 2009, Dragan et al. 2010, Lin and Zhan 2012, Metin et al. 2013, Saraswathi and Viswanath 2012). Incorporation of small amounts of zeolites leads to an improvement of the composite properties, such as mechanical strength and thermal and chemical stability. On the other hand, the biopolymer matrix arises with additional features like biodegradability, biocompatibility, antibacterial activity, and chelating properties. The properties of the biopolymer–zeolite composites can be also controlled by micro-/nanostructural parameters such as dimension, shape, distribution, volume fraction, alignment, and packing arrangement of zeolites (Caro et al. 2000, Dogan 2012, Vu et al. 2002). Development and tailoring of biopolymer–zeolite composites offer the possibility to promote their use in many fields of applications like automotive, aerospace, building, electrical, optoelectronic, biomedical, and environmental protection.

Major challenges in design and fundamental understanding of biopolymer–zeolite composites are related to their complex structure, to the uniform distribution of zeolites within biopolymer matrix, and to the relationship between dispersion and optimal properties of biocomposites. Several strategies have been studied to achieve well-dispersed zeolites in biopolymer matrix, including blending biopolymers and zeolites in solvent, solvent casting method (often with surface functionalization and/or sonication pretreatment), and in situ polymerization (Bastani et al. 2013, Chmielewská et al. 2011, Dragan and Dinu 2009, Dragan et al. 2010, Mintova and Valtchev 1996, Nigiz et al. 2012, Vu et al. 2002). Blending biopolymers and zeolites in solvent resulted in a better dispersion. The biopolymer–zeolite composites have been mainly evaluated for their performance in environmental pollutant removal and in separations of aqueous–organic mixtures (Bhat and Aminabhavi 2007a, Dogan 2012, Dragan et al. 2011, Humelnicu et al. 2011, Ji et al. 2012, Kittur et al. 2005, Nešić et al. 2013, Saraswathi and Viswanath 2012, Spiridon et al. 2013, Wan Ngah et al. 2013, Xie et al. 2013).

In this chapter, the synthesis, characterization, and applications for the removal of heavy metal ions and for separations of aqueous–organic mixtures will be highlighted for chitosan (CS), alginate (Alg), and cellulose, as biopolymer matrix, embedded with synthetic or natural zeolites. Removal of dyes by CS–zeolite composites will be also discussed. The sorption capacities and the pervaporation (PV) separation performances of biopolymer–zeolite composites have been compared with those of the raw zeolite, pristine biopolymer, or other biopolymer-based composites. This chapter also includes the mechanisms that might be involved during the sorption process.

6.2 CHITOSAN–ZEOLITE COMPOSITES

A strong interest has been lately addressed to biopolymers, as adsorbents coming from renewable resources, because they are a more cost-effective alternative to the existing adsorbents like activated carbon (AC) and synthetic ion exchangers

(Crini 2005, Wan Ngah et al. 2011). Among biopolymers, CS, the linear cationic polysaccharide composed of β-(1 $\rightarrow$ 4)-2-amino-2-deoxy-D-glucopyranose and β-(1 $\rightarrow$ 4)-2-acetamido-2-deoxy-D-glucopyranose units, randomly distributed along the polymer chain, has attracted numerous scientists due to its good complexing ability with metal ions. CS-based composite materials such as CS–perlite composites (Hasan et al. 2007), CS–cellulose composites (Sun et al. 2009, Zhou et al. 2004), CS–cotton composites (Qu et al. 2009), CS–bentonite composites (Wan Ngah et al. 2010), CS–montmorillonite composites (Nešić et al. 2012), CS–montmorillonite–magnetite composites (Chen et al. 2013), and CS–g-poly(acrylic acid)–attapulgite composites (Wang et al. 2009) have been designed by different strategies in order to remove metal ions and dyes from wastewaters.

Nowadays, other naturally occurring low-cost materials such as zeolites are intensively investigated as adsorbents in separation processes (Hernández-Montoya et al. 2013, Sanghi and Verma 2013, Wang and Peng 2010, Zou et al. 2011). Zeolites are crystalline microporous aluminosilicates with ion exchange properties suitable for a wide range of applications in catalysis and separation of liquids and gases. Among natural zeolites, clinoptilolite (CPL) is one of the most frequently studied (Aytas et al. 2004, Bedelean et al. 2006, Günay et al. 2007, Hernández-Montoya et al. 2013, Kilincarslan and Akyil 2005, Qiu et al. 2009, Sprynskyy et al. 2006, Zou et al. 2011). It was demonstrated that the cation exchange capability of CPL depends on the pretreatment method and that conditioning improves its ion exchange ability and removal efficiency (Aytas et al. 2004, Günay et al. 2007, Kilincarslan and Akyil 2005, Qiu et al. 2009, Sprynskyy et al. 2006, Zou et al. 2011). However, the cation exchange capacity of the natural CPL is relatively low, and therefore surfactant-modified zeolites or biopolymer–zeolite composite materials with chelating-ion exchange properties have been synthesized to increase the adsorption capacity for environmental pollutants (Bondarev et al. 2011, Chao and Chen 2012, Dinu and Dragan 2010a,b, Dogan 2012, Dragan et al. 2010, Ji et al. 2012, Xie et al. 2013, Zhan et al. 2013).

6.2.1 SYNTHESIS AND CHARACTERIZATION OF CHITOSAN–ZEOLITE COMPOSITES

The first report on ionic composites, as beads, based on CS and CPL, a widespread natural zeolite, was presented by Dragan and colleagues (Dragan and Dinu 2009, Dragan et al. 2010). The zeolite sample used comes from volcanic tuffs containing 60%–70% CPL, cropped out in Măcicaş area (Cluj County, Romania), and has the following ideal composition: $(NaKCa_{0.5})_{5.4}(Al_{5.4}Si_{30.6}O_{72})\cdot 20H_2O$ (Si/Al = 5.7). CPL particles with sizes between 0.032 and 0.05 mm were used for the preparation of composites. The cross-linked ionic composites consist of CPL particles dispersed (embedded) in the CS matrix, the whole system being stabilized by an ionotropic gelation of CS with sodium tripolyphosphate (TPP) in "tandem" with a covalent cross-linking with epichlorohydrin (ECH). The schematic representation of the formation of CS–CPL composites is depicted in Figure 6.1.

As Figure 6.1 shows, in the first step, the multivalent anion TPP forms ionic cross-links by the interaction with the amino groups of CS, locking the CPL microparticles in the composite microspheres. However, the composite microspheres stabilized only by the ionotropic gelation do not preserve their integrity when they are used in acidic

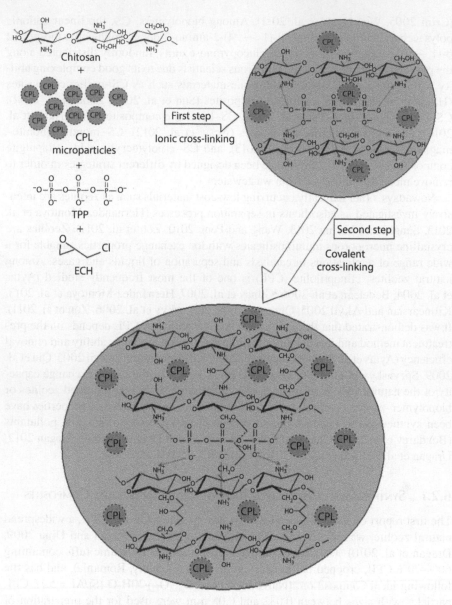

FIGURE 6.1 Schematic representation of the formation of CS–CPL composite through ionic and covalent cross-links.

or basic environment (Crini 2005). Therefore, the covalent cross-linking with ECH has been used to prepare CS–CPL composites with a high chemical stability. ECH was preferred as a cross-linker because further hydrophilic groups are generated (OH) (see Figure 6.1, second step), which could contribute to the increase of the chelating performances of the whole system. Ionic composite materials with different amounts of CPL were prepared. The ionic composites as beads were characterized by scanning electron microscopy (SEM), energy-dispersive x-ray analysis (EDX), x-ray diffraction (XRD), Fourier transform infrared spectroscopy (FTIR), and thermogravimetric analysis (TGA).

The first information on the presence of CPL in different ratios in the CS–CPL composites as well as on the sizes of beads in dry state and on the elemental composition of the CS–CPL composites was given by SEM–EDX. A dense and uniform morphology without pores exhibited the cross-linked CS beads without zeolite, while the CS–CPL composites presented different morphologies. The presence of CPL was observed as discrete zones for the sample with 9.09 wt.% of CPL and even agglomerates for the composites with the highest content of loaded CPL, that is, 33.3 wt.% of CPL (Dragan et al. 2011). The average size of the microspheres in dry state was evaluated from SEM images, and the value obtained was around 800 μm. The EDX spectra of CS–CPL composites were consistent with the presence of Al and Si from CPL and C, N, and O from CS.

The XR diffractograms of cross-linked CS and CPL showed a low crystallinity for the cross-linked CS and a high degree of crystallinity for CPL. When CS and CPL were mixed in different proportions to prepare CS–CPL composites, the crystallinity and the number of definite peaks increased with the increase of CPL percentage loaded in composites. At the same time, by analyzing the numerical registration data, a smooth left shift of the peaks position of about 0.1° has been observed, which corresponds to an increase in the interplanar distances. This small change would support a physical interaction between the continuous phase of the cross-linked CS and the dispersed phase of the CPL microparticles.

FTIR spectroscopy was performed to identify any changes in the structure of CS–CPL composites compared with the cross-linked CS. The main characteristic bands of CPL and CS were found in the FTIR spectra of CS–CPL composites. For example, the peak at 608 cm^{-1}, assigned to the vibration of the external linkage of the tetrahedra, and the peak at 467 cm^{-1}, which resulted from the stretching vibrations of Al–O bonds, were observed in all composites, their intensity increasing with the increase of the loaded CPL content. The band at 795 cm^{-1}, assigned to Si–O–Si bonds, was visible only in composites with a high content of CPL (i.e., 27.27 and 33.3 wt.%). Increasing the CPL content, some of the characteristic bands of CS were either red shifted or blue shifted (Dragan et al. 2010). Also, the band at 1032 cm^{-1}, assigned to the stretching vibration of the C–O bonds in anhydroglucose ring, was diminished with the increase of CPL content.

The synthesis strategy mentioned earlier for the preparation of CS–CPL composite has been recently used by Metin and colleagues to design CS–zeolite composite beads for the removal of the acidic dye (Metin et al. 2013). The zeolite sample with an average particle size less than 33 μm was provided from the Bigadic region of Turkey. The CS–zeolite biocomposite beads were characterized by SEM–EDX, FTIR, and TGA. CS–CPL composites have been also prepared by encapsulation of zeolites from volcanic tuffs cropping out in Marsid area (Romania), in CS matrix (Bondarev et al. 2011).

CS–zeolite A composite without any cross-linking, CS–zeolite A composite cross-linked with ECH, and CS–zeolite A composite cross-linked with ECH and TPP have been recently reported by the group of Wan Ngah (Wan Ngah et al. 2012a, 2013).

Zeolite A is a synthetic sodium aluminum silicate and has been used as inorganic filler due to its interesting structure that has the ability for adsorption of specific molecules. Based on the FTIR, TGA, and XRD analysis, a bridging mechanism for the formation of the double cross-linked CS–zeolite A composite was proposed. Moreover, it was found that the double cross-linked CS–zeolite A composite presents improved stability in distilled water, acetic acid, and NaOH comparative to the other two samples (Wan Ngah et al. 2013).

CS–zeolite composites, as beads, have been also synthesized in an alkaline medium adjusted to pH 9 by an aqueous solution of NaOH. A natural zeolite originated from a mineral deposit in Jinyun County, China, has been used to prepare the composite materials (Lin and Zhan 2012). The CS–zeolite composites showed enhanced sorption capacity for organic pollutants. The predicted maximum monolayer humic acid adsorption capacity for CS–zeolite composite at pH 7 and 30°C was found to be 74.1 mg/g.

CS-based composite membranes with improved PV separation characteristics have been generated by incorporating synthetic zeolites into CS (Kittur et al. 2005) or by embedding silicalite zeolite in CS cross-linked by glutaraldehyde (GA) (Patil and Aminabhavi 2008). The membranes were characterized by FTIR, ion exchange capacity measurements, XRD, and SEM. In a recent study, CS–zeolite A composite materials as films, with enhanced sorption capacity for anionic dyes, have been also reported (Nešić et al. 2013). The biological and antimicrobial activities of CS–zeolite A composites have been also investigated (Yu et al. 2013a).

6.2.2 APPLICATIONS OF CHITOSAN–ZEOLITE COMPOSITES

6.2.2.1 Removal of Heavy Metal Ions

One of the most important environmental problems throughout the world is hazardous heavy metal pollution of wastewaters. The sources of heavy metal ions in the wastewaters include mines and industries like paints, pigments, metal fabrication, batteries, corrosion control, fertilizers, and textiles. Because of their high solubility in the aquatic environments, heavy metals can be absorbed by living organisms. If the heavy metals are ingested beyond the permitted concentration, they can cause serious health disorders (Fu and Wang 2011). Therefore, it is necessary to treat the wastewaters contaminated with metal ions prior to their discharge into the environment.

Chemical precipitation, membrane separation, ion exchange, evaporation, and electrolysis are conventional methods usually used to remove heavy metal ions from industrial effluents (Fu and Wang 2011). They are often ineffective, especially in removing heavy metal ions from dilute solutions. Among the conventional techniques commonly used in the removal of heavy metal ions, the adsorption process is mainly preferred either when the enrichment of trace metal amounts or a high selectivity for a certain metal ion is required (Dinu and Dragan 2008, Dinu et al. 2009, Fu and Wang 2011, Wang and Peng 2010).

It is essential to note that the selection of the most suitable treatment technique of wastewater contaminated with heavy metals depends on the initial metal concentration, the components of the wastewater, operational cost, plant flexibility and reliability, and environmental impact. Therefore, a central task of many groups of researchers is finding cheap adsorbents with high efficiency and selectivity for heavy metal ions. Adsorbents coming from renewable resources like polysaccharides (especially CS) or natural zeolites are lately preferred due to their biocompatibility and their origin from unlimited resources. Numerous studies have been devoted to evaluation of adsorption properties of new composite materials based on CS and zeolites for heavy metal ions (Bondarev et al. 2011, Dinu and Dragan 2010a,b, Dragan and Dinu 2009, Dragan et al. 2010, 2011, Humelnicu et al. 2011, Wan Ngah et al. 2012a,b, 2013, Xie et al. 2013). Incorporation of zeolites in a CS network is an effective method to design composite systems with improved sorption properties. In fact, by embedding CPL microparticles in a matrix of cross-linked CS, it was found that the sorption properties of the whole system for Cu^{2+} ions were enhanced (Dragan and Dinu 2009, Dragan et al. 2010, 2011). From the studies on the influence of zeolite content on the adsorption capacity of the CS–CPL composites for Cu^{2+} ions, an abrupt increase of the adsorption capacity of CS–CPL composites compared with cross-linked CS was observed, starting with 9.09 wt.% up to about 20 wt.% of CPL loaded in the composite. The increase of the adsorption capacity of the CS–CPL composites was explained by a synergy of both components, the presence of CPL microparticles leading to the increase of the accessibility of metal ions at the functional groups of the CS network. Therefore, the CS–CPL composite loaded with 20 wt.% of CPL was selected for a comparative evaluation of the Cu^{2+}, Co^{2+}, and Ni^{2+} ion removal as a function of some parameters, like the number of consecutive sorption–desorption cycles, the initial pH, the contact time, the initial metal ion concentration, and the temperature (Dinu and Dragan 2010a,b, Dragan et al. 2011). The reusability of the CS–CPL composite compared with that of the cross-linked CS without CPL for the sorption of Cu^{2+} is depicted in Figure 6.2a. The effect of pH on the sorption capacity of CS–CPL composite for Cu^{2+}, Co^{2+}, and Ni^{2+} ions is shown in Figure 6.2b.

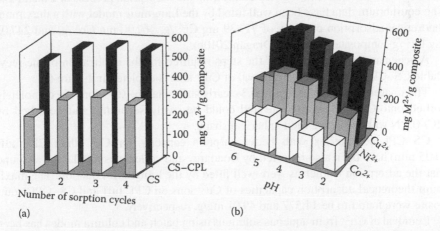

FIGURE 6.2 (a) Effect of the sorption cycles on Cu^{2+} ion retention on cross-linked CS and on CS–CPL composites loaded with 20 wt.% of CPL, at pH 5. (b) Effect of solution pH on the sorption capacity of CS–CPL composite loaded with 20 wt.% of CPL for Cu^{2+}, Ni^{2+}, and Co^{2+} ions.

Figure 6.2a shows that in the first cycle of adsorption, an increase of the adsorption capacity can be observed in the case of CS–CPL composite compared with cross-linked CS, this fact being attributed to the increase of accessibility of Cu^{2+} ions at the functional groups of the CS matrix due to the presence of CPL microparticles. Desorption of the Cu^{2+} ions from the composite after the first cycle was achieved with 0.1 M HCl in about 20 min. The eluent used for the regeneration of the spent sorbent was 0.1 M NaOH. The regenerated composite was reused in another cycle of adsorption. In the second cycle of adsorption, a significant increase of the sorption capacity can be seen, for both sorbents. The increase of the adsorption capacity in the second cycle of adsorption, that is, after the sorbent regeneration, was attributed to the increase of the number of amine groups available for Cu^{2+} binding by the complete removal of the TPP anions involved in the ionic gelation step (see Figure 6.1, first step). The stabilization of the CS–CPL composite by the covalent cross-linking was demonstrated by the values of the adsorption capacity, which were almost constant even after the fourth cycle of adsorption (Figure 6.2a, Dragan et al. 2010).

The pH of the metal ion solution plays an important role in the whole adsorption process. As can be seen from Figure 6.2b, the amount of M^{2+} adsorbed by the CS–CPL composite slowly increased when the pH of the M^{2+} solution increased from 2 to 5, the optimum adsorption pH being located at 5. At low pH, most of the amino groups of CS in the composite were in the form of NH_3^+, the electrostatic repulsion between M^{2+} and NH_3^+ ions preventing the adsorption of M^{2+} ions onto the composite. At pH > 5, the M^{2+} retention decreased because a small amount of M^{2+} started to deposit as $M(OH)_2$.

The kinetics data were analyzed by pseudo-first-order, pseudo-second-order, and intraparticle diffusion equations. It was found that the adsorption kinetics were well described by the pseudo-second-order equation, supporting that the chemisorption would be the rate-determining step controlling the adsorption process of a metal ion. The equilibrium data have been well fitted by the Langmuir model with a maximum theoretical adsorption capacity of 719.39 mg Cu^{2+}/g, 467.90 mg Co^{2+}/g, and 247.03 mg Ni^{2+}/g composite (Dinu and Dragan 2010b).

A schematic representation of the steps required for the application of the recyclable CS–CPL composite for removal of Cu^{2+} ions is depicted in Figure 6.3.

The images presented in Figure 6.3 clearly show that the Cu^{2+} ions are completely sorbed onto CS–CPL composite and could be easily separated and desorbed by HCl 0.1 N without losing their spherical shape.

CS–CPL composites with better sorption capacity for Cu^{2+} than CPL tuff (0.05 mm) have been also reported by Bondarev and colleagues (2011). It was found that the adsorption isotherms were well fitted by the Langmuir equation. The maximum theoretical adsorption capacities of Cu^{2+} ions on CPL tuff and CS–CPL composite were found to be 14.577 and 49.01 mg/g, respectively.

Removal of Cu^{2+} from aqueous solutions using batch and column modes has been also performed on CS–zeolite A composites prepared in alkaline medium (Wan Ngah et al. 2012b). It was found that the optimum pH, stirring rate, and stirring time were 3.0, 300 rpm, and 60 min, respectively. The Cu^{2+} uptake rate was analyzed using the pseudo-first-order and pseudo-second-order kinetic models, and it showed

FIGURE 6.3 Schematic presentation by optical images of the sorption–desorption process of Cu²⁺ ions on/from recyclable CS–CPL composites, as beads.

better regression value for the pseudo-first order, that is, physical sorption mechanism. The sorption isotherm was well described by the Langmuir model, and the maximum adsorption capacity was 25.88 mg/g. Based on the bed depth service time (BDST) model, the critical bed depth was found to be 3.44 cm, and the breakthrough curves were well fitted to the Clark model with decreasing rate of mass transfer as the bed height increased, being an indication of the limitation of the availability of the number of sites of adsorbent for interactions with metal ions. From the desorption studies, Cu²⁺ was only able to be recovered by ethylenediaminetetraacetic acid; however, the removal percentage was quite low, and therefore, regeneration of the CS–zeolite A composite was not favored. Therefore, new CS–zeolite A composites have been synthesized by modification of the type of cross-linking (Wan Ngah et al. 2013). CS–zeolite A composite cross-linked with ECH, CS–zeolite A composite cross-linked with ECH and TPP, and CS–zeolite A composite without any cross-linking have been designed, and their adsorption properties for Cu²⁺ ions have been investigated. The maximum adsorption capacities based on the Langmuir model for CS–zeolite A composite cross-linked with ECH, CS–zeolite A composite cross-linked with ECH and TPP, and CS–zeolite A composite without any cross-linking were 51.32, 14.75, and 25.61 mg Cu²⁺/g, respectively. Experimental data for the CS–zeolite A composites were well fitted by the pseudo-second-order kinetic model, indicating the chemisorption as the rate controlling step of the sorption process. Moreover, a kinetic model consisting of pseudo-second-order adsorption and pseudo-first-order desorption was used to describe the binding behavior of the adsorption/desorption process in Cu²⁺-loaded CS–zeolite A composites. The results showed that the kinetic model was well fitted to the experimental data showing that $-CH_2OH$ groups of CS act as major binding sites for Cu^{2+} ions.

Nowadays, removal of radioactive ions from the wastewaters is also a huge problem, these radioisotopes being extremely dangerous for the environment and human health because of their high toxicity, even at very low concentrations, and long half-lives. Therefore, adsorption features of UO_2^{2+} and Th^{4+} ions from simulated radioactive solutions onto a CS–CPL composite, as beads, have been investigated compared with cross-linked CS without zeolite (Humelnicu et al. 2011). The adsorption process of both radiocations obeyed the pseudo-second-order kinetics, supporting the chemisorption as the rate-determining step. The equilibrium data obtained for the adsorption of radiocations onto the CS–CPL composite well fitted in the Sips model with a maximum theoretical adsorption capacity of 438.55 mg Th^{4+}/g composite and 536.35 mg UO_2^{2+}/g composite. The higher retention capacity against UO_2^{2+} compared with Th^{4+} ions was attributed to the higher level of hydration of Th^{4+} ions in solution, that is, the adsorption increasing with the decrease of the hydrated radiocation radius.

Table 6.1 summarizes the experimental conditions and the values of the theoretical adsorption capacities found for the removal of UO_2^{2+} and Th^{4+} ions from simulated radioactive solutions on different sorbents.

According to the values listed in Table 6.1, the adsorption capacities and removal efficiencies of the UO_2^{2+} and Th^{4+} radiocations found for CS-based composites are higher than those of different types of zeolites, and this recommends the CS-based composites as a better alternative for the sorption of Th^{4+} and UO_2^{2+} radiocations. It should be also noted that the sorption capacities and the removal efficiencies of zeolites depend on the source of zeolite as well as on the experimental conditions used to carry out the sorption (sorbent dosage, temperature, initial solution pH).

To check if the CS–CPL composite can be useful for selective separation of metal cations, sorption studies were performed under competitive conditions, that is, using binary and quaternary mixtures of UO_2^{2+} or Th^{4+} with Cu^{2+}, Fe^{2+}, and Al^{3+} (Humelnicu et al. 2011). In binary systems, it was found that the presence of Al^{3+} cations led to the lowest sorption amounts, for both UO_2^{2+} and Th^{4+} radiocations, that is, 373.4 and 302.63 mg/g, respectively. The results obtained as adsorbed amount of each cation from quaternary mixtures are presented in Figure 6.4.

The overall adsorption tendency of the CS–CPL composite toward UO_2^{2+} and Th^{4+} radiocations, under competitive conditions, follows the following order: $Cu^{2+} > UO_2^{2+} > Fe^{2+} > Al^{3+}$ (Figure 6.4a) and $Cu^{2+} > Th^{4+} > Fe^{2+} > Al^{3+}$ (Figure 6.4b), respectively.

An important characteristic of the composite sorbents is the rate of desorption of the ion adsorbed. The eluents used for desorption of Th^{4+} and UO_2^{2+} from the CS–CPL composite were 0.1 M HCl, 0.1 M HNO_3, and 0.1 M Na_2CO_3, respectively. The results indicated 0.1 M Na_2CO_3 as the best desorption agent for the UO_2^{2+} ions with a desorption level around 92%, while for the desorption of the Th^{4+} ions, the best results were obtained with 0.1 M HCl solution, the desorption level being around 85% (Humelnicu et al. 2011). The results of desorption tests showed that the sorption process is reversible and also the possibility to reuse the composite sorbent taking into account the high recovery level of both radiocations.

TABLE 6.1
Adsorption Capacities and Experimental Conditions of UO_2^{2+} and Th^{4+} Radiocation Removal Using Different CS-Based Composites and Zeolites

Sorbent	Initial Metal Conc. (mg/L)	Dose of Sorbent (g)	pH	Adsorption Capacity (mg/g)		Removal Efficiency (%)		Reference
				UO_2^{2+}	Th^{4+}	UO_2^{2+}	Th^{4+}	
CS–perlite composite	10	1	5	149.25	—	95	—	Hasan et al. (2007)
CS–CPL composite[a]	10–100	0.04	5.5	536.35	438.55	92	93	Humelnicu et al. (2011)
CS–poly(acrylamide) composite	10–80	0.1	5	72.9	118.32	NA	NA	Akkaya and Ulusoy (2008)
Magnetic CS composite	10–100	0.004	5.5	666.67	312.50	89	87	Hritcu et al. (2012)
CPL[b]	0.025–0.175	0.1	5	1.23	—	88	—	Aytas et al. (2004), Kilincarslan and Akyil (2005)
Raw zeolite[c]	40	10–20	6	11.13	—	94	—	Zou et al. (2011)
Manganese oxide-coated zeolite[c]	25–400	5	4	15.1	—	NA	—	Han et al. (2007)
Synthetic zeolite NaA	80	2.5	2	1.69	—	67.2	—	Nibou et al. (2011)

Note: NA, not available.

a CPL, Măcicaş Area, Cluj County, Romania.

b CPL, Balikesir deposits in Turkey.

c Raw zeolite, Xinyang City in China.

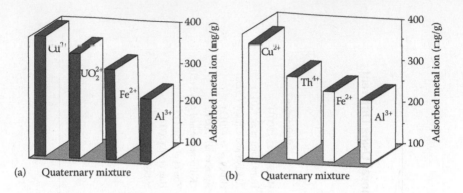

FIGURE 6.4 Adsorbed metal ion from quaternary mixtures containing UO_2^{2+}, Cu^{2+}, Fe^{2+}, and Al^{3+} ions (a) and Th^{4+}, Cu^{2+}, Fe^{2+}, and Al^{3+} ions (b).

6.2.2.2 Separation of Solvent Mixtures by Chitosan–Zeolite Composites

Organic–inorganic composite membranes received considerable attention, especially zeolite-filled membranes being widely studied due to their better PV performances over the pure polymeric membranes (Caro et al. 2000, Kittur et al. 2005, Libby et al. 2003, Nawawi et al. 1997, Patil and Aminabhavi 2008, Sun et al. 2008, Wang et al. 2010, Yuan et al. 2007). PV is a membrane-based technique in which the membrane functions as a selective barrier for the mixture to be separated. PV is a promising process in the chemical industry for the separation of azeotropic mixtures due to the low energy consumption and mild working conditions. Kittur et al. investigated the PV separation of water–isopropanol mixtures on zeolite NaY-filled CS membranes (Kittur et al. 2005). The experimental results showed that the membrane containing 30 wt.% of zeolite NaY presents the highest separation selectivity of 2620, with a flux of 11.50×10^{-2} kg/m² h at 30°C, and 5 wt.% of water in the feed, suggesting that the membrane could be effectively used to break the azeotropic point of water–isopropanol mixture. In Figure 6.5, the flux and separation characteristics of some CS-based membranes for water–isopropanol mixtures are presented (Kittur et al. 2005, Nawawi et al. 1997).

As Figure 6.5a shows, the separation factors of CS–zeolite NaY membranes are reasonably higher than those of other CS-based membranes. However, the pure CS membrane, CS–polysulfone membrane, and CS cross-linked with hexamethylene diisocyanate (HMDI) membrane exhibited significantly higher values of the permeation flux (Figure 6.5b), compared with CS–zeolite NaY membranes. This behavior was correlated with the thickness of the membranes, which was about 10, 20, and 40 µm for pure CS, CS–polysulfone, and CS–zeolite NaY membranes, respectively.

Zeolite-filled CS membranes were also investigated for their PV separation characteristics over that of pristine CS membrane for toluene–methanol and toluene–ethanol mixtures (Patil and Aminabhavi 2008). The composite membranes were cross-linked with GA. It was found that the zeolite-filled CS membranes were more toluene selective than alcohol selective. Toluene permeated preferentially with a selectivity of 264 and fluxes of 0.019–0.027 kg/m² h for toluene–methanol mixtures.

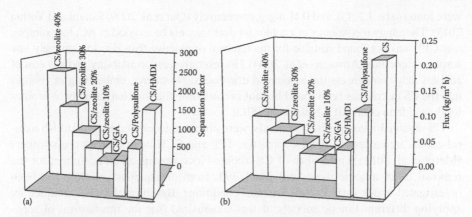

FIGURE 6.5 Comparative PV characteristics of some CS-based composite membranes for water–isopropanol mixtures: separation factor (a) and permeation flux (b).

Selectivity of 301 with fluxes ranging from 0.019 to 0.026 kg/m^2 h was observed for toluene–ethanol mixtures.

CS membranes filled by glycidoxypropyltrimethoxysilane (GPTMS)-modified zeolite were prepared and evaluated for methanol permeability compared with that of Nafion®117, pure CS, and CS–GPTMS membranes (Wang et al. 2010). Compared with the Nafion®117, the pure CS membrane showed much lower methanol permeability, which suggests that the CS is one of the excellent methanol barrier polymers. The CS membranes filled by modified zeolite showed much lower methanol permeability than the pure CS membrane. The decrease of the methanol permeability was attributed to two reasons: (1) the introduced zeolite particles increase the diffusion path length for the methanol molecules (Libby et al. 2003), and (2) the incorporated zeolite particles could cause the local rigidity of the CS matrix and compress the volumes between the CS chains (Yuan et al. 2007). The selectivity of CS–GPTMS–zeolite membranes is comparable with that of Nafion®117 at 2 M methanol concentration and much higher at 12 M methanol concentration. Considering the low methanol permeability, high selectivity, environmental benignity, low cost, as well as facile preparation, it may be concluded that the CS membranes filled with GPTMS-modified zeolite exhibit promising application potential as a proton exchange membrane for the direct methanol fuel cell (Wang et al. 2010).

6.2.2.3 Removal of Dyes by Chitosan–Zeolite Composites

Pollution caused by dyes is a common problem faced by many industrial countries. Release of colored wastewaters in effluents can affect the photosynthetic activity in aquatic life because of the reduced light penetration, and therefore, their decontamination is strongly required. The most commonly used adsorbent for the removal of dyes is AC manufactured from carbonaceous materials such as wood, coal, and waste agricultural biomass (Furlan et al. 2010, Sanghi and Verma 2013). However, because of the high cost of conventionally used AC, alternative sorbents have been investigated by employing natural zeolites, as inexpensive biosorbents. For example, the maximum adsorption capacities of the natural zeolite for Safranine T, Acid Red, and Amido Black 10B dyes were found to be 1.2, 1.0, and 0.11 mg/g, respectively (Qiu et al. 2009, Sanghi and Verma 2013). The removal efficiency of zeolites for dyes may not be as good as AC; for example, raw CPL was not found suitable for the removal of reactive dyes due to extremely low sorption capacities (Armagan et al. 2004). However, the easy availability and low cost of zeolites may compensate the associated drawbacks. Nowadays, embedding of zeolites into the CS matrix has been found helpful in increasing the sorption capacity of zeolites for dyes (Metin et al. 2013, Nešić et al. 2013).

CS–zeolite biocomposites, as beads, were prepared by embedding a Turkish natural zeolite into a cross-linked CS matrix. TPP and ECH were used as cross-linkers (Metin et al. 2013). The ability of CS–zeolite biocomposite as an adsorbent for the removal of an anionic dye, Acid Black 194, from an aqueous solution, has been investigated under various experimental conditions. Based on the data obtained by applying different kinetic models, it was established that the mechanism of sorption of Acid Black 194 on CS–zeolite biocomposites involved three steps: diffusion of dye molecules through the solution onto the surface of the adsorbent, adsorption of dye molecules on the surface of the biocomposite through molecular interactions, and diffusion of dye molecules from the surface to the interior of the adsorbent. Also, it was observed that the adsorption capacity increased when initial dye concentration or temperature increased. The maximum adsorption capacity of the dye onto CS–zeolite biocomposites was calculated as 2140 mg/g at 4000 mg/L initial dye concentration (Metin et al. 2013). The negative value of $\Delta G°$ and the value of enthalpy less than 84 kJ/mol showed that the adsorption process is spontaneous and physical. A new biocomposite based on CS modified by zeolite A, as adsorbent for an anionic dye, Bezactive Orange 16, has been also reported (Nešić et al. 2013). The results showed that the sorption capacity increased with the increase of dye concentration and with increase of pH up to 6. In alkaline solutions, the adsorption capacity decreased. Analysis of kinetic parameters by applying four kinetic models showed that the adsorption of Bezactive Orange 16 onto CS–zeolite A followed the pseudo-second-order model. According to the Langmuir model, the CS–zeolite A composites presented a maximum adsorption capacity of about 305.8 mg/g in acidic pH, being higher than those reported for sludge, AC, and CS–montmorillonite composites, which were equal to 97, 159, and 279 mg/g, respectively (Furlan et al. 2010, Nešić et al. 2012, Sanghi and Verma 2013).

6.3 ALGINATE–ZEOLITE COMPOSITES

Alg, a natural polysaccharide extracted from brown seaweeds, is a very promising biosorbent, being preferred over other materials because of its features such as biodegradability, hydrophilic properties, natural origin, abundance, and presence of binding sites due to its carboxylate functions. Alg is composed of β-D-mannuronate and α-L-guluronate, linked by β-1,4- and α-1,4-glycosidic bounds. The carboxylate groups of the Alg provide the ability to form biodegradable gels in the presence of multivalent cations, more specifically with calcium ions, via ionic interactions. In the environmental field, Alg beads are widely used for the removal of heavy metal ions from aqueous solution (Ibanez and Umetsu 2002). On the other hand, Alg beads containing different components like maghemite (Idris et al. 2012), humic acid (HA)

(Pandey et al. 2003), AC (Park et al. 2007), or zeolite (Chmielewská et al. 2011, Choi et al. 2009, Dogan 2012), which have the role to enhance the sorption capacity of the whole system, are widely investigated. Cross-linked composite membranes based on Alg and natural or synthetic zeolites have been also studied for the PV separation of water–isopropanol, water–tetrahydrofuran (THF), water–acetic acid, water–1, 4-dioxane, water–dimethylformamide (DMF), and water–ethanol mixtures (Bhat and Aminabhavi 2007a, 2009, Kahya et al. 2011, Kariduraganavar et al. 2004, Kittur et al. 2004, Nigiz et al. 2012). Recently, Alg–rectorite composites (Yang et al. 2012), Alg–halloysite composites (Liu et al. 2012), and Alg–magnetic ferrite composites (Mahmoodi 2013) have been reported as efficient sorbents for the removal of cationic dyes such as Basic Blue 9, methylene blue, Basic Blue 41, and Basic Red 18. So far, the scientific literature gives no data on the removal of dyes by Alg–zeolite composites.

6.3.1 Synthesis and Characterization of Alginate–Zeolite Composites

Alg-based composites, as beads, have been generated by embedding a synthetic zeolite type A and AC into cross-linked Alg matrix. A solution of $CaCl_2$ has been used to form the insoluble gel spheres (Choi et al. 2009). Alg–CPL composites have been also reported, by incorporation of a Slovakian natural zeolite into Alg cross-linked using Fe(III) and Ca(II) chlorides (Chmielewska et al. 2011). The small-angle x-ray scattering (SAXS) analysis revealed information about the chemical composition and porosity of the Alg–CPL composite and of the zeolite sample. According to the SAXS porosity calculations, the CPL and iron alginate (FeAlg)–CPL composites exhibited porosity around 28% and 37%, respectively. The internal morphology of cross-linked Alg, CPL sample, FeAlg–CPL, and calcium alginate (CaAlg)–CPL composites was observed by SEM, scanning tunneling microscopy, and atomic force microscopy. The synthesized FeAlg–CPL composite displayed a rough, nearly rope-like surface morphology being a profound change compared with the original tablet-like Slovakian zeolite morphology. Alg-based composites, as beads, have been also prepared by embedding Turkish CPL and HA into cross-linked Alg matrix (Dogan 2012). The entrapment of CPL and HA into the Alg matrix was proved by FTIR, SEM, TGA, and XRD. The Turkish CPL sample exhibited cauliflower morphology, while that of the corresponding composite was different, the presence of CPL being observed as discrete zones and even agglomerates formation.

Alg-based composites, as membranes, were synthesized by solution casting method using synthetic zeolites or natural zeolite microparticles and sodium alginate (NaAlg). The membranes were cross-linked with GA (Bhat and Aminabhavi 2007a, 2009, Kahya et al. 2011, Kittur et al. 2004, Nigiz et al. 2012). Zeolite K-LTL prepared by hydrothermal route with different ratios between SiO_2/Al_2O_3, zeolite 4A type, and zeolite NaY type was used as synthetic zeolites. CPL microparticles with a ratio between SiO_2 and Al_2O_3 equal to 6 and sizes obtained by sieving of about 13 and 38 μm were provided by Etibank A.Ş. and Enli Madencilik A.S. (Bigadiç, Balıkesir, Turkey) and were utilized as natural zeolites for composite designing. The SEM images showed the uniform distribution of CPL and zeolite K-LTL microparticles in the NaAlg-based composites. The uniform dispersion of zeolite microparticles in the NaAlg matrix is important to enhance the flux and selectivity during PV separation experiments of solvent mixtures.

6.3.2 APPLICATIONS OF ALGINATE–ZEOLITE COMPOSITES

6.3.2.1 Removal of Heavy Metal Ions

Alg-synthetic zeolite–AC composites, as beads, were used for the removal of mixed contaminants containing both organic and inorganic compounds (Choi et al. 2009). The adsorption of Zn^{2+} and toluene as target contaminants onto the Alg-synthetic zeolite–AC composites was investigated by performing both equilibrium and kinetic batch tests. Equilibrium tests showed that adsorption of both contaminants followed Langmuir isotherm and the composites were capable of removing Zn^{2+} and toluene with a maximum theoretical sorption capacity of 4.3 and 13.0 mg/g, respectively. The values obtained were higher compared with those reported for AC, zeolite, and Alg–AC composites (Choi et al. 2009, Fu and Wang 2011, Hernandez-Montoya et al. 2013, Wang and Peng 2010). It was further revealed from kinetic tests that removal efficiencies of Zn^{2+} and toluene were 54% and 86%, respectively, for the initial solution concentrations of 250 mg/L. This indicates that the Alg-synthetic zeolite–AC composites can be used as promising adsorbents for simultaneous removal of organic and inorganic compounds from wastewater.

Composite-adsorbent materials based on CaAlg, CPL, and HA were used for the removal of Cd^{2+}, Hg^{2+}, and Pb^{2+} (Dogan 2012). The sorption properties of CaAlg–CPL–HA were compared with those of HA, CPL, CaAlg, CaAlg–HA, and CaAlg–CPL. The kinetic data obtained for the Alg, Alg–HA, Alg–CPL, and Alg–HA–CPL adsorbents were evaluated using pseudo-first-order model and pseudo-second-order model. It was found that the pseudo-second-order kinetic model well described the experimental data for Cd^{2+}, Hg^{2+}, and Pb^{2+} ions on all adsorbents investigated. The equilibrium results were fitted to Langmuir, Freundlich, Dubinin–Radushkevich, and Temkin isotherm models. The Freundlich and Langmuir models well described the sorption of Pb^{2+} and Hg^{2+} ions, while Dubinin–Radushkevich model well fitted the equilibrium data for Cd^{2+} removal.

Removal of nitrate, sulfate, and Zn^{2+} ions from aqueous solutions through adsorption onto FeAlg–CPL and CaAlg–CPL composites was studied using an equilibrium batch technique (Chmielewská et al. 2011). The equilibrium adsorption data for the sorption of nitrate, sulfate, and Zn^{2+} ions were well fitted by Freundlich and Langmuir isotherms. The highest sorption capacity was provided by the CaAlg–CPL composite toward Zn^{2+} ions, while the FeAlg–CPL composite exhibited high sorption capacity for nitrate and sulfate ions.

Table 6.2 summarizes the adsorption capacities and the experimental conditions of the Alg–zeolite composites and of the raw zeolite for removal of some heavy metal ions, like Zn^{2+}, Hg^{2+}, Cd^{2+}, and Pb^{2+}, from aqueous solution.

Comparison of the sorption capacities of the Alg–zeolite composite adsorbents with those of the natural zeolite, presented in Table 6.2, clearly shows that the Alg–zeolite composite-based adsorbents are more effective than CPL for the removal of Zn^{2+}, Hg^{2+}, and Pb^{2+} ions. The use of various isotherm models indicated that many of the adsorption processes were described in terms of the Langmuir approach (Bondarev et al. 2011, Chmielewská et al. 2011, Choi et al. 2009, Hernández-Montoya et al. 2013). This general model describes a situation where at low concentrations in solution, strong uptake of the

TABLE 6.2
Experimental Conditions, Adsorption Capacities, and Isotherm Models That Will Describe the Sorption Process of the Alg–Zeolite Composites and of the Raw Zeolite for Zn^{2+}, Hg^{2+}, Cd^{2+}, and Pb^{2+} Ions

Sorbent	Metal ions	C_i^a (mg/L)	m^b (g)	pH	Temperature (°C)	q_{max}^c (mg/g)	Isotherm	Reference
CaAlg–zeolite A–AC	Zn^{2+}	50–1000	20	6–7	32	4.3	Langmuir	Choi et al. (2009)
CaAlg–Slovakian CPL	Zn^{2+}	100–400	0.5	6.7	22	70	Langmuir	Chmielewská et al. (2011)
FeAlg–Slovakian CPL	Zn^{2+}	100–400	0.5	6.7	22	50	Langmuir	Chmielewská et al. (2011)
Slovakian CPL	Zn^{2+}	100–400	0.5	6.7	22	20	Langmuir	Chmielewská et al. (2011)
Mexican CPL	Zn^{2+}	20–250	0.02	5	30	6.47	Langmuir	Hernández-Montoya et al. (2013)
CaAlg–Turkish CPL	Hg^{2+}	50–200	0.5	3.5–4	25	74	Langmuir and Freundlich	Dogan (2012)
CaAlg–HA–Turkish CPL	Hg^{2+}	50–200	0.5	3.5–4	25	53	Langmuir and Freundlich	Dogan (2012)
CaAlg–Turkish CPL	Cd^{2+}	50–200	0.5	3.5–4	25	4.4	Dubinin–Radushkevich	Dogan (2012)
CaAlg–HA–Turkish CPL	Cd^{2+}	50–200	0.5	3.5–4	25	3.6	Dubinin–Radushkevich	Dogan (2012)
Romanian CPL	Cd^{2+}	30–100	0.25	5	22	13.6	Langmuir	Bondarev et al. (2011)
Mexican CPL	Cd^{2+}	20–250	0.02	5	30	3.70	Langmuir	Hernández-Montoya et al. (2013)
Ukrainian CPL	Cd^{2+}	80	0.5	6.2	25	4.22	Langmuir and Freundlich	Sprynskyy et al. (2006)
CaAlg–Turkish CPL	Pb^{2+}	50–200	0.5	3.5–4	25	117	Langmuir and Freundlich	Dogan (2012)
CaAlg–HA–Turkish CPL	Pb^{2+}	50–200	0.5	3.5–4	25	111	Langmuir and Freundlich	Dogan (2012)
Turkish CPL	Pb^{2+}	50–400	0.5	4.5	22	80.9	Sips and Toth	Günay et al. (2007)
Mexican CPL	Pb^{2+}	20–250	0.02	5	30	42.8	Langmuir	Hernández-Montoya et al. (2013)
Ukrainian CPL	Pb^{2+}	800	0.5	6.2	25	27.7	Langmuir and Freundlich	Sprynskyy et al. (2006)

Note: AC, activated carbon; HA, humic acid.

a C_i, initial metal concentration.

b m, mass of sorbent.

c q_{max}, maximum theoretical sorption capacity.

metal ion on the adsorbent is observed, but beyond a certain concentration, a plateau is reached in metal ion uptake, which corresponds to the maximum binding capacity, q_{max} (column 7, Table 6.2). As can be seen from Table 6.2, the adsorption processes were also described by the Dubinin–Radushkevich model, Sips and Toth model, or Freundlich model (Dogan 2012, Günay et al. 2007).

6.3.2.2 Separation of Solvent Mixtures

Membrane-separation techniques, with an easy operation and high energy savings, are greatly appreciated in a variety of applications in the medical, food, industrial, energy, and environment fields (Kittur et al. 2004). Nowadays, the research studies on the PV process have been concentrated on the separation of water–alcohol, water–acetic acid, water–THF, or water–DMF systems using composite membranes based on Alg and zeolites (Bhat and Aminabhavi 2007a,b, 2009, Kahya et al. 2011, Kittur et al. 2004, Nigiz et al. 2012). For example, the PV dehydration of isopropanol, 1,4-dioxane, and THF was tested on NaAlg–zeolite K-LTL composite membranes as a function of membrane thickness and feed compositions (Bhat and Aminabhavi 2007a). It was found that the permeation flux and selectivity of the composite membrane were improved compared with that of the pristine NaAlg membrane due to the addition of zeolite K-LTL particles. Molecular sieving effect created by the uniform distribution of zeolite K-LTL particles and its hydrophilic nature in addition to its interaction with hydrophilic NaAlg was responsible for the appreciable increase of the membrane composite performance. Efficiency of the composite membranes in removing water from the mixed aqueous media was more significant for isopropanol and 1,4-dioxane than for THF. Furthermore, the composite membranes could withstand the repetitive cyclic PV runs on the laboratory level module.

The PV separation of water–DMF mixtures on cross-linked composite membranes based on NaAlg and CPL has been also studied (Kahya et al. 2011). The effects of feed composition and operating temperature on the permeation rates and the separation factors have been investigated. It was observed that the flux values of composite membranes were much higher, whereas the selectivities were lower than those of pristine NaAlg membranes. NaAlg–zeolite 4A and NaAlg–CPL composite membranes were also studied for separation properties of ethanol–water mixtures (Nigiz et al. 2012). The effects of zeolite type, zeolite loading, zeolite particle size, and feed composition on PV performance were evaluated. It was observed that the flux and selectivity increased with the increase of zeolite content up to 15% in the membrane composites, while an increase of the water concentration in the feed mixtures conducted to the decrease of the selectivity. Also, it was seen that the synthetic zeolite 4A was more proper than CPL for designing composite membranes with improved PV features. In order to increase PV performance, "priming" procedure was applied to NaAlg–zeolite 4A composite membrane loaded with 10 wt.% of zeolite 4A. "Priming" procedure consists of covering the NaAlg–zeolite 4A composite membrane with a thin polymer film in order to eliminate the interfacial voids observed between the polymer matrix and the zeolite particles. It was found that the selectivity of modified composite membrane was eight times higher than that of pristine composite membrane. However, the results reported were not sufficient to recommend these membranes for commercial use.

The PV performances of different NaAlg–zeolite composite membranes for some aqueous–organic mixtures compared to other membranes are presented in Table 6.3.

As can be seen in Table 6.3, the selectivity of NaAlg-based composite membranes prepared with different amounts of zeolite for PV separation of water–THF, water–acetic acid, and water–ethanol mixtures was improved compared with that of other Alg-based composite membranes. The best results for PV separation of water–acetic acid and water–ethanol mixtures have been obtained on NaAlg–zeolite 4A composite membranes at 30°C and 10% of water in the feed (Bhat and Aminbavi 2009), while NaAlg–zeolite K-LTL (10 wt.%) composite membrane presented high selectivity for PV separation of water–THF mixtures at 30°C and 6.7% of water in the feed (Bhat and Aminbavi 2007a).

6.4 CELLULOSE–ZEOLITE COMPOSITES

Cellulose, the most widely available and renewable biopolymer in nature, is a very promising raw material available at low cost for the preparation of various functional materials. Each glucose unit in the cellulose backbone contains three hydroxyl groups that can undergo acetyl substitution. Cellulose acetate (CA) is frequently used as a polymer matrix to build composite materials because it can be easily molded into different forms, such as membranes, fibers, and spheres and by its hydrophilic nature contributes to an improvement of the accessibility of aqueous solutions to the surface of composite materials (Chen et al. 2004, Ji et al. 2012, 2013, Liu and Bai 2006, Ma et al. 2008, Nataraj et al. 2011, Zafar et al. 2012, Zhou et al. 2004).

To enhance the sorption performances of the cellulose-based polymeric membranes, inorganic compounds such as zeolites have been incorporated into the polymer matrix (Agarwal et al. 2012, Bastani et al. 2013, Dogan and Hilmioglu 2010a,b, Ji et al. 2012, 2013, Mintova and Valtchev 1996, Vu et al. 2002). Moreover, zeolites have a high mechanical strength and good thermal and chemical stability, and the membranes incorporating these fillers can be used over a wide range of operating conditions (Bastani et al. 2013, Caro et al. 2000).

Modified cellulose–zeolite composites have been proposed for numerous applications such as medical antibacterial materials, deodorizers, absorbent pads, sanitary napkins, gas separators, adsorbents for heavy metal ions, and organic pollutants like phenols, dyes, and PV separation membranes for solvent mixtures.

6.4.1 SYNTHESIS AND CHARACTERIZATION OF CELLULOSE–ZEOLITE COMPOSITES

The preparation of a soft membrane with molecular sieving properties is attractive from both practical and fundamental standpoints. One general approach is to combine a crystalline, porous solid such as zeolites, with a flexible, cheap, and abundant organic matrix such as natural cellulose. The early reports on cellulose–zeolite composite materials indicated for composite preparation the use of adhesive polymers, "electret" technology, or in situ zeolite crystallization (Caro et al. 2000, Mintova and Valtchev 1996, Mintova et al. 1996, Vu et al. 2002, Zafar et al. 2012). The adhesive polymer method was found to present several disadvantages, the most important being the weak zeolite–cellulose interaction and the random distribution

TABLE 6.3

Comparison of PV Performance of the NaAlg–Zeolite Membranes with Other Membranes for Aqueous–Organic Mixtures

Membrane Type	Temperature (°C)	Water in Feed wt.%	J^a (kg/m²h)	α^b	Reference
Water + THF					
NaAlg–zeolite K-LTL (10 wt.%)	30	6.7	0.12	3081	Bhat and Aminabhavi (2007a)
NaAlg–HEC (10 wt.%)	30	10	0.183	1516	Bhat and Aminabhavi (2007b)
Zeolite A–ceramic support	45	6.1	0.023	2000	Urtiaga et al. (2003)
Zeolite NaY–Al$_2$O$_3$ support	60	10	0.15	230	Li et al. (2001)
Zeolite ZSM-5–Al$_2$O$_3$ support	60	6.7	0.01	3.4	Li et al. (2001)
Water + acetic acid					
Pure Alg	30	5	2.24	28	Kittur et al. (2004)
NaAlg–zeolite NaY (30 wt.%)	30	5	3.80	42	Kittur et al. (2004)
NaAlg–zeolite 4A (10 wt.%)	30	10	0.188	991	Bhat and Aminabhavi (2009)
Cross-linked NaAlg with HDM	70	15	0.262	161	Wang (2000)
NaAlg–PAAm-g-GG	30	10	0.029	15	Bhat and Aminabhavi (2007b)
Water + ethanol					
NaAlg–zeolite 4A (10 wt.%)	30	10	0.137	1334	Bhat and Aminabhavi (2009)
NaAlg–cellulose	60		0.068	1175	Yang et al. (2000)
NaAlg–PHEMA (20 wt.%)	30	4	0.132	571	Bhat and Aminabhavi (2007b)
NaAlg–AlPO$_4^{-5}$ (20 wt.%)	30	4	0.104	980	Bhat and Aminabhavi (2007b)
NaAlg–CPL (10 wt.%)	25	2	0.123	43	Nigiz et al. (2012)
NaAlg–zeolite 4A (10 wt.%)	25	2	0.106	396	Nigiz et al. (2012)
NaAlg–zeolite MCM41 (20 wt.%)	30	10	0.645	1090	Bhat and Aminabhavi (2007b)

Note: HEC, hydroxyethylcellulose; HDM, hexanediamine; CPL, caprolactam; GG, guargum; PHEMA, poly(hydroxyethylmethacrylate).

a Permeation flux.

b Separation factor.

of zeolite crystallites on the cellulose surface. In addition, incorporating an adhesive (a third chemical entity) in the zeolite–cellulose composite could be deemed inacceptable for certain applications, especially if the polymer is unstable or soluble in the filtration medium (Mintova and Valtchev 1996, Zafar et al. 2012). The so-called "electret" method requires the use of high voltages, which might damage substances and pose an occupational risk. The third synthetic approach normally yields byproducts and materials that are unstable on washing with boiling water. Therefore, other methods have been proposed to prepare zeolite–cellulose composites (Agarwal et al. 2012, Dogan and Hilmioglu 2010a, Ji et al. 2012, Vu et al. 2002). One method consists of pretreatment of natural, low-lignin cellulose fibers with NaOH, KOH, or Na_2SiO_3, and anchoring preformed L- and Y-zeolite crystallites onto the cellulose surface (Vu et al. 2002). Characterization experiments, including SEM, diffuse reflectance infrared spectroscopy, XRD, and Brunauer–Emmett–Teller (BET)-specific surface area measurements, were carried out to monitor how the properties of the zeolite–cellulose composite materials have been changed under different synthesis conditions. It was found that the cellulose-based composites loaded with 20% L- or Y-zeolites were stable and resistant toward leaching of the zeolite phase in aqueous media for several hours (Vu et al. 2002).

Cellulose–zeolite membranes have been also synthesized using zeolite 13X with an average particle size of 310 nm and zeolite 4A with an average particle size of 270 nm and linter pulp, as cellulose sample, dissolved in N-methylmorpholine-N-oxide/H_2O mixture (Dogan and Hilmioglu 2010a). The resulting composite membranes were characterized by FTIR, TGA, SEM, and transmission electron microscopy (TEM). Furthermore, in order to eliminate the interfacial voids, the zeolite-filled regenerated cellulose membranes were coated with CS (Dogan and Hilmioglu 2010b). Electrospinning technique has been also used to generate cellulose–polyethylene terephthalate (PET) blend nanofibers, which were then coated with two zeolites, that is, Linde type-A (LTA) and, respectively, mordenite. The internal morphology of the composites was confirmed using SEM (Agarwal et al. 2012). CA–zeolite composite fibers were also synthesized by wet spinning method (Ji et al. 2012, 2013). SEM micrographs showed that the CA–zeolite fibers possessed highly porous and spongelike structures.

In Figure 6.6, SEM images of some cellulose–zeolite composites prepared using pretreated cellulose and different types of zeolites are selected (Agrawal et al. 2012, Dogan and Hilmioglu 2010a,b, Vu et al. 2002).

As images from Figure 6.6a and b show, the zeolites Y and LTA were successfully loaded on cellulose fibers treated with NaOH and on cellulose–PET electrospinned fibers. On the other hand, image C confirms that the zeolite was distributed evenly throughout the regenerated cellulosic membranes filled with 50 wt.% of zeolite 4A, with no apparent clustering. Furthermore, the image from Figure 6.6c clearly shows that the zeolites were embedded in the membrane matrix with small voids around them, meaning that the adherence between zeolite particles and the polymer was poor. Therefore, regenerated cellulosic membranes filled with zeolite 4A were coated with CS, the image from Figure 6.6d indicating the successful coating of CS on the zeolite-filled cellulose substrate.

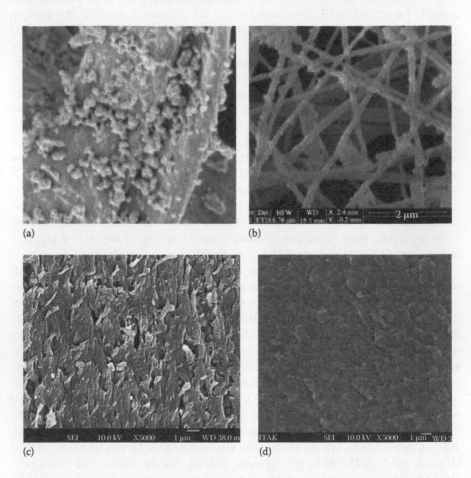

(a)   (b)

(c)   (d)

FIGURE 6.6 SEM images of zeolite–cellulose composites prepared with different types of zeolites. (a) Zeolite Y loaded on NaOH–cellulose fibers. (Reprinted from *Microporous Mesoporous Materials*, 55, Vu, D., Marquez, M., and Larsen, G., A facile method to deposit zeolites Y and L onto cellulose fibers, 93–101, Copyright 2002, with permission from Elsevier.) (b) Zeolite LTA loaded on electrospinned cellulose–PET fibers. (Adapted from Agarwal, S. et al., *J. Inorg. Mater.*, 27, 332, 2012.) (c) Regenerated cellulosic membranes filled with 50 wt.% of zeolite 4A. (Reprinted from *Vacuum*, 84, Dogan, H. and Hilmioglu, N.D., Zeolite-filled regenerated cellulose membranes for pervaporative dehydration of glycerol, 1123–1132, Copyright 2010, with permission from Elsevier.) (d) Regenerated cellulosic membranes filled with zeolite 4A and coated with CS. (Reprinted from *Desalination*, 258, Dogan, H. and Hilmioglu, N.D., Chitosan coated zeolite filled regenerated cellulose membrane for dehydration of ethylene glycol-water mixtures by pervaporation, 120–127, Copyright 2010, with permission from Elsevier.)

6.4.2 Applications of Cellulose–Zeolite Composites

6.4.2.1 Cellulose–Zeolite Composites for Heavy Metal Ion Removal

Heavy metals, such as Cd^{2+}, Co^{2+}, Cu^{2+}, Ni^{2+}, Pb^{2+}, and Hg^{2+}, are common water pollutants. Adsorption is one of the most used techniques for the removal of heavy metals from wastewater, because it offers flexibility in design and operation and, in many cases, the treated effluents are free of color and odor and, therefore, suitable for reuse. In addition, because adsorption is usually reversible, the regeneration of the adsorbent with resultant economy of operation is possible (O'Connell et al. 2008). Adsorbents produced by direct modification of cellulose showed good binding capacities of heavy metal ions from aqueous solution (Nataraj et al. 2011, O'Connell et al. 2008, Yu et al. 2013b). Moreover, by incorporation of zeolites on CA (Ji et al. 2012), it was found that the equilibrium of sorption of Cu^{2+} and Ni^{2+} ions could be attained very quickly in about 20 min. The highly porous and spongelike structure exhibited by CA–zeolite fibers (Figure 6.7a) allowed rapid passage and diffusion of heavy metal ions into the internal pores for contact with the adsorptive sites of the zeolite particles. Also, desorption of metal ions was accomplished with HCl solution, and the results showed that the adsorbents could be reused at least five times without significant loss in adsorption performance. The maximum theoretical adsorption capacities of the CA–zeolite fibers for Cu^{2+} and Ni^{2+} ions at 298 K were found to be 28.57 and 16.95 mg/g, respectively (Ji et al. 2012).

The removal of Cu^{2+} by CA–zeolite fibers was also studied in a packed bed upflow column (Ji et al. 2013). It was found that the Cu^{2+} removal depended strongly on the pH of solution, bed depth, flow rate, and initial concentration. The experimental data were well fitted by the Thomas model and the experimental data showed good agreement with the data predicted by the BDST model.

To examine the practical application of the CA–zeolite fibers, the treatment of industrial effluents was also investigated in the fixed bed column and the reusability of the CA–zeolite fibers column was also assessed. The breakthrough curves for removal of Cu^{2+} from the electroplating wastewater are shown in Figure 6.7b (Ji et al. 2013).

Cu^{2+} uptake for industrial effluent was found to be 20.7 mg/g, whereas for synthetic solution, it was 21.7 mg/g. The presence of organic pollutants in the solution decreased the Cu^{2+} uptake capacity of the CA–zeolite fibers. As shown in Figure 6.7b, the breakthrough time and the exhaustion time decreased with successive runs. In fact, complete desorption was not possible because the electrostatic and complexation reactions occurred between the sorbent and Cu^{2+} ions (Ji et al. 2013). It was found that about 2.1 L of industrial effluent could be treated in the three cycles of adsorption, and the concentrations of Cu^{2+} decreased from 20 to 0.1 mg/L.

The sorption of Cu^{2+} and Cd^{2+} ions has been also followed on ethyl cellulose–CPL composites, as beads (Bondarev et al. 2011). According to the Langmuir model, the maximum theoretical adsorption capacities of the ethyl cellulose–CPL composites for Cu^{2+} and Cd^{2+} ions were found to be 10.1 and 13.62 mg/g, respectively. It may be concluded that, in the case of each cationic species, significant variation in adsorption levels was observed depending on the cellulose modification method as well as on the nature of the zeolite.

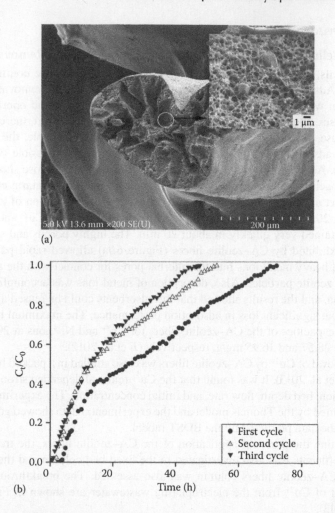

FIGURE 6.7 (a) SEM images of CA–zeolite fiber. (b) Breakthrough curves for Cu²⁺ adsorption onto CA–zeolite fiber from industrial effluent during different regeneration cycles. Sorption conditions: initial metal ion concentration 19 mg/L, bed depth 15 cm, flow rate 5.4 mL/min, and pH 5.5. (Reprinted from *Colloids Surf. A*, 434, Ji, F., Li, C., Xu, J., and Liu, P., Dynamic adsorption of Cu(II) from aqueous solution by zeolite-cellulose acetate blend fiber in fixed-bed, 88–94, Copyright 2013, with permission from Elsevier.)

6.4.2.2 Cellulose–Zeolite Composites for Separation of Solvent Mixtures

Zeolite-filled cellulose membranes have been also investigated for the PV separation of solvent mixtures (Bhat and Aminabhavi 2009, Caro et al. 2000, Dogan and Hilmioglu 2010a,b, Kahya et al. 2011, Ma et al. 2008, Nigiz et al. 2012, Saraswathi and Viswanath 2012, Wang et al. 2010). For instance, the use of regenerated cellulosic membranes filled with zeolite 13X or zeolite 4A for the dehydration of water–ethylene glycol mixtures has been reported (Dogan and Hilmioglu 2010b). It was demonstrated that 20 wt.% zeolite 13X-filled cellulose membrane showed higher selectivity than cellophane- and pristine-regenerated cellulosic membrane.

On the other hand, 30 wt.% zeolite 4A-filled membrane exhibited higher selectivity and lower flux than pristine-regenerated cellulosic membrane. This behavior was attributed to the pore sizes of zeolites 13X, which were higher than those of zeolite 4A, and thus facilitated the permeation of water molecules. However, the regenerated cellulose membranes including different zeolite content gave low separation factor value in the separation of water from ethylene glycol (Dogan and Hilmioglu 2010b). High separation factor values were obtained with the CS-coated zeolite-filled cellulose membranes because coating of zeolite-filled membrane with CS eliminated the nonselective voids (Dogan and Hilmioglu 2010a). Moreover, the CS-coated zeolite-filled cellulose membranes showed commercially acceptable selectivity and higher flux than pristine-regenerated cellulose membranes.

Composite membranes obtained by incorporating zeolite NaY type in NaAlg–(carboxymethyl cellulose [CMC]-*g*-AAm) membranes with significant improvement in the separation by PV of water–isopropanol mixtures have been recently reported (Saraswathi and Viswanath 2012). An increase in the zeolite content in the membrane resulted in simultaneous increases in both the permeation flux and selectivity. The possibility of achieving infinite selectivity for the feed mixture containing 10 wt.% of water was also demonstrated. These composite membranes are an attractive alternative over other types of membranes because it provided better PV separation performance over the NaAlg–(CMC-*g*-AAm) membrane, 30 wt.% zeolite NaY-filled NaAlg membrane, or 10 wt.% Na⁺MMT zeolite-filled NaAlg membrane (Bhat and Aminabhavi 2007b, Kariduraganavar et al. 2004).

6.5 CONCLUSIONS AND PERSPECTIVES

In this chapter, a comprehensive overview of recent developments and progress on biopolymer–zeolite composites was presented. It was demonstrated that biopolymer–zeolite composites exhibit higher performance in the adsorption of environmental pollutants, like heavy metal ions and dyes, and in PV separation of solvent mixtures compared with natural or synthetic zeolites or pristine biopolymers. CS–zeolite composites were found to be very efficient for the removal of cationic ions from aqueous solution, while alginate–zeolite and cellulose–zeolite composites exhibited enhanced PV separation characteristics for water–THF, water–ethanol, and water–isopropanol mixtures. On the other hand, the removal of dyes has been reported only on CS–zeolite composites. So, in the future, the alginate–zeolite and cellulose–zeolite composites should also be evaluated for dye removal. The sorption tests using biopolymer–zeolite composites were mainly carried out on batch mode, but the research should be also directed to design and carry out some pilot-plant scale studies to check their feasibility at the industrial level. More importantly, real wastewater should be tested instead of simulated wastewater.

This chapter also revealed the work done on improving the selectivity and permeability of polymeric membranes in PV separation of aqueous–organic mixtures by fabricating biopolymer–zeolite membranes. However, better prediction of membrane performance under industrially relevant feed conditions is also important as the real conditions are usually fluctuated over time and membrane properties are a strong function of these operating parameters.

These issues indicate that biopolymer–zeolite composites still have a long way to go to fully exploit their potential for the removal of environmental pollutants and for the separation of aqueous–organic mixtures. Nevertheless, the biopolymer–zeolite composites are economically feasible because they are easy to prepare and involve inexpensive chemical reagents, which are coming from renewable resources.

ACKNOWLEDGMENT

This work was supported by a grant from the Ministry of National Education, CNCS-UEFISCDI, project number PN-II-ID-PCE-2011-3-0300.

ABBREVIATIONS

AC	Activated carbon
Alg	Alginate
CA	Cellulose acetate
CMC	Carboxymethyl cellulose
CPL	Clinoptilolite
CS	Chitosan
DMF	Dimethylformamide
ECH	Epichlorohydrin
EDX	Energy-dispersive x-ray analysis
FTIR	Fourier transform infrared spectroscopy
GA	Glutaraldehyde
GG	Guar gum
GPTMS	Glycidoxypropyltrimethoxysilane
HA	Humic acid
HEC	Hydroxyethylcellulose
HMDI	Hexamethylene diisocyanate
PAAm	Poly(acrylamide)
PET	Polyethylene terephthalate
PHEMA	Poly(hydroxyethylmethacrylate)
PV	Pervaporation
SAXS	Small-angle x-ray scattering
SEM	Scanning electron microscopy
TEM	Transmission electron microscopy
TGA	Thermogravimetric analysis
THF	Tetrahydrofuran
TPP	Sodium tripolyphosphate
XRD	X-ray diffraction

REFERENCES

Agarwal, S., S. Sundarrajan, and S. Ramakrishna. 2012. Functionalized cellulose: PET polymer fibers with zeolites for detoxification against nerve agents. *J. Inorg. Mater.* 27: 332–336.
Akkaya, R. and U. Ulusoy. 2008. Adsorptive features of chitosan entrapped in polyacrylamide hydrgel for Pb^{2+}, UO_2^{2+}, and Th^{4+}. *J. Hazard. Mater.* 151: 380–388.

Armagan, B., M. Turan, and M.S. Celik. 2004. Equilibrium studies on the adsorption of reactive azo dyes into zeolite. *Desalination* 170: 33–39.

Aytas, S.O., S. Akyil, and M. Eral. 2004. Adsorption and thermodynamic behavior of uranium on natural zeolite. *J. Radioanal. Nucl. Chem.* 260: 119–125.

Bastani, D., N. Esmaeili, and M. Asadollahi. 2013. Polymeric mixed matrix membranes containing zeolites as filler for gas separation applications: A review. *J. Ind. Eng. Chem.* 19: 375–393.

Bedelean, H., M. Stanca, A. Măicăneanu, and S. Burca. 2006. Zeolitic volcanic tuffs from Măcicaş (Cluj County), natural raw materials used for NH_4^+ removal from wastewaters, *Geologia* 52: 43–49.

Bhat, S.D. and T.M. Aminabhavi. 2007a. Zeolite K-LTL-loaded sodium alginate mixed matrix membranes for pervaporation dehydration of aqueous–organic mixtures. *J. Membr. Sci.* 306: 173–185.

Bhat, S.D. and T.M. Aminabhavi. 2007b. Pervaporation separation using sodium alginate and its modified membranes—A review. *Sep. Purif. Rev.* 36: 203–229.

Bhat, S.D. and T.M. Aminabhavi. 2009. Pervaporation-aided dehydration and esterification of acetic acid with ethanol using 4A zeolite-filled cross-linked sodium alginate-mixed matrix membranes. *J. Appl. Polym. Sci.* 113: 157–168.

Bondarev, A., S. Mihai, O. Pantea, and S. Neagoe. 2011. Use of biopolymers for the removal of metal ion contaminants from water. *Macromol. Symp.* 303: 78–84.

Caro, J., M. Noack, P. Kölsch, and R. Schäfer. 2000. Zeolite membranes—State of their development and perspective. *Microporous Mesoporous Mater.* 38: 3–24.

Chao, H.P. and S.H. Chen. 2012. Adsorption characteristics of both cationic and oxyanionic metal ions on hexadecyltrimethylammonium bromide-modified NaY zeolite. *Chem. Eng. J.* 193–194: 283–289.

Chen, D., W. Li, Y. Wub, Q. Zhu, Z. Lu, and G. Du. 2013. Preparation and characterization of chitosan-montmorillonite magnetic microspheres and its application for the removal of Cr (VI). *Chem. Eng. J.* 221: 8–15.

Chen, Z., M. Deng, Y. Chen, G. He, M. Wu, and J. Wang. 2004. Preparation and performance of cellulose acetate-polyethyleneimine blend microfiltration membranes and their applications. *J. Membr. Sci.* 235: 73–86.

Chmielewská, E., I. Sabová, H. Peterlik, and A. Wu. 2011. Batch-wise adsorption, SAXS and microscopic studies of zeolite pelletized with biopolymeric alginate. *Braz. J. Chem. Eng.* 28: 63–71.

Choi, J.W., K.S. Yang, D.J. Kim, and C.E. Lee. 2009. Adsorption of zinc and toluene by alginate complex impregnated with zeolite and activated carbon. *Curr. Appl. Phys.* 9: 694–697.

Crini, G. 2005. Recent developments in polysaccharide-based materials used as adsorbents in wastewater treatment. *Prog. Polym. Sci.* 30: 38–70.

Dinu, M.V. and E.S. Dragan. 2008. Heavy metals adsorption on some iminodiacetate chelating resins as a function of the adsorption parameters. *React. Funct. Polym.* 68: 1346–1354.

Dinu, M.V. and E.S. Dragan. 2010a. Adsorption of heavy metals on ionic composites based on chitosan. *Bull. Inst. Pol. Iasi.* Tom LVI(LX): 171–178.

Dinu, M.V. and E.S. Dragan. 2010b. Evaluation of Cu^{2+}, Co^{2+} and Ni^{2+} ions removal from aqueous solution using a novel chitosan-clinoptilolite composite: Kinetics and isotherms. *Chem. Eng. J.* 160: 157–163.

Dinu, M.V., E.S. Dragan, and A.W. Trochimczuk. 2009. Sorption of Pb(II), Cd(II) and Zn(II) by iminodiacetate chelating resins in non-competitive and competitive conditions. *Desalination* 249: 374–379.

Dogan, H. 2012. Preparation and characterization of calcium alginate-based composite adsorbents for the removal of Cd, Hg, and Pb ions from aqueous solution. *Toxicol. Environ. Chem.* 94: 482–499.

Dogan, H. and N.D. Hilmioglu. 2010a. Zeolite-filled regenerated cellulose membranes for pervaporative dehydration of glycerol. *Vacuum* 84: 1123–1132.

Dogan, H. and N.D. Hilmioglu. 2010b. Chitosan coated zeolite filled regenerated cellulose membrane for dehydration of ethylene glycol-water mixtures by pervaporation. *Desalination* 258: 120–127.

Dragan, E.S. and M.V. Dinu. 2009. Removal of copper ions from aqueous solution by adsorption on ionic hybrids based on chitosan and clinoptilolite. *Ion Exchange Lett.* 2: 15–18.

Dragan, E.S., M.V. Dinu, and M. Mihai. 2011. Chapter 6: Separations by multicomponent ionic systems based on natural and synthetic polycations. In Sengupta, A.K. (ed.), *Ion Exchange and Solvent Extraction: A Series of Advances*, pp. 233–291. CRC Press, Boca Raton, FL.

Dragan, E.S., M.V. Dinu, and D. Timpu. 2010. Preparation and characterization of novel composites based on chitosan and clinoptilolite with enhanced adsorption properties for Cu^{2+}. *Bioresour. Technol.* 101: 812–817.

Fu, F. and Q. Wang. 2011. Removal of heavy metal ions from wastewaters: A review. *J. Environ. Manag.* 92: 407–418.

Furlan, F.R., L.G. De Melo da Silva, A.F. Morgado, A.A.U. de Souza, and S.M.A.G. Ulson de Souza. 2010. Removal of reactive dyes from aqueous solutions using combined coagulation-flocculation and adsorption on activated carbon. *Resour. Conserv. Recycl.* 54: 283–290.

Günay, A., E. Arslankaya, and I. Tosun. 2007. Lead removal from aqueous solution by natural and pretreated clinoptilolite: Adsorption equilibrium and kinetics. *J. Hazard. Mater.* 146: 362–371.

Han, R., W. Zou, Y. Wang, and L. Zhu. 2007. Removal of U(VI) from aqueous solutions by manganese oxide coated zeolite: Discussion of adsorption isotherms and pH effect. *J. Environ. Radioact.* 93: 127–143.

Hasan, S., T.K. Ghosh, M.A. Prelas, D.S. Viswanath, and V.M. Boddu. 2007. Adsorption of uranium on a novel bioadsorbent-chitosan-coated perlite. *Nucl. Technol.* 159: 59–71.

Hernández-Montoya, V., M.A. Pérez-Cruz, D.I. Mendoza-Castillo, M.R. Moreno-Virgen, and A. Bonilla-Petriciolet. 2013. Competitive adsorption of dyes and heavy metals on zeolitic structures. *J. Environ. Manag.* 116: 213–221.

Hritcu, D., D. Humelnicu, G. Dodi, and M.I. Popa. 2012. Magnetic chitosan composite particles: Evaluation of thorium and uranyl ion adsorption from aqueous solutions. *Carbohydr. Polym.* 87: 1185–1191.

Humelnicu, D., M.V. Dinu, and E.S. Dragan. 2011. Adsorption characteristics of UO_2^{2+} and Th^{4+} ions from simulated radioactive solutions onto chitosan-clinoptilolite sorbents. *J. Hazard. Mater.* 185: 447–455.

Ibanez, J.P. and Y. Umetsu. 2002. Potential of protonated alginate beads for heavy metals uptake. *Hydrometallurgy* 64: 89–99.

Idris, A., N. Suriani, M. Ismail, N. Hassan, E. Misran, and A.F. Ngomsik. 2012. Synthesis of magnetic alginate beads based on maghemite nanoparticles for Pb(II) removal in aqueous solution. *J. Ind. Eng. Chem.* 18: 1582–1589.

Ji, F., C. Li, B. Tang, J. Xu, G. Lu, and P. Liu. 2012. Preparation of cellulose acetate-zeolite composite fiber and its adsorption behavior for heavy metal ions in aqueous solution. *Chem. Eng. J.* 209: 325–333.

Ji, F., C. Li, J. Xu, and P. Liu. 2013. Dynamic adsorption of Cu(II) from aqueous solution by zeolite-cellulose acetate blend fiber in fixed-bed. *Colloids Surf. A: Physicochem. Eng. Asp.* 434: 88–94.

Kahya, S., O. Şanlıb, and E. Çamurlu. 2011. Crosslinked sodium alginate and sodium alginate-clinoptilolite (natural zeolite) composite membranes for pervaporation separation of dimethylformamide-water mixtures: A comparative study. *Des. Water Treat.* 25: 297–309.

Kariduraganavar, M.Y., A.A. Kittur, S.S. Kittur, S.S. Kulkarni, and K. Ramesh. 2004. Development of novel pervaporation membranes for the separation of water–isopropanol mixtures using sodium alginate and NaY zeolite. *J. Membr. Sci.* 238: 165–175.

Kilincarslan, A. and S. Akyil. 2005. Uranium adsorption characteristic and thermodynamic behavior of clinoptilolite zeolite. *J. Radioanal. Nucl. Chem.* 264: 541–548.

Kittur, A.A., S.S. Kulkarni, M.I. Aralaguppi, and M.Y. Kariduraganavar. 2005. Preparation and characterization of novel pervaporation membranes for the separation of water–isopropanol mixtures using chitosan and NaY zeolite. *J. Membr. Sci.* 247: 75–86.

Kittur, A.A., S.M. Tambe, S.S. Kulkarni, and M.Y. Kariduraganavar. 2004. Pervaporation separation of water–acetic acid mixtures through NaY zeolite-incorporated sodium alginate membranes. *J. Appl. Polym. Sci.* 94: 2101–2109.

Li, S.G., V.A. Tuan, R.D. Noble, and J.L. Falconer. 2001. Pervaporation of water-THF mixtures using zeolite membranes. *Ind. Eng. Chem. Res.* 40: 4577–4585.

Libby, B., W.H. Smyrl, and E.L. Cussler. 2003. Polymer–zeolite composite membranes for direct methanol fuel cells. *AIChE J.* 49: 991–1001.

Lin, J. and Y. Zhan. 2012. Adsorption of humic acid from aqueous solution onto unmodified and surfactant-modified chitosan/zeolite composites. *Chem. Eng. J.* 200–202: 202–213.

Liu, C. and R. Bai. 2006. Adsorptive removal of copper ions with highly porous chitosan-cellulose acetate blend hollow fiber membranes. *J. Membr. Sci.* 284: 313–322.

Liu, L., Y. Wan, Y. Xie, R. Zhai, B. Zhang, and J. Liu. 2012. The removal of dye from aqueous solution using alginate-halloysite nanotube beads. *Chem. Eng. J.* 187: 210–216.

Ma, X., C. Hu, R. Guo, X. Fang, H. Wu, and Z. Jiang. 2008. HZSM5-filled cellulose acetate membranes for pervaporation separation of methanol-MTBE mixtures. *Sep. Purif. Technol.* 59: 34–42.

Mahmoodi, N.M. 2013. Magnetic ferrite nanoparticle–alginate composite: Synthesis, characterization and binary system dye removal. *J. Taiwan Inst. Chem. Eng.* 44: 322–330.

Metin, A.U., H. Çiftçi, and E. Alver. 2013. Efficient removal of acidic dye using low-cost biocomposite beads. *Ind. Eng. Chem. Res.* 52: 10569–10581.

Mintova, S. and V. Valtchev. 1996. Deposition of zeolite A on vegetal fibers. *Zeolites* 16: 31–34.

Mintova, S., V. Valtchev, B. Schoeman, and J. Sterte. 1996. Preparation of zeolite Y-vegetal fiber composite materials. *J. Porous Mater.* 3: 143–150.

Nataraj, S.K., S. Roy, M.B. Patil, M.N. Nadagouda, W.E. Rudzinski, and T.M. Aminabhavi. 2011. Cellulose acetate-coated α-alumina ceramic composite tubular membranes for wastewater treatment. *Desalination* 281: 348–353.

Nawawi, M., M. Ghazali, and R.Y.M. Huang. 1997. Pervaporation dehydration of isopropanol with chitosan membranes. *J. Membr. Sci.* 124: 53–62.

Nešić, A.R., S.J. Veličković, and D.G. Antonović. 2012. Characterization of chitosan- montmorillonite membranes as adsorbents for Bezactiv Orange V-3R dye. *J. Hazard. Mater.* 209–210: 256–263.

Nešić, A.R., S.J. Veličković, and D.G. Antonović. 2013. Modification of chitosan by zeolite A and adsorption of Bezactive Orange 16 from aqueous solution. *Composites Part B: Eng.* 53: 145–151.

Nibou, D., S. Khemaissia, S. Amokrane, M. Barkat, S. Chegrouche, and A. Mellah. 2011. Removal of UO_2^{2+} onto synthetic NaA zeolite. Characterization, equilibrium and kinetic studies. *Chem. Eng. J.* 172: 296–305.

Nigiz, F.U., H. Dogan, and N.D. Hilmioglu. 2012. Pervaporation of ethanol-water mixtures using clinoptilolite and 4A filled sodium alginate membranes. *Desalination* 300: 24–31.

O'Connell, D.W., C. Birkinshaw, and T.F. O'Dwyer. 2008. Heavy metal adsorbents prepared from the modification of cellulose: A review. *Bioresour. Technol.* 99: 6709–6724.

Pandey, A.K., S.D. Pandey, V. Misra, and S. Devi. 2003. Role of humic acid entrapped calcium alginate beads in removal of heavy metals. *J. Hazard. Mater.* 98: 177–181.

Park, H.G., T.W. Kim, M.Y. Chae, and I.K. Yoo. 2007. Activated carbon-containing alginate adsorbent for the simultaneous removal of heavy metals and toxic organics. *Process Biochem* 42: 1371–1377.

Patil, M.B. and T.M. Aminabhavi. 2008. Pervaporation separation of toluene-alcohol mixtures using silicalite zeolite embedded chitosan mixed matrix membranes. *Sep. Purif. Technol.* 62: 128–136.

Qiu, M., C. Qian, J. Xu, J. Wu, and G. Wang. 2009. Studies on the adsorption of dyes into clinoptilolite. *Desalination* 243: 286–292.

Qu, R., C. Sun, F. Ma, Y. Zhang, C. Ji, Q. Xu, C. Wang, and H. Chen. 2009. Removal and recovery of Hg(II) from aqueous solution using chitosan-coated cotton fibers. *J. Hazard. Mater.* 167: 717–727.

Sanghi, R. and P. Verma. 2013. Decolorisation of aqueous dye solutions by low-cost adsorbents: A review. *Color Technol.* 129: 85–108.

Saraswathi, M. and B. Viswanath 2012. Separation of water–isopropyl alcohol mixtures with novel hybrid composite membranes. *J. Appl. Polym. Sci.* 126: 1867–1875.

Spiridon, O.B., E. Preda, A. Botez, and L. Pitulice. 2013. Phenol removal from wastewater by adsorption on zeolitic composite. *Environ. Sci. Pollut. Res.* 20: 6367–6381.

Sprynskyy, M., B. Buszewski, A.P. Terzyk, and J. Namiesnik. 2006. Study of the selection mechanism of heavy metal (Pb^{2+}, Cu^{2+}, Ni^{2+}, and Cd^{2+}) adsorption on clinoptilolite. *J. Colloid Interface Sci.* 304: 21–28.

Sun, H., L. Lu, X. Chen, and Z. Jiang. 2008. Surface modified zeolite chitosan membranes for pervaporation dehydration of ethanol. *Appl. Surf. Sci.* 254: 5367–5374.

Sun, X., B. Peng, Y. Ji, J. Chen, and D. Li. 2009. Chitosan(chitin)-cellulose composite biosorbents prepared using ionic liquid for heavy metal ions adsorption, *AIChE J.* 55: 2062–2069.

Urtiaga, A., E.D. Gorri, C. Casado, and I. Ortiz. 2003. Pervaporative dehydration of industrial solvents using a zeolite NaA commercial membrane. *Sep. Purif. Technol.* 32: 207–213.

Vu, D., M. Marquez, and G. Larsen. 2002. A facile method to deposit zeolites Y and L onto cellulose fibers. *Microporous Mesoporous Mater.* 55: 93–101.

Wang, S. and Y. Peng. 2010. Natural zeolites as effective adsorbents in water and wastewater treatment. *Chem. Eng. J.* 156: 11–24.

Wang, X., Y. Zheng, and A. Wang. 2009. Fast removal of copper ions from aqueous solution by chitosan-g-poly(acrylic acid)-attapulgite composites. *J. Hazard. Mater.* 168: 970–977.

Wang, X.P. 2000. Modified alginate composite membranes for the dehydration of acetic acid. *J. Membr. Sci.* 170: 71–79.

Wang, Y., Z. Jiang, H. Li, and D. Yang. 2010. Chitosan membranes filled by GPTMS-modified zeolite beta particles with low methanol permeability for DMFC. *Chem. Eng. Proc.: Proc. Intensif.* 49: 278–285.

Wan Ngah, W.S., N.F.M. Ariff, and M.A.K.M. Hanafiah. 2010. Preparation, characterization, and environmental application of crosslinked chitosan-coated bentonite for tartrazine adsorption from aqueous solutions. *Water Air Soil Pollut.* 206: 225–236.

Wan Ngah, W.S., L.C. Teong, and M.A.K.M. Hanafiah. 2011. Adsorption of dyes and heavy metal ions by chitosan composites: A review. *Carbohydr. Polym.* 83: 1446–1456.

Wan Ngah, W.S., L.C. Teong, R.H. Toh, and M.A.K.M. Hanafiah. 2012b. Utilization of chitosan–zeolite composite in the removal of Cu(II) from aqueous solution: Adsorption, desorption and fixed bed column studies. *Chem. Eng. J.* 209: 46–53.

Wan Ngah, W.S., L.C. Teong, R.H. Toh, and M.A.K.M. Hanafiah. 2013. Comparative study on adsorption and desorption of Cu(II) ions by three types of chitosan–zeolite composites. *Chem. Eng. J.* 223: 231–238.

Wan Ngah, W.S., L.C. Teong, C.S. Wong, and M.A.K.M. Hanafiah. 2012a. Preparation and characterization of chitosan–zeolite composites. *J. Appl. Polym. Sci.* 125: 2417–2425.

Xie, J., C. Li, L. Chi, and D. Wu. 2013. Chitosan modified zeolite as a versatile adsorbent for the removal of different pollutants from water. *Fuel* 103: 480–485.

Yang, G., L. Zhang, T. Peng, and W. Zhong. 2000. Effects of Ca^{2+} bridge cross-linking on structure and pervaporation of cellulose/alginate blend membranes. *J. Membr. Sci.* 175: 53–60.

Yang, L., X. Ma, and N. Guo. 2012. Sodium alginate-Na^+-rectorite composite microspheres: Preparation, characterization, and dye adsorption. *Carbohydr. Polym.* 90: 853–858.

Yu, L., J. Gong, C. Zeng, and L. Zhang. 2013a. Preparation of zeolite-A-chitosan hybrid composites and their bioactivities and antimicrobial activities. *Mater. Sci. Eng. C* 33: 3652–3660.

Yu, X., S. Tong, M. Ge, L. Wu, J. Zuo, C. Cao, and W. Song. 2013b. Adsorption of heavy metal ions from aqueous solution by carboxylated cellulose nanocrystals. *J. Environ. Sci.* 25: 933–943.

Yuan, W., H. Wu, B. Zheng, X. Zheng, Z. Jiang, X. Hao, and B. Wang, 2007. Sorbitol plasticized chitosan-zeolite hybrid membrane for direct methanol fuel cell. *J. Power Sources* 172: 604–612.

Zafar, M., M. Ali, S.M. Khan, T. Jamil, and M.T.Z. Butt 2012. Effect of additives on the properties and performance of cellulose acetate derivative membranes in the separation of isopropanol-water mixtures. *Desalination* 285: 359–365.

Zhan, Y., J. Lin, and J. Li 2013. Preparation and characterization of surfactant-modified hydroxyapatite-zeolite composite and its adsorption behavior toward humic acid and copper(II). *Environ. Sci. Pollut. Res. Int.* 20: 2512–2526.

Zhou, D., L. Zhang, J. Zhou, and S. Guo. 2004. Cellulose-chitin beads for adsorption of heavy metals in aqueous solution. *Water Res.* 38: 2643–2650.

Zou, W., H. Bai, L. Zhao, K. Li, and R. Han. 2011. Characterization and properties of zeolite as adsorbent for removal of uranium(VI) from solution in fixed bed column. *J. Radioanal. Nucl. Chem.* 288: 779–788.

7 Metal-Impregnated Ion Exchanger for Selective Removal and Recovery of Trace Phosphate

Sukalyan Sengupta and Arup K. SenGupta

CONTENTS

7.1 INTRODUCTION

Eutrophication or the overenrichment of global water bodies has grown from a rising concern to a major water-quality-impairment parameter over the past few decades. This phenomenon is mainly attributable to the influx of nitrogen (N) and phosphorus (P) into the water bodies from a variety of sources ranging from detergents to fertilizers to atmospheric deposition. Eutrophication is a key driver in a number of environmental problems including reduced light penetration resulting in sea-grass mortality, increases in harmful algal blooms, and hypoxic and anoxic conditions. It is generally agreed upon that P is the rate-limiting nutrient in freshwater ecosystems, that is, the level of P in freshwater bodies governs the growth of aquatic organisms. Many streams and lakes in the United States have very little capacity to assimilate P loading during the "critical" warm and dry summer period without significant water quality degradation. Large diurnal swings in pH and dissolved oxygen may occur as excessive amounts of nutrients are metabolized by aquatic plants and algae. The range of these swings is often measured to exceed the state water quality criteria established to protect fish and other aquatic organisms in their various life stages. Therefore, the amount of phosphorus currently entering these waters exceeds the seasonal loading capacity and must be reduced if these water quality problems are to be resolved (Goldman et al. 1973, Rhyther and Dunstan 1971, Valiela 1995).

In the United States, an EPA (1997) study found that 36% of the lakes are eutrophic. In more focused studies, the problem has been found to be even more intensive. For example, over 1000 water bodies in Idaho, Oregon, and Washington are identified as being impaired due to excessive phosphorus (P) loading and are included on state Clean Water Act lists for water quality problems (EPA 2007). Control of point sources of P has been a major focus of environmental regulatory agencies and has resulted in stricter regulations for wastewater treatment.

The primary source of P in the environment is anthropogenic, present as both point sources (synthetic detergent in domestic wastewater) and nonpoint sources (agricultural runoff). The approximate contributions of P from major sources to domestic wastewater is estimated as 0.6 kg P/cap/year from human wastes, 0.3 kg P/cap/year from laundry detergents with no restrictions on P content, and 0.1 kg P/cap/year from household detergents and other cleaners (Sedlak 1991). Without significant commercial or industrial loads, the influent concentration of total phosphorus in wastewater treatment plant (WWTP) effluent may range from 6 to 8 mg/L as P. There has been a concerted effort to reduce the amount of P in laundry detergents, and now, there is growing momentum to eliminate P from dishwashing detergents. But even if these efforts are totally successful, the P concentration in WWTP effluent would reduce only to 4–5 mg/L P (USGS 1999). Unfortunately, this is much higher than the natural capacity of rivers and lakes to assimilate P. Studies (Heathwaite and Sharpley 1999, Seviour et al. 2003, Sharpley et al. 1994, 2003) indicate that lake water concentrations of P above 0.02 ppm generally accelerate eutrophication. Therefore, federal and state regulators have been applying strict P effluent limits. For instance, the Everglades Forever Act mandates a total P concentration of 10 ppb (Florida Everglades Forever Act 1994); additionally, a recent plan establishes a systemic P reduction in effluent WWTP streams emptying into the Spokane River, with an ultimate mandate of 10 ppb (Hansen 2006). While the 10 ppb limit is the strictest and rare, many WWTPs have been forced to retrofit or modify existing processes to meet limits such as 0.07 mg/L (Durham Advanced Wastewater Treatment Facility, Tigard, OR, and Rock Creek AWTF in Hillsboro, OR) or 0.1 mg/L (effluent total P limit for all major POTWs discharging to Assabet River in MA). And there are many other WWTPs that are currently regulated to meet discharge standard of 500 ppb but have been notified that the limits will be reduced to 100 ppb. Thus, WWTPs have to prepare themselves for stricter P discharge limits in the range of 10–100 ppb.

Another important aspect of the phosphorus issue is the potential for its decrease in worldwide supplies. Phosphorus is a nonrenewable source and is primarily available from phosphate rock, high-quality deposits of which are controlled by just five countries. Some predictions suggest (Herring and Fantel 1993) that if the current trend of phosphate rock mining is continued, the stock is predicted to dwindle in next 50 years. Thus, it is imperative that phosphorus recycle and reuse become an integral part of a WWTP process.

Techniques for phosphorus removal from wastewater are in general biological and physical–chemical. It is readily recognized that traditional biological nutrient removal (BNR) and precipitation–sorption processes are unable to reduce phosphate concentrations below 100 μg/L as P (Cooper et al. 1993, Jenkins et al. 1971, Jenkins and Hermanowicz 1991, Kuba et al. 1993). Moreover, these methods suffer from the drawbacks of excessive sludge production, high operating costs, and inability to attain complete nutrient removal due to thermodynamic and kinetic limitations.

7.2 ANION EXCHANGER DEVELOPMENT

With the constraints of the biological and chemical phosphorus removal processes and the complexity involved in the recovery of phosphorus from the sludge produced in the processes, physical–chemical adsorption, or sorption as it is referred to in literature, can be a viable solution to the challenge faced by the global wastewater industry. Many natural and engineered sorbent media have been investigated in recent years, granular-activated alumina, zirconium oxide, and iron oxide being a few well-studied inorganic sorbents in this regard (Chubar et al. 2005, Genz et al. 2004, Seida and Nakano 2002, Tanada et al. 2003, Zeng et al. 2004). However, these metal oxide particles lack the mechanical strength and abrasion resistance properties for prolonged operation in fixed-bed units consequently forming fines leading to increased head loss in the system (Blaney et al. 2007). The metal oxides referred to earlier are also not amenable to efficient regeneration and are hence recommended for single-use applications (Genz et al. 2004).

While polymeric materials offer the requisite durability and mechanical strength, they lack the specific phosphate selectivity in the presence of competing sulfate ions ubiquitously present in both point and nonpoint sources of phosphorus. Earlier endeavors in this area include the REM NUT® process (Liberti et al. 1981) and the development of a copper-loaded polymeric ligand exchanger (PLE) (SenGupta and Zhao 2000, Zhao and SenGupta 1996, 1998). In spite of their advantages and promising pilot-scale applications, they never reached full-scale application due to technoeconomic inefficiencies (Petruzzelli et al. 2004).

It has been well demonstrated through research in recent past that the oxides of polyvalent metals, namely, Fe(III), Ti(IV), and Zr(IV), exhibit strong ligand sorption properties (Lewis acids) through the formation of inner-sphere complexes (Dzombak and Morel 1990, Stumm and Morgan 1995). Of them, hydrated iron(III) oxide [FeOOH] or hydrated ferric oxide (HFO) is benign, inexpensive, and stable over a wide pH range. Many previous studies confirmed that HFO or α-goethite has high sorption affinity for phosphates (Golterman 1995, Hiemstra and van Riemsdjik 1999, Seida and Nakano 2002), and their selective sorption is attributed to the ligand exchange in the coordination sphere of structural iron atoms.

In view of the economic constraints faced by the earlier PLEs and the preferential sorption affinity for phosphate exhibited by the HFO particles, attempts have been made to disperse or dope HFO nanoparticles within various support media, such as alginate, zeolite, cation exchange resin, and activated carbon (Cumbal et al. 2003, DeMarco et al. 2003, Huang and Vane 1989, Katsoyiannis and Zouboulis 2002, Onyango et al. 2003). A new hybrid anion exchange resin was developed by impregnating HFO nanoparticles within polymeric anion exchanger and it displayed significantly increased sorption capacity (Blaney et al. 2007, Cumbal and SenGupta 2005) and is commercially available as hybrid anion exchanger (HAIX) or LayneRT™.

Two combinations of transition metals and/or metal oxides were loaded onto the DOWEX™ M4195 (referred to as DOW) resin to compare their performance with the ArsenX. Table 7.1 shows the properties of each PLE, and Figure 7.1 shows the polymeric backbone onto which HFO nanoparticles are impregnated. Details of PLE synthesis are available in Sengupta and Pandit (2011) and are not repeated here.

The configurations chosen and the justification of their choice are as follows:

1. *Only HFO*—Since DOW is a free-form resin, it can be presumed that there would be no phosphate removal by ion exchange. The removal would be solely due to the ligand sorption onto HFO. This would give an estimate of relative distribution of the phosphate removal by ligand sorption and ion exchange in HAIX.

   For the impregnation of HFO nanoparticles onto the resin, a weighed mass of the dried virgin resin (10.0 g) was added to 1.0 L of 1% solution of $FeCl_3 \cdot 6H_2O$ (2.0 g/L as Fe) at pH $\approx$ 2.0 (to ensure that no $Fe(OH)_3$ precipitates out) and was vigorously stirred for 1 h by a magnetic stirrer at 750 rpm. The pH of the solution was then stepwise increased to around 8 uniformly

TABLE 7.1
Properties of Iron Oxide: Impregnated PLEs

Characteristics	HAIX	DOW-HFO	DOW-HFO-Cu
Structure	Macroporous polystyrene–divinylbenzene	Macroporous polystyrene–divinylbenzene	Macroporous polystyrene–divinylbenzene
Appearance	Brown spherical beads	Tan to dark-brown opaque beads	Tan to dark-brown opaque beads
Functional group	Quaternary ammonium	Bis-picolylamine	Bis-picolylamine
Iron content	75–90 mg as Fe/g resin	45–60 mg as Fe/g resin	40–50 mg as Fe/g resin
Bulk density	790–840 g/L	673 g/L	673 g/L
Particle size	300–1200 μm	297–841 μm	297–841 μm

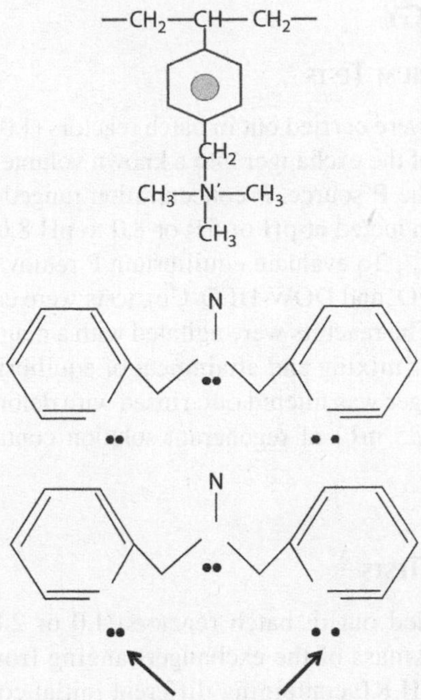

FIGURE 7.1 Polymeric backbone onto which HFO nanoparticles are impregnated. Top: HAIX; Middle: DOW-HFO; Bottom: DOW-HFO-Cu.

paced over 3 h, by dropwise addition of 1.0 N sodium hydroxide [NaOH] solution to form the FeOOH precipitate as per the following equation:

$$Fe^{3+} + 3OH^- \rightarrow FeOOH + H_2O \qquad (7.1)$$

The mixing speed was reduced to around 200 rpm and was kept for 24 h to optimize the contact of the resin beads with freshly formed HFO nanoparticles. After 24 h, the resultant solution was repeatedly rinsed with deionized water to remove excess precipitate. The resin was then oven dried at 45°C ± 2°C for 24 h and stored in moisture-free glass vials. The average HFO impregnation was found to vary between 45 and 60 mg as Fe/g of resin. This resin is referred to as DOW resin loaded with HFO (DOW-HFO) this point forward for typographical convenience.

2. *Both HFO nanoparticles and Cu²⁺*—As both HFO and Cu²⁺ have shown high selectivity for PO_4^{3-}, this would allow demonstrating whether they can indeed be loaded in conjunction and if possible what would be the removal capacity. To load the DOW resin with both HFO and Cu²⁺, DOW-HFO was taken as the starting material. The DOW-HFO was loaded with Cu²⁺ as per the process descriptions given in Sengupta and Pandit (2011). Anticipated loading capacity (40–50 mg Fe/g of resin) was observed. This loaded resin would be referred to as DOW resin loaded with HFO and copper ("DOW-HFO-Cu") further in this text for ease of notation.

7.3 METHODOLOGY

7.3.1 Batch Equilibrium Tests

Batch equilibrium tests were carried out in batch reactors (1.0 or 2.0 L glass beakers) by adding known mass of the exchanger into a known volume of a synthetic solution. Na_2HPO_4 was used as the P source. P concentration ranged from 0.5 to 1.0 (mg/L as P). The tests were conducted at pH of 5.0 or 8.0 to pH 8.0 to assess the effect of pH on the uptake of PO_4^{3-}. To evaluate equilibrium P removal capacity of the three resins (HAIX, DOW-HFO, and DOW-HFO-Cu), tests were carried out at room temperature ($20°C \pm 2°C$). The reactors were agitated with a magnetic stirrer at 400 rpm for 48 h to ensure proper mixing and attainment of equilibrium. After equilibrium was attained, the exchanger was filtered out, rinsed with deionized water, and stirred with a known volume (25 mL) of regenerant solution containing 2.5% NaCl and 2.0% NaOH.

7.3.2 Batch Kinetic Tests

Kinetic tests were carried out in batch reactors (1.0 or 2.0 L glass beakers) by adding a predetermined mass of the exchanger ranging from 0.5 to 1.0 g into 1.0 or 2.0 L of solution at pH 8.0 containing different initial concentrations of orthophosphate phosphorous ranging from 10.0 to 45.0 mg as P/L. To ensure the attainment of uniform solute–sorbent contact and to eliminate the diffusion resistance in the liquid film (i.e., the Nernst film), the solution was agitated on a magnetic stirrer at 400 rpm. Thus, intraparticle diffusion was the rate-limiting step under the experimental conditions, and the intraparticle diffusivity ($\bar{D}$) was determined. The diameter of each exchanger particle was obtained from the parent resin supplier's brochure. The average particle size for HAIX was 750 µm (0.75 mm); for DOW-HFO and DOW-HFO-Cu, it was 570 µm (0.57 mm). Samples (6 mL each) were collected at every 30 min interval for 6.0 h and then again after 20.0, 24.0, and 48.0 h. Uptake of orthophosphate species or other solutes by the sorbents at different times was determined from mass balance calculations on the influent and effluent solutions.

7.3.3 Fixed-Bed Column Runs

Fixed-bed column runs were carried out in Adjusta-Chrom® #11 (Ace Glass Inc., Vineland, NJ) glass columns 300 mm long with 10 mm inner diameter. The influent was pumped into the column in a downflow direction using synchronous pumps, FMI Lab Pump, Model QSY (Fluid Metering Inc., Syosset, NY). Effluent sample from the column was collected by a Spectra/Chrom® CF-1 fraction collector (spectrum chromatography). The flow was adjusted to maintain constant empty bed contact time (EBCT) for all the column runs at 3.0 min. An online pH meter was also connected in the system to monitor the effluent pH. Regeneration of the bed was performed similarly by passing the regenerant in a downflow direction and the spent regenerant samples were also collected in a similar fashion. The regenerated bed was rinsed with deionized water till the effluent pH was <8.5. The regenerated bed was then used for another exhaustion run.

7.3.4 REGENERATION, RECYCLING OF SPENT REGENERANT, AND RECOVERY OF PHOSPHATE

Based on earlier studies (Pandit 2010), a 2.5% NaCl + 2.0% NaOH solution was chosen as the regenerant for all the regeneration runs. The sorbent media was regenerated with ≈20 bed volumes (BVs) of the regenerant solution. Following analysis of orthophosphate phosphorous in the spent regenerant samples, $Ca(NO_3)_2 \cdot 4H_2O$ or a mixture of $MgSO_4 \cdot 7H_2O$ and NH_4Cl was added to the spent regenerant to precipitate $Ca_3(PO_4)_2$ or $Mg(NH_4)PO_4$ (struvite), the latter being a slow-release fertilizer. The precipitate was allowed to settle for a couple of hours and the supernatant was filtered through 0.45 μm filter and analyzed for different solutes. The precipitate was mildly rinsed with deionized water and dried in an oven at 85°C, and an elemental analysis using energy-dispersive x-ray (EDX) was performed on it using an Oxford Instruments, Inc. (High Wycombe, United Kingdom), INCA EDX system in conjunction with a scanning electron microscope (SEM) (JSM 5610 from JEOL, Inc., Peabody, MA) at an accelerating voltage of 15 kV. Following the precipitation of orthophosphate phosphorous from the spent regenerant, 1.5% NaOH was added to it to compensate for the loss in OH⁻ ions in the regenerant due to regeneration of the media and possible precipitation and this mixture was again used as a regenerant for the next cycle.

7.3.5 ANALYTICAL PROCEDURE

All the chemicals were analyzed as per the Standard Methods of Water and Wastewater Treatment (APHA, AWWA, and WEF 1998). All anions such as PO_4^{3-}, Cl^-, NO_3^-, and SO_4^{2-} were analyzed using a DIONEX ion chromatograph (Model—ICS 900) coupled with an AS40 autosampler. The iron content in the resins was determined using a HACH spectrophotometer (Model—DV 4000 UV/Vis) by the FerroVer® method. Before analysis, the samples were digested as per the Standard Methods of Water and Wastewater Treatment (APHA, AWWA, and WEF 1998).

7.4 RESULTS

7.4.1 FIXED-BED COLUMN RUN: EXHAUSTION

Figure 7.2 shows the P breakthrough curve for the three anion exchangers studied in this study when the influent synthetic solution was identical. The SO_4^{2-} and P concentration being 130.0 and 4.25 mg/L, respectively, this is a case where P is a trace species. P breakthrough is the earliest for DOW-HFO. Since this is a free-form resin, it has no exchange capacity. Therefore, the only mechanism of P removal is the Lewis acid–base (LAB) inner-sphere complex formed between HPO_4^{2-} (since the pH is 8) and $FeOH_2^+$. The P breakthrough curve for HAIX closely follows that of DOW-HFO. In the case of HAIX, P removal can occur by ion exchange (of the

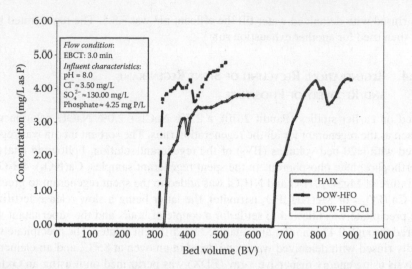

FIGURE 7.2 Comparative breakthrough profile of P with different sorbents.

preloaded Cl⁻ with HPO_4^{2-}) and LAB between HPO_4^{2-} and $FeOH_2^+$. The following affinity sequence is generally obtained for strong-base anion exchange resins (the parent resin of HAIX) and has been demonstrated in previous works (Blaney et al. 2007, Zhao and SenGupta 1996):

$$SO_4^{2-} > PO_4^{3-} > NO_3^- > Br^- > NO_2^- > Cl^-$$

Thus, for the case of anion exchange where SO_4^{2-} is present at a much higher concentration than P, it provides stronger coulombic interaction (CI) and almost completely prevails over P for the preloaded Cl⁻ sites. P removal is almost completely due to LAB interaction between HPO_4^{2-} and $FeOH_2^+$. The role of SO_4^{2-} vis-à-vis P being a trace species is discussed in the following section, but it will suffice to state here that competition by SO_4^{2-} does not impair the performance of HAIX. For the third exchanger, DOW-HFO-Cu, the capacity of the sorbent media is enhanced and the breakpoint of P is observed much later than (about 200 BVs) HAIX or DOW-HFO. In this particular scenario, P was removed by LAB interaction with both the dispersed HFO nanoparticles and the Cu^{2+} ion immobilized on the surface. For the Cu^{2+} sites on the exchanger, LAB + CI for HPO_4^{2-} is higher than CI alone for SO_4^{2-}; therefore, SO_4^{2-} competition is avoided and this exchanger provides the highest P uptake. These experimental data validate that if two different ligand exchangers can successfully be hosted on a polymeric backbone, increased sorption capacity is attained, with little or no compromise on the individual capacity of the ligand exchangers.

The role played by SO_4^{2-} concentration in the case of P removal by HAIX was further evaluated. Figure 7.3 provides the P breakthrough curve for a HAIX column that was fed an influent that varied only in SO_4^{2-} concentration (161.6 and 245.9 mg/L).

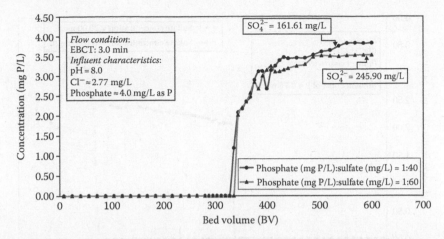

FIGURE 7.3 Comparison of breakthrough profiles of P for HAIX for different background concentrations of SO_4^{2-} (P is present as trace species compared with SO_4^{2-}).

Since the P:SO_4^{2-} is 40.4 or 61.5, P can be considered as a trace species in both cases. Figure 7.3 shows that the P breakthrough behavior remained the same for both the cases and it can be inferred that as long as P is a trace species compared to SO_4^{2-} (that is ubiquitous in wastewater effluents), P uptake capacity of HAIX is not affected by SO_4^{2-} concentration. The typical SO_4^{2-} and P concentration in the secondary effluent of a municipal wastewater treatment plant being in this range, little or no competition is expected from the increase in concentration of SO_4^{2-}.

However, when the fixed-bed column was operated with a much lower background concentration of SO_4^{2-}, at P:SO_4^{2-} = 1.0:5.00, there was significant improvement in the performance of the column. The breakthrough BV for P almost doubled, even though there was some competition from the SO_4^{2-}, which is evident from its later breakpoint in the effluent graph and chromatographic elution of P. Figures 7.4 and 7.5 illustrate the P breakthrough behaviors obtained from the HAIX-loaded fixed-bed column runs for high and low background concentration of SO_4^{2-}. The sudden drop in P concentration at ~700 BV as observed in Figure 7.4 was due to an interruption of flow, and the nature of the curve is further elucidated in Section 5.4.

In all the fixed-bed column runs performed, the pump used to drive the inflow through the fixed bed needed to be turned off (once or more, depending on the length of the run) for a brief period to prevent overheating. This observation was consistent in all cases. However, this interruption test had no effect on the ultimate removal capacity. The removal of P is attributed to CI + LAB interaction, with the latter being the major player, whereas the removal of SO_4^{2-} by HAIX is purely due to CI and LAB interaction has no role to play. Thus, when the background concentration of SO_4^{2-} is very high, the ion exchange sites get exhausted rapidly with SO_4^{2-} and its breakthrough occurs almost instantaneously. P being present as a trace species does not pose a competition to the SO_4^{2-} for CI removal, but gets exchanged due to LAB interaction with HFO till the exhaustion of its (HFO) capacity. On the other hand,

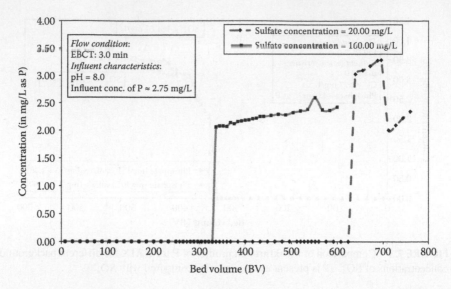

FIGURE 7.4   Comparative breakthrough profile of P for HAIX at low and high background concentration of SO_4^{2-}.

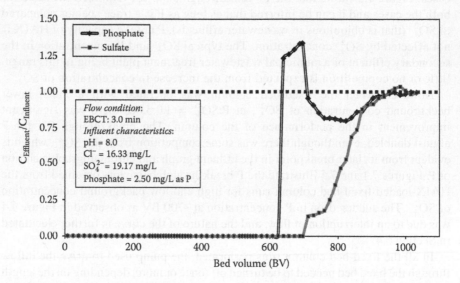

FIGURE 7.5 $C_{Effluent}/C_{Influent}$ graph of P and SO_4^{2-} for HAIX at low background concentration SO_4^{2-}.

when the background concentration of SO_4^{2-} is low, both P and SO_4^{2-} compete for the ion exchange sites, in addition to P removal by LAB interaction. Since SO_4^{2-} is more preferred by the CI exchange mechanism, P undergoes a brief chromatographic elution before the breakpoint of SO_4^{2-}, where SO_4^{2-} replaces P from the exchange sites and gets removed from the solution phase. However, P removal due to LAB interaction is not compromised at any stage. Hence, it can be inferred that P removal by HAIX is attributed to both CI and LAB exchange mechanism at low background concentration of SO_4^{2-}, while at high background concentration of SO_4^{2-}, it is entirely attributable to LAB interaction and is independent of SO_4^{2-} concentration.

7.4.2 Fixed-Bed Column Run: Regeneration, Recovery, and Recycling

Excellent regeneration of the exchanger was observed in all cases. Figure 7.6a through c shows the regeneration profile for the three exchangers studied for a regenerant of 2.5% NaCl + 2% NaOH.

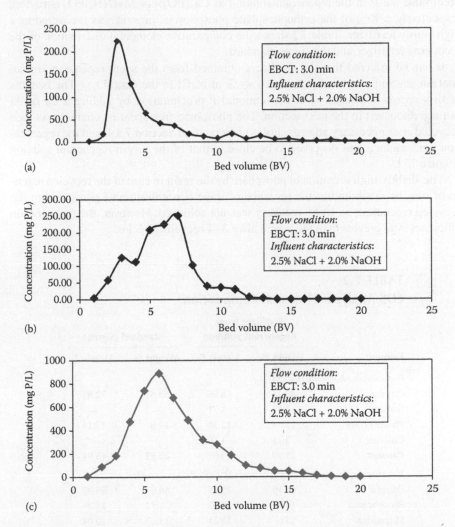

FIGURE 7.6 Typical regeneration profile of P for (a) HAIX, (b) DOW-HFO, and (c) DOW-HFO-Cu.

Recovery of more than 90.0% of the sorbed P was attained within 10 BVs in all cases, even after 10 cycles of exhaustion and regeneration run. Desorption of the SO_4^{2-} was also remarkable and >90.0% ± 5.00% of the sorbed SO_4^{2-} was desorbed within 15 BVs in all cases. The mass of SO_4^{2-} in the exchanger in increasing order of magnitude was DOW-HFO < DOW-HFO-Cu < HAIX, as may be expected from the earlier discussion. Also, there was no compromise on the performance of the regenerated resin with that of the virgin resin. A decrease of only about 1.5% in the breakpoint of P was observed after 10 cycles of exhaustion–regeneration of the same resin bed.

After the addition of $Ca(NO_3)_2 \cdot 4H_2O$ or a mixture of $MgSO_4 \cdot 7H_2O$ and NH_4Cl to precipitate the P in the regenerant solution as $Ca_3(PO_4)_2$ or $Mg(NH_4)PO_4$ (struvite), respectively, >90% of the orthophosphate phosphorous present was recovered as a high-purity fertilizer. Table 7.2 shows the comparative elemental distribution of the recovered fertilizer and the sigma standard.

It can be inferred that the fertilizers obtained from the spent regenerant do not contain any impurity except for ≈3.6 wt.% of NaCl in the $Ca_3(PO_4)_2$. The benefits of this approach compared to conventional P precipitation by adding a Ca or Al salt are discussed in the next section. The phosphate-free spent regenerant was then recycled after necessary adjustments as elaborated in Section 7.3.4, and the regeneration performance was observed to be close to that of the virgin regenerant solution (Figure 7.7).

The slightly high retention of phosphate by the resin in case of the recycled regenerant may be attributed to the presence of trace concentration of phosphate in the recycled regenerant as 100% recovery was not achieved. However, the regeneration efficiency was greater than 90% even after 3–4 recycling cycles.

TABLE 7.2
EDS Analysis: Elemental Composition

Element	Fertilizer from Regenerant Solution		Standard (Sigma)	
	Weight %	Atomic %	Weight %	Atomic %
I—Calcium phosphate ($Ca_3(PO_4)_2$)				
Oxygen	49.35	68.96	53.57	72.87
Sodium	1.82	1.71	—	—
Phosphorous	17.74	12.36	19.6	13.18
Chlorine	3.11	1.89	—	—
Calcium	27.99	15.07	26.82	13.94
II—Magnesium ammonium phosphate ($Mg(NH_4)PO_4$): Struvite				
Oxygen	66.19	75.56	56.67	65.92
Phosphorous	16.99	10.02	22.12	13.29
Magnesium	13.6	10.21	13.12	10.04
Nitrogen	3.23	4.21	8.1	10.76

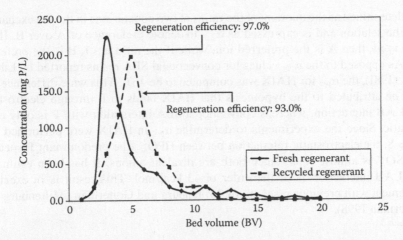

FIGURE 7.7 Comparison of the regeneration efficiency of the fresh and recycled regenerant for HAIX column.

7.5 DISCUSSION

7.5.1 PHOSPHOROUS RECOVERY

Chemical precipitation of phosphorus in wastewater by addition of aluminum or ferric salt is a conventional process for the removal of phosphorus, but many studies (Banu et al. 2008, Szabo et al. 2008, Takacs et al. 2006) have shown the critical role of pH and alkalinity, and the process may need addition of an external buffer to raise the pH/alkalinity before attempting precipitation of Al (or Fe) salt. Moreover, it is clear (Debarbadillo et al. 2010, Esvelt et al. 2010) that to achieve effluent P in the range of <50 µg/L, the molar ratio of Al (or Fe)/P required is >>1, even in the range of 200. The technique employed here applied molar Ca (or Mg)/P of 1.0. The main features of the chemical addition step are as follows: (1) the regenerant solution already is highly alkaline (pH > 12) and (2) the regenerant solution volume is much lower (1/20 of the volume of wastewater treated). Moreover, after filtering the Ca (or Mg) salt from the regenerant, the regenerant solution is reused for the next cycle, as explained in Section 7.3.4. Thus, although the process described in this communication uses chemicals, the quantity needed is much lower than alternative chemical precipitation processes, and the solution volume on which the chemicals are applied is only about 5% of the wastewater treated.

7.5.2 QUANTIFYING THE RELATIVE CONTRIBUTION OF DIFFERENT MECHANISMS IN PHOSPHATE SORPTION

Conventional strong-base anion exchange resins have poor P selectivity compared to SO_4^{2-}. Liberti et al. have reported the $\alpha_{P/S}$ for different SBA resins; the maximum value is 0.25 (Liberti et al. 1981). The separation factor or α can be defined as the preference of the ion exchanger for one of the two counterions (Helfferich 1995).

It is determined as the quotient of the ratios of two counterions in the ion exchanger and the solution and is expressed as $\alpha_{A/B}$ to denote preference of A over B. If the ratio is >1, then A is the preferred ion, while if the ratio is <1, B is the preferred ion. As opposed to the $\alpha_{P/S}$ values for conventional SBA resins reported by Liberti et al. (1981), the $\alpha_{P/S}$ for HAIX was computed to be 46.0. This wide difference can only be attributed to the hypothesis that HAIX binds to P through electrostatic and LAB interaction, whereas conventional SBA interaction with P is only electrostatic. Since the experiments to determine $\alpha_{P/S}$ in HAIX were performed at a pH ≈ 8, the electrostatic interaction between HPO_4^{2-} (the predominant P species) and SO_4^{2-} is almost equal since both are divalent anions. It has been estimated that LAB interaction is of the order of –3 kcal/mol. This result is in excellent agreement with previous research by Miltenburg and Golterman (Miltenburg and Golterman 1998).

7.5.3 KINETIC MODELING OF SORPTION

Fixed-bed sorption mechanism is often governed by the kinetics of the sorption process. Generally, the process is controlled by the mass transfer (intraparticle and/or film diffusion mass transfer) rate. From the results of interruption test, it was inferred that intraparticle diffusion was the rate-limiting step in the sorbent's sorption of phosphate. This kinetic behavior has been validated for selective sorption of trace solutes in absence of chemical reactions by many researchers. Similar interruption data and interpretation in favor of such an inference for selective ion exchangers and sorbents are available in open literature (Gjerde and Fritz 1978, Helfferich 1995). Figure 7.8 shows the breakthrough profile emphasizing the interruption test.

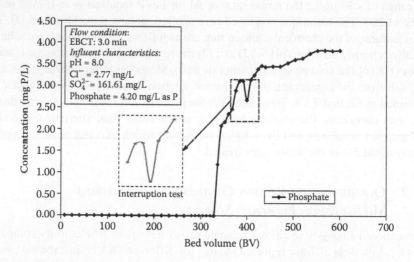

FIGURE 7.8 Breakthrough profile of phosphate showing interruption test.

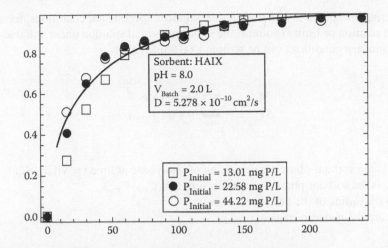

FIGURE 7.9 Intraparticle diffusion model plot with experimental data.

During one of the fixed-bed column run experiment with HAIX, the flow of the influent was intentionally stopped after 392 BVs with the effluent phosphate concentration of 3.117 mg P/L, for approximately 3.0 h. After that, the original flow condition was restored and the effluent phosphate concentration reached 3.25 mg P/L at 416 BV, that is, within 24 BVs (as shown in Figure 7.9).

7.5.3.1 Determination of Intraparticle Diffusivity

Results of batch kinetic study for phosphate–chloride binary system were studied to determine the intraparticle diffusivity, D. Restricting the discussion to cases where the diffusion is radial, the equation for a constant diffusion coefficient can be expressed as

$$\frac{\partial C}{\partial t} = D\left(\frac{\partial^2 C}{\partial r^2} + \frac{2}{r}\frac{\partial C}{\partial r}\right) \tag{7.2}$$

where
  D is the intraparticle diffusivity
  C is the concentration
  r is the radial coordinate (distance from the bead center)
  t is the time

On substituting u = Cr, where u is the mass flux entering the spherical bead, we get

$$\frac{\partial u}{\partial t} = D\frac{\partial^2 u}{\partial r^2} \tag{7.3}$$

Assuming phosphate to be a trace species under experimental conditions, for a well-stirred solution of limited volume, the exact analytical solution under suitable initial and boundary conditions can be written as (Crank 1975)

$$\frac{M_t}{M_\infty} = 1 - \sum_{n=0}^{\infty} \frac{6\alpha(\alpha+1)e^{\frac{-Dq_n^2 t}{a^2}}}{9+9\alpha+q_n^2\alpha^2} \tag{7.4}$$

where
  M_t is the sorbent-phase concentration of phosphate at time $t = V(C_{P,0}-C_{P,t})$
  M_∞ is the sorbent-phase d at equilibrium $= V(C_{P,0}-C_{P,\infty})$
  a is the radius of the bead
  q_ns are the nonzero roots of

$$\tan q_n = \frac{3q_n}{3+\alpha q_n^2} \tag{7.5}$$

The parameter α is expressed in terms of the final fractional uptake of the solute by the sphere by the relation

$$\frac{M_\infty}{VC_0} = \frac{1}{1+\alpha} \tag{7.6}$$

where
  V is the batch volume
  C_0 is the initial aqueous-phase concentration of phosphate

The batch kinetic tests were carried out for three different concentrations of phosphate, namely, 13.01, 22.58, and 44.22 mg P/L. The average value of α obtained was 1.661. The values of q_ns were interpolated from the values provided in open literature (Crank 1975). The first six values of q_n were only considered as it converges within that range. The other terms being known, the intraparticle diffusivity was obtained by fitting the batch kinetic data to Equation 7.4 as shown in Figure 7.9. The mean effective intraparticle diffusivity was found to be 5.278×10^{-10} cm²/s (with standard deviation $= 0.011 \times 10^{-10}$ cm²/s). Excellent fit was observed for all three initial concentrations of phosphate. The solid line in the figure shows the best-fit line. The observed values were a magnitude higher than phosphate exchange in OH⁻ type strong basic anion exchanger (Galinada and Yoshida 2004).

7.6 CONCLUSIONS

Selective removal of trace phosphate from ubiquitous background presence of other competing anions present in wastewater is a major challenge faced by the industry. Chemical precipitation is limited by the production of excessive sludge and biological removal processes are restricted by the inability to remove trace concentration. Also, both these processes suffer from inadequate recovery of the removed phosphate, which is a nonrenewable resource. The overall goal of this study was to estimate the removal capacity of phosphate by three iron-oxide-impregnated polymers, HAIX, DOW-HFO, and DOW-HFO-Cu, at different background concentrations of competing anions and to assess the recovery of the phosphate as a solid-phase fertilizer. Laboratory experiments were carried out to investigate the key features of the fixed-bed sorption process, which is capable of removing trace concentrations of phosphate, to recover the phosphate, and to reuse the spent regenerant to minimize the production of any solid or liquid stream. Theoretical analysis and experimental evidence were also provided to demonstrate the underlying principles of sorption and kinetics of the process.

The major conclusions that can be drawn from this study are synopsized as follows:

1. HAIX, DOW-HFO, and DOW-HFO-Cu showed high selectivity towards phosphate when compared to competing anions, especially sulfate.
2. LAB interaction (i.e., formation of coordination bond between the anionic ligand and the central metal atom forming inner-sphere complexes) accompanied by the electrostatic attraction (i.e., ion-pair formation) is the core mechanism leading to the high sorption affinity for HAIX and DOW-HFO-Cu, whereas for DOW-HFO, LAB interaction is the only mechanism of phosphate sorption.
3. All three materials studied, HAIX, DOW-HFO, and DOW-HFO-Cu, are amenable to efficient regeneration. Single-step regeneration with 2.5% sodium chloride and 2.0% sodium hydroxide consistently recovered more than 95.0% of sorbed phosphate within 10 BVs. Only minor capacity drop (about 1.5%) was observed after 10 cycles of exhaustion–regeneration.
4. The spent regenerant may be reused after supplementing for the hydroxide lost in regeneration. The regeneration efficiency was unaltered compared to virgin regenerant.
5. Phosphate can be recovered from the spent regenerant as calcium phosphate or magnesium ammonium phosphate (struvite) upon addition of calcium nitrate or a combination of ammonium chloride and magnesium sulfate, respectively.
6. No significant bleeding of the iron (from HAIX) or copper (from DOW-HFO-Cu) was found in the regenerant.
7. Intraparticle diffusion is the rate-limiting step for phosphate sorption by HAIX.

HAIX, or LayneRT™, the commercial form, can potentially be applied to wastewater treatment facilities as a tertiary treatment step to cope up with the newer and stricter regulations regarding discharge limit of phosphate. In spite of putting up a stellar performance in terms of removal, DOW-HFO-Cu was not evaluated further due to its commercial nonavailability and high cost of the parent resin, which might be deterrent for its full-scale application.

ABBREVIATIONS

CI	Coulombic interaction
DOW-HFO	DOW resin loaded with hydrated ferric oxide
DOW-HFO-Cu	DOW resin loaded with hydrated ferric oxide and copper
EBCT	Empty bed contact time
EDX	Energy-dispersive x-ray
HAIX	Hybrid anion exchanger
HFO	Hydrated ferric oxide
LAB	Lewis acid–base
PLE	Polymeric ligand exchanger
SEM	Scanning electron microscope
WWTP	Wastewater treatment plant

REFERENCES

APHA, AWWA, and WEF. 1998. *Standard Methods for the Examination of Water and Wastewater* (20th edn.). Baltimore, MD: American Public Health Association, American Water Works Association, Water Environment Federation.

Banu, R.J., K.U. Do, and I.T. Yeom. 2008. Phosphorus removal in low alkalinity secondary effluent using alum. *Int. J. Environ. Sci. Technol.* 5: 93–98.

Blaney, L.M., S. Cinar, and A.K. SenGupta. 2007. Hybrid anion exchanger for trace phosphate removal from water and wastewater. *Water Res.* 41: 1603–1613.

Chubar, N.I., V.A. Kanibolotskyy, V.V. Strelko, G.G. Gallios, V.F. Samanidou, and T.O. Shaposhnikova. 2005. Adsorption of phosphate ions on novel inorganic ion exchangers. *Colloids Surf. A* 255: 55–63.

Cooper, P., T. Dee, and G. Yang. 1993. Nutrient removal-methods of meeting the EC urban wastewater directive. In Paper presented at *the Fourth Annual Conference on Industrial Wastewater Treatment*, Esher, Surrey, March 10, 1993.

Crank, J. 1975. *The Mathematics of Diffusion* (2nd edn.). Oxford, U.K.: Oxford University Press.

Cumbal, L., J. Greenleaf, D. Leun, and A.K. SenGupta. 2003. Polymer supported inorganic nanoparticles: Characterization and environmental applications. *React. Funct. Polym.* 54: 167–180.

Cumbal, L. and A.K. SenGupta. 2005. Arsenic removal using polymer-supported hydrated iron(III) oxide nanoparticles: Role of donnan membrane effect. *Environ. Sci. Technol.* 39: 6508–6515.

Debarbadillo, C., G. Shellswell, W. Cyr, B. Edwards, R. Waite, B. Sabherwal, J. Mullan, and R. Mitchell. 2010. Development of full-scale sizing criteria from tertiary pilot testing results to achieve ultra-low phosphorus limits at Innisfil, Ontario. *Proceedings of the 83rd Annual Water Environment Federation Technical Exhibition and Conference*, New Orleans, LA, October 2–6, 2010.

DeMarco, M.J., A.K. SenGupta, and J.E. Greenleaf. 2003. Arsenic removal using a polymeric/inorganic hybrid sorbent. *Water Res.* 37: 164–176.

Dzombak, D.A. and F.M. Morel. 1990. *Surface Complexation Modeling: Hydrous Ferric Oxide*. Hoboken, NJ: John Wiley & Sons, Inc.

Esvelt, L., M. Esvelt, B. Walker, and L. Hendron. 2010. Pilot studies for reducing RPWRF effluent TP for discharge to the Spokane river. *Proceedings of the 83rd Annual Water Environment Federation Technical Exhibition and Conference*, New Orleans, LA, October 2–6, 2010.

Florida Everglades Forever Act. 1994. *Florida Department of Environmental Protection–Office of Ecosystem Projects*. Tallahassee, FL.

Galinada, W.A. and H. Yoshida. 2004. Intraparticle diffusion of phosphates in OH-type strongly basic ion exchanger. *AIChE J.* 50: 2806–2815.

Genz, A., A. Kornmüller, and M. Jekel. 2004. Advanced phosphorus removal from membrane filtrates by adsorption on activated aluminium oxide and granulated ferric hydroxide. *Water Res.* 38: 3523–3530.

Gjerde, D.T. and J.S. Fritz. 1978. *Ion Chromatography*. New York: A. Hüthig.

Goldman, J.C., K.R. Tenore, and H.I. Stanley. 1973. Inorganic nitrogen removal from wastewater: Effect on phytoplankton growth in coastal marine waters. *Science* 180: 955–956.

Golterman, H. 1995. Theoretical aspects of the adsorption of ortho-phosphate onto iron-hydroxide. *Hydrobiologia* 315: 59–68.

Hansen, B. 2006. Long-term plan seeks to reduce phosphorus in Spokane river. *Civil Eng.* 76: 24–25.

Heathwaite, L. and A. Sharpley. 1999. Evaluating measures to control the impact of agricultural phosphorus on water quality. *Water Sci. Technol.* 39: 149–155.

Helfferich, F.G. 1995. *Ion Exchange*. Mineola, NY: Dover Publications Inc.

Herring, J.R. and R.J. Fantel. 1993. Phosphate rock demand into the next century: Impact on world food supply. *Nat. Resour. Res.* 2: 226–246.

Hiemstra, T. and W. van Riemsdjik. 1999. Surface structural ion adsorption modeling of competitive binding of oxyanions by metal (Hydr)oxides. *J. Colloid Interface Sci.* 210: 182–193.

Huang, C.P. and L.M. Vane. 1989. Enhancing As^{5+} removal by a Fe^{2+}-treated activated carbon. *Water Pollut. Control Fed. J.* 61: 1596–1603.

Jenkins, D. and S.W. Hermanowicz.1991. Principles of chemical phosphorus removal. In: Sedlak, R.I. (ed.), *Phosphorus and Nitrogen Removal from Municipal Wastewater: Principles and Practice* (2nd edn.). Chelsea, MI: Lewis Publishers.

Jenkins, O., J.F. Fergusson, and A.B. Menar. 1971. Chemical processes for phosphate removal. *Water Res.* 5: 369–387.

Katsoyiannis, I.A. and A.I. Zouboulis. 2002. Removal of arsenic from contaminated water sources by sorption onto iron-oxide-coated polymeric materials. *Water Res.* 36: 5141–5155.

Kuba, T., G.J.F. Smolders, M.C.M. van Loosdrecht, and J.J. Heijnen. 1993. Biological phosphorus removal from wastewater by anaerobic-anoxic sequencing batch reactor. *Water Sci. Technol.* 27: 241–252.

Liberti, L., G. Boari, D. Petruzzelli, and R. Passino. 1981. Nutrient removal and recovery from wastewater by ion exchange. *Water Res.* 15: 337–342.

Miltenburg, J. and H. Golterman. 1998. The energy of the adsorption of o-phosphate onto ferric hydroxide. *Hydrobiologia* 36: 93–97.

Onyango, M.S., H. Matsuda, and T. Ogada. 2003. Sorption kinetics of arsenic onto iron-conditioned zeolite. *J. Chem. Eng. Jpn.* 36: 477–485.

Pandit, A. 2010. Selective removal and recovery of phosphate from wastewater. MS thesis, University of Massachusetts Dartmouth, North Dartmouth, MA.

Rhyther, J.H. and W.M. Dunstan. 1971. Nitrogen, phosphorus, and eutrophication in the coastal marine environment. *Science* 171: 1008–1013.

Sedlak, R. (ed.). 1991. *Phosphorus and Nitrogen Removal from Municipal Wastewater: Principles and Practice*. Boca Raton, FL: Lewis Publishers.

Seida, Y. and Y. Nakano. 2002. Removal of phosphate by layered double hydroxides containing iron. *Water Res.* 36: 1306–1312.

SenGupta, A.K. and D. Zhao. 2000. Selective removal of phosphates and chromates from contaminated water by ion exchange. U.S. Patent 6136199.

Sengupta, S. and A. Pandit. 2011. Selective removal of phosphorus from wastewater combined with its recovery as a solid-phase fertilizer. *Water Res.* 45: 3318–3330.

Seviour, R.J., T. Mino, and M. Onuki. 2003. The microbiology of biological phosphorus removal in activated sludge systems. *FEMS Microbiol. Rev.* 27: 99–127.

Sharploy, A.N., S.C. Chapra, R. Wedepohl, J.T. Sims, T.C. Daniel, and K.R. Reddy. 1994. Managing agricultural phosphorus for protection of surface waters: Issues and options. *J. Environ. Qual.* 23: 437–451.

Sharpley, A.N., T. Daniel, T. Sims, J. Lemunyon, R. Stevens, and R. Parry. 2003. *Agricultural Phosphorus and Eutrophication* (2nd edn.). United States Department of Agriculture, Agricultural Research Service.

Stumm, W. and J.J. Morgan. 1995. *Aquatic Chemistry: Chemical Equilibria and Rates in Natural Waters* (3rd edn.). New York: John Wiley & Sons.

Szabo, A., I. Takacs, S. Murthy, G.T. Daigger, I. Liksco, and S. Smith. 2008. Significance of design and operational variables in chemical phosphorus removal. *Water Environ. Res.* 80: 407–416.

Takacs, I., S. Murthy, S. Smith, and M. McGrath. 2006. Chemical phosphorus removal to extremely low levels: Experience of two plants in the Washington, DC area. *Water Sci. Technol.* 53: 21–28.

Tanada, S., M. Kabayama, N. Kawasaki, T. Sakiyama, T. Nakamura, and M. Araki. 2003. Removal of phosphate by aluminum oxide hydroxide. *J. Colloid Interface Sci.* 257: 13–140.

USEPA. 1997. *Chesapeake Bay Nutrient Reduction Program and Future Directions.* Anapolis, MD: USEPA Chesapeake Bay Program.

USEPA. 2007. Advanced wastewater treatment to achieve low concentration of phosphorus. EPA 910-R-07-002. Seattle, WA.

USGS. 1999. *Phosphorus in a Ground-Water Contaminant Plume Discharging to Ashumet Pond, Cape Cod, Massachusetts.* Northborough, MA: USGS.

Valiela, I. 1995. *Marine Ecological Processes.* New York: Springer-Verlag.

Zeng, L., X. Li, and J. Liu. 2004. Adsorptive removal of phosphate from aqueous solutions using iron oxide tailings. *Water Res.* 38: 1318–1326.

Zhao, D. and A.K. SenGupta. 1996. Selective removal and recovery of phosphate in a novel fixed-bed process. *Water Sci. Technol.* 33: 139–147.

Zhao, D. and A.K. SenGupta. 1998. Ultimate removal of phosphate from wastewater using a new class of polymeric ion exchangers. *Water Res.* 32: 613–1625.

8 Molecularly Imprinted Polymers for Water Polishing

Joanna Wolska and Marek Bryjak

CONTENTS

8.1 INTRODUCTION

Recently, it has been reported that many chemicals released to the environment can disrupt the endocrine system of wildlife and humans. A large number of nonsteroidal chemicals, referred to as xenoestrogens, can bind to estrogen receptor and evoke biological responses. The key factor causing interaction of xenoestrogens with the binding pocket of the receptors has chemical similarity to estradiol (usually a phenolic A-ring) (Witorsch 2002). Xenoestrogens appear in natural resources such as food, water, or soil in very low concentrations ranging from ng/L to µg/L (Staples et al. 2000). For illustration, the amount of some xenoestrogens in the surface waters is shown in Table 8.1.

The low concentration of xenoestrogens does not mean these compounds are not harmful to the endocrine system. Even at nanogram level, they can disturb hormone activities and can cause deterioration of sperm quality among males, abnormal or

TABLE 8.1

Concentrations of Some Xenoestrogens in the Surface Water

Country	4-*Tert*-octylphenol (mg/m³)	4-Nonylphenol (mg/m³)	Bisphenol A (mg/m³)
Denmark	<0.1	<0.1–0.29	<0.001–0.44
Germany	0.0004–0.0036 or <0.01–0.189	0.001–0.221 or <0.01–0.485	0.009–0.776 or <0.05–0.272
Netherlands	0.05–6.3	<−0.11–4.1	0.0088–1
England and Wales	<1 or <0.01–13	<0.2–180 or <0.03–5.2	—
USA	—	<0.11–0.64	<1–8

Source: Dudziak, M. and Bodzek, M., *Ochrona Środowiska*, 31, 9, 2009.

delayed development of reproductive organs, and increased risks of prostate cancer, testicular cancer, breast cancer, and endometriosis with the associated infertility in females (Chahoud et al. 2001, Shin et al. 2001, Zhao et al. 2004). Thus, the highest concern is the development of methods for separation of these compounds from aqueous solutions (Wang et al. 2011).

Numerous substances such as herbicides, pesticides, fungicides, plasticizers, some monomers, phenylphenols, bisphenols, alkylphenols, dioxins, chlorophenols, and phthalates by their action through endocrine-disruptive mechanisms (Coughlin et al. 2011, Witorsch 2002) have been named EDCs.

8.2 WATER POLLUTANTS: ENDOCRINE-DISRUPTING COMPOUNDS

Since the early 1950s, abnormalities in reproduction and developmental systems as well as disorders in immunodeficiency or malfunctioning of brain have been reported all over the world. In the past three decades, evidences of previously mentioned phenomena have been found in human and wildlife populations and their intensity has been related to the presence of EDCs.

EDCs are the species that appear in the nanogram level, but even so small dosage can vitally affect the hormonal system of higher organisms including mammalians. Many EDCs have natural origin but some of them are the civilization products: pharmaceuticals or personal care products. As they appear almost everywhere, they are continuously introduced to the environment from different sources and are found in aquatic systems worldwide. Conventional water treatment processes are inefficient for the removal of EDCs, and it is required to find more efficient methods for water purification. The frequently used sorption of organic compounds on activated carbon particles, in the form of powdered or granulated beads, is mostly efficient in the removal of hydrophobic contaminants. The problem gets serious when natural organic matter or surfactants appear in treated water. They adsorb preferentially on the surface of sorbents and block the adsorption centers of activated carbon. Oxidation processes

(Higuchi et al. 2002) can affect other organic compounds being in water causing less effective EDC degradation. Finally, the nanofiltration and/or reverse osmosis is not enough specifically for EDCs and, furthermore, they are high-energy-demanding processes. The use of MIPs for EDC removal opens the new perspectives for water polishing technology. The footprints prepared in a polymer matrix with its high affinity to particular endocrine disruptor make the possibility to use highly selective materials for the removal of low concentrated but extremely harmful pollutant.

8.3 PREPARATION OF MOLECULARLY IMPRINTED MATERIALS

8.3.1 PARTICLE IMPRINTING METHODS

As the imprinting matrix, the high cross-linked organic polymers are synthesized usually by free radical copolymerization of functional monomers and cross-linking agent in the presence of porogen. The reaction is conducted in an appropriate solvent that contains the template molecules. This results in preparation of porous material containing specific binding sites for the template molecules (Sellergren 2001).

Various requirements for preparation of MIPs have extensively been developed, and many polymerization methods to synthesize the requested products have been introduced. The most frequently applied methods are the following:

- Mass polymerization (block polymerization)
- Suspension polymerization
- Multistep swelling polymerization
- Two-step process with membrane emulsification followed by suspension polymerization
- Precipitation polymerization
- Surface-imprinted polymerization
- In situ polymerization
- Polymerization of reactive surfactants

Each method has its own advantages and disadvantages. The main features are juxtaposed in Table 8.2.

8.3.1.1 Block Polymerization

This method is the most frequently used for MIP preparation. Using it, one can obtain block of polymers in the shapes related to the shape of a polymerization vessel. After polymerization, the obtained block has to be crushed, grounded, and sieved to get particles of required sizes. Due to its simplicity, the method is attractive and popular for new composition of reactive mixture and templates. It is very fast and simple and its optimization is relatively straightforward—it does not require special operator skills or sophisticated equipment. However, the method has its own disadvantages—ground particles have polydisperse size and irregular shape and some of the footprints are destroyed during the cracking treatment. Furthermore, grinding and sieving cause the loss of a large quantity of material, which sometimes is about 50%–70% of the initial amount of bulk polymer (Yan and Row 2006).

TABLE 8.2
Short Characteristics of MIP Polymerization Methods

Polymerization	Advantages	Disadvantages
Mass	Simple and universal	Need to grind the product
	Does not require additional skills and special equipment	Irregular shape of particles
		Low productivity
Suspension in water	Spherical beads	Complicated interface system
	High repeatability	Adverse effect of water
	Simple scaling up	Need of special surfactants
		Difficult to imprint with hydrophilic functional monomers
Suspension in perfluorocarbon	Applicable to most imprinting systems	Specialized surfactants needed
	Particle size adjustable	For some applications, fluorinated compounds are dangerous
Multistep	Monodisperse beads of controlled diameter	Complicated process
	Excellent column packing for high-performance liquid chromatography (HPLC)	Not compatible to noncovalent imprinting systems
		Difficult choice of reaction conditions
		Low loading capacity
		Required of aqueous emulsions
Two-step method	Monodisperse microspheres	Adverse effect of water
		Careful selection of reaction parameters
Precipitation	Monodisperse and clean microspheres	High consumption of template
	High productivity	Large dilution factor
Surface imprinted	General compatibility with imprinting	Complicated process
	Monodisperse product	Time consuming
	Thin layers of polymer	Low loading capacity and grafting yield
In situ polymerization	One step, in situ polymerization	Extensive optimization required for each new template system
	Cost efficient, good porosity	

Sources: Ye, L. et al., *Anal. Chim. Acta*, 435, 187, 2001; Yan, H. and Row, K.H., *Int. J. Mol. Sci.*, 7, 155, 2006.

Radical polymerization process is usually exothermic. Due to that behavior, it is a danger of overheating the prepolymerization complexes that leads to bad structuring of obtained footprints. Wang et al. (2011) using block polymerization of methacrylic acid (MAA) and ethylene glycol dimethacrylate (EGDMA) obtained MIPs with malachite green as template and used them for the separation of this component from seawater and seafood samples. Byun et al. (2013) obtained materials selective for bisphenol A (BPA) or 2,4-dichlorophenoxyacetic acid by bulk polymerization of MMA, methyl methacrylate (MMA), and EGDMA. Wolska and Bryjak (2012) used block polymerization of styrene (S) and divinylbenzene (DVB) to obtain MIPs for the removal of BPA and phthalates.

8.3.1.2 Suspension Polymerization

Many applications request the use of fine polymeric particles. This requirement led
to the synthesis of MIP materials by suspension polymerization, which eliminates
the problem of temperature stabilization of the reaction mixture that appears in block
polymerization. The product of suspension polymerization does not require mechanical
ical crushing of materials and offers spherical beads. Suspension polymerization is
the process of heterogeneous radical reaction when mechanical stirring of reactants
allows obtaining droplets of monomer phase, in the first stage, and particles as the
final product, in the second stage. This process is usually carried out in water phase
that affects the polymerization efficiency, and furthermore, the selectivity of the
MIPs could be decreased. The water is thought to weaken the interaction between
the template molecule and monomer due to its high polarity.

Joshi et al. (1999) synthesized MIPs by suspension polymerization with the
template of BPA. As monomers, they used phenyl methacrylate, p-cumylphenol
methacrylate and MAA, and EGDMA as cross-linker. They checked the effect of
porogen on sorption and selectivity of the sorbent. Kawaguchi et al. (2005) prepared
molecularly imprinted beads by suspension polymerization from 4-vinylpyridine
and EGDMA. As templates, they used BPA, 4-$tert$-buthylphenol (BP), or bisphenol
A-d_{16} (BPA-d_{16}). Walsh et al. (2011) obtained MIPs from MAA and EGDMA.

In order to avoid adverse effects of water, Mayes and Mosbach (1996) applied
liquid perfluorocarbon as dispersing phase. As a monomer mixture, they used MAA
and EGDMA in different ratios and three types of porogens: chloroform, acetone, or
toluene. Sometimes, perfluorocarbon has been replaced by mineral oil (Kempe and
Kempe 2004, 2006). In all these approaches, it was noted that suspension polymerization
ization gave spherical polymer beads with high repeatability.

8.3.1.3 Multistep Swelling Polymerization

In recent years, new techniques to produce molecularly imprinted polymeric microspheres
spheres with very narrow size distribution have been searched for. One of the relatively
new methods is multistep swelling polymerization. Although this process can give fine
spherical MIP particles, it is complicated enough, and due to the use of aqueous phase,
water molecules can interfere with imprinting process and decrease the selectivity of
obtained particles (Yan and Row 2006). Multistep swelling polymerization was firstly
described by Hosoya et al. (1996). Using multistep polymerization, the following templates
plates have been imprinted: (S)-α-methylbenzylamine (Hosoya et al. 1996), S-naproxen
(Haginaka et al. 1999a, Haginaka and Sanbe 2001), S-ibuprofen (Haginaka et al. 1999b),
and BPA (Sanbe and Haginaka 2002). As a seed particle, they used polystyrene particles
with the diameter of 1 μm that were saturated with functional monomers, cross-linking
agent, and Porofor and polymerized inside polymer matrix.

8.3.1.4 Membrane Emulsification Followed by Suspension Polymerization

For the production of molecularly imprinted polymeric microspheres, the membrane
emulsification process followed by suspension polymerization was also adapted. The
key step of the method is the formation of emulsion with controlled and uniform size
distribution. It is achieved by injection of disperse phase through pores of membrane into

immiscible liquid that usually contains emulsion stabilizers. The diameter of dispersed droplets depends on physical properties of both liquids, pore structure of membranes, and hydrodynamic conditions of the emulsification process. As membrane, some specially prepared metal films with regularly scattered pores, porous glasses, or polymers are used. In the second step, the elusion droplets are polymerized and fine polymer particles are obtained. Compared with conventional suspension polymerization, this method is more efficient and allows the obtaining of high-quality products. It is easily scalable and, by an appropriate selection of process parameters, allows preparation of materials with requested properties (Omi et al. 1997, Sugiura et al. 2002). A two-step process for MIP preparation was described by Wolska and Bryjak (2012). The authors imprinted the following EDC templates: dimethyl phthalate, diethyl phthalate, dibutyl phthalate, or BPA in the polymer matrix of S and DVB.

8.3.1.5 Precipitation Polymerization

The method can give monodisperse microspheric beads with high surface area. The process involves coagulation of the nanogelled beads and their growth due to capture of oligomers from the surrounding solution. The size of the polymeric beads and their porosity can be altered by selection of polymerization parameters. The precipitation polymerization is very effective and it does not need to use any surfactants; however, its main drawback is high consumption of toxic solvents.

Zhang et al. (2006) prepared MIPs for BPA removal by precipitation polymerization of 4-vinylpyridine and trimethylolpropane trimethacrylate (TRIM) in the solution of toluene and acetonitrile. Pierto et al. (2011) prepared MIPs with 17β-estradiol template from copolymers of MAA and p-vinylbenzoic acid with EGDMA as cross-linking agent and acetonitrile as solvent. The obtained particles were tested for the removal of estriol, norethisterone, estrone, BPA, and other EDCs. Ye et al. (2001) obtained MIP particles by precipitation polymerization of MAA and TRIM with theophylline or 17β-estradiol as templates. Yoshimatsu et al. (2007) obtained microspheres and nanospheres of copolymers MAA/DVB, MAA/TRIM, and MAA/DVB/TRIM with (R,S)-propranolol as the template.

8.3.1.6 Surface Imprinting Polymerization

This method involves grafting the thin layers of MIP onto preformed beads. Porous silica was used as the core structure mostly. This technique is very attractive, especially for larger molecules or microorganisms that are difficult to imprint into the polymeric matrix due to the problems with removal of such templates from materials (Sellergren 2001, Yan and Row 2006). Hua et al. (2009) obtained surface imprinting hydrogel on silica microspheres for selective recognition of bovine serum albumin. They carried out covalent immobilization of a water-soluble UV-sensitive initiator onto the surface of silica beads that initiated polymerization of N-[3-(dimethylamino)propyl]methacrylamide and N-isopropylacrylamide functional monomers. Zhu et al. (2011) obtained the surface MIP layers on the silica beads, with imprints of imidazole. Kitahara et al. (2010) synthesized surface-imprinted microspheres selective for cholesterol, when the grafted layers were formed from copolymer of S and DVB. Chang et al. (2010) applied reversible addition–fragmentation chain transfer (RAFT) polymerization protocol for imprinted layers of methacrylamide and DVB. The same authors prepared surface-imprinted magnetic particles for the removal of chlorinated phenols (Chang et al. 2012), using MAA and DVB monomers to form the surface layer. Lee and Kim (2009) grafted EGDMA and MAA on carbon nanotubes, with theophylline as the template.

8.4   MIP APPLICATIONS FOR EDC REMOVAL

There are a very limited number of contributions that describe the use of MIP materials for the removal of endocrine disruptors. Usually, this process is carried out by the sorption of EDCs on activated carbon or by their catalytic/photolytic oxidation (Delgado et al. 2012, Kruithof and Martijn 2013, Margot et al. 2013). Sometimes, active sludge is used to transform endocrine disruptors to less harmful components (Husain and Qayyun 2013, Ifelebuegu 2011). Some works were also conducted on nanofiltration or reverse osmosis processes (Breaken and van der Bruggen 2009, Yangali-Quitanilla et al. 2011). It seems that described in the following hybrid system, which employs sorption of EDCs on MIP particles with membrane separation of formed complexes, can be considered as an alternative for methods applied so far.

8.4.1   SORPTION-MEMBRANE FILTRATION HYBRIDS

The recent achievements in the materials science offer new types of selective sorbents in the form of organic or inorganic particles that are able to chelate species, adsorb them in the imprinted footprints, or coordinate with functional groups. It is expected that the particles of binding agents attract target compounds and, as larger species, are easily separated by membrane filtration. In the second step of the process, the separated complexes are split to the components and the binding agent can be recirculated to the sorption unit. This approach allows the removal of some species that can appear in the solution at very low quantities. The combination of affinity of the advanced binding agents with particle separation on membranes reveals many advantages comparing to the conventionally used fixed-bed systems. The main benefit of the complexation-enhanced hybrid process is the high separation efficiency and lower operational costs as compared with other classical separations. First, when the binding agent appears at large dispersion, the process can run in a semihomogenous medium. Second, some synergetic effects may intensify the process nearby the membrane surface when binding agent is concentrated (effect of concentration polarization or deposition on the surface) (Koltuniewicz et al. 2004).

The integrated system includes usually two separation elements: sorption and desorption units. In the first one, the target component is bound to the coupling agent, and as a large complex, it is retained by the membrane filtration. Small molecules are allowed to pass through the membrane pores. The complex is decomposed in the second unit by the action of splitting compounds. Coupling agent is stopped by the porous membrane and directed to the first unit, while freed targeted species are collected in permeate. The layout of the hybrid system is shown in Figure 8.1.

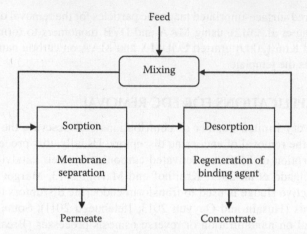

FIGURE 8.1 A scheme of a hybrid process. (From Kabay, N. and Bryjak, M.: Hybrid processes combining sorption and membrane filtration. In *Encyclopedia of Membrane Science and Technology*. E. Hoek and V.V. Tarabara, eds. pp. 1–21. 2013. Copyright Wiley-VCH Verlag GmbH & Co. KGaA. Reproduced with permission.)

The essence of the hybrid process is to complex target molecules with binding agent and to separate complexes by means of membrane filtration. The type of membrane employed depends on the size of separated complexes. Hence, the following hybrid processes can be specified (Kabay and Bryjak 2013):

- Small-organics-enhanced nanofiltration—binding agent less than 1 nm
- Polymer-enhanced ultrafiltration—binding agent several nanometers
- Micellar-enhanced ultrafiltration—binding agent several nanometers
- Colloid-enhanced ultrafiltration—binding agent less than micrometer
- Suspension-enhanced microfiltration—binding agent several micrometers

MIP spheres can be used for the last type of hybrid process. When the suspension-enhanced microfiltration process is designed, the sorption kinetics as well as dynamics of the hybrid process should be considered. For simplicity, it can be assumed that uptake of EDCs on the MIP particles is very fast and the bottlenecking step of the hybrid process is the rate of coupling agent replacement in the sorption loop. Hence, one can make mass balance for the hybrid system (see Figure 8.2).

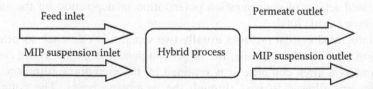

FIGURE 8.2 Fluxes layout in the hybrid systems. (From Kabay, N. and Bryjak, M.: Hybrid processes combining sorption and membrane filtration. In *Encyclopedia of Membrane Science and Technology*. E. Hoek and V.V. Tarabara, eds. pp. 1–21. 2013. Copyright Wiley-VCH Verlag GmbH & Co. KGaA. Reproduced with permission.)

Assume that the system is fed with two streams: the solution of the EDCs at a concentration of C_f and volume flux of J_f and sorbent suspension stream described by flux of J_s^i and X^i concentration of MIP particles that carry some amounts of EDCs (q^i). Two streams leave the hybrid system: permeate, with flux of J_p, and EDC content of C_p, with loaded suspension of MIP, with flux J_s^o, MIP concentration X^o, and load of q^o. Hence, one can write the following equation to balance EDC concentration:

$$J_f C_f + J_s^i X^i q^i = J_p C_p + J_s^o X^o q^o \qquad (8.1)$$

Simplifying Equation 8.1, one can express the concentration of EDCs in permeate as (Kabay and Bryjak 2013)

$$C_p = C_f - \frac{J_s^i}{J_f} X^i (q^o - q^i) \qquad (8.2)$$

Hence, to reduce the concentration of endocrine disruptor in permeate, it is recommended to increase the rate of MIP suspension replacement, to use higher concentration of MIP particles in the sorption loop, and to use as clean as possible MIP sorbent.

The case of BPA removal illustrates the previously presented relations. The sorbent was prepared by membrane emulsification followed by polymerization of methyl methacrylate and EGDMA suspension (Wolska and Bryjak 2014). It was found that the obtained sorbent was able to remove different amounts of BPA that was related to the concentration of MIP in the suspension mostly. For particular concentration of BPA in feed solution, the efficiency for its removal raised from 40% for 0.1 g/L MIP concentration to 54% for 0.5 g/L and to almost 80% for 2 g/L concentration of molecularly imprinted particles. That relation verifies Equation 8.2: the removal ratio gets higher when more sorbent flows through sorption loop.

Similar effects were noted for the removal of other water contaminants by sorption-membrane filtration system (Bryjak et al. 2009, Guler et al. 2011, Kabay et al. 2008). Hence, it can be easily predicted that the use of hybrid system for the removal of endocrine disruptors from aqueous solutions is worthy of evaluation and water obtained by that method can be free of EDC substances. However, the main obstacle for wider use of hybrid system in water reclamation technology is the worry of its price. The cost of the hybrid process is composed of the cost of sorbent preparation and the cost of sorbent regeneration (Al Abdulgader et al. 2013). When some, not too expensive, methods for preparation of fine MIP particles are evolving, but the methods of sorbent regeneration are still under study. Some preliminary works on that subject can be found in literature (Fernandez-Alvarez et al. 2009). However, the limited number of such data shows that the suspension-enhanced microfiltration hybrid system cannot be offered to commercialization now and is still at the beginning of its development. Still the integrated methods of membrane bioreactor followed by ultrafiltration and nanofiltration/reverse osmosis processes (Lee et al. 2008) are considered as the best methods for EDC removal.

8.5 CONCLUDING REMARKS

Among different methods for the removal of trace amounts of EDCs, the suspension-enhanced microfiltration seems to have great potential. The critical point of it is to establish technology for the production of fine and monodisperse molecularly imprinted particles. The developing method of membrane emulsification followed by suspension polymerization could fulfill such request. The production of hundreds of kilograms of particles per hour, technically plausible to reach now, should give the green light for wider use of the hybrid system. In the case of process unit, it is possible to adopt the submerge membrane bioreactor, the complete installation offered now for waste water treatment plant. In such unit, water polishing can be run in the large extent and trace amounts of EDCs removed effectively. It seems that the last element is to develop an effective method for sorbent regeneration. After that, the hybrid system should pass some pilot tests, and after getting the positive results, the suspension-enhanced micro-filtration method can be considered to be useful for water cleaning technology.

ABBREVIATIONS

BP	4-*tert*-buthylphenol
BPA	bisphenol A
BPA-d$_{16}$	bisphenol A-d$_{16}$
C_f	EDCs concentration in feed
C_p	EDCs concentration in permeate
DVB	divinylbenzene
EDCs	endocrine-disrupting compounds
EGDMA	ethylene glycol dimethacrylate
J_f	volume flux of feed
J_p	volume flux of permeate
J_s^i	inlet suspension volume flux
J_s^o	outlet suspension volume flux
MAA	methacrylic acid
MIP(s)	molecularly imprinted polymer(s)
MMA	methyl methacrylate
q^i	load of sorbent with EDCs at hybrid system inlet
q^o	load of sorbent with EDCs at hybrid system outlet
RAFT	reversible addition–fragmentation chain transfer
S	styrene
TRIM	trimethylolpropane trimethacrylate
X^i	particles concentration at hybrid system inlet
X^o	particles concentration at hybrid system outlet

REFERENCES

Al Abdulgader, H., V. Kochkodan, and N. Hilal. 2013. Hybrid ion exchange—Pressure driven membrane processes in water treatment: A review. *Sep. Purif. Technol.* 116: 253–264.

Breaken, L. and B. van der Bruggen. 2009. Feasibility of nanofiltration for the removal of endocrine disrupting compounds. *Desalination* 240: 127–131.

Bryjak, M., J. Wolska, I. Soroko, and N. Kabay. 2009. Adsorption–membrane filtration process in boron removal from first stage seawater RO permeate. *Desalination* 24: 127–132.

Byun, H.-S., D.-S. Yang, and S.-H. Cho. 2013. Synthesis and characterization of high selective molecularly imprinted polymers for bisphenol A and 2,4-dichlorophenoxyacetic acid by using supercritical fluid technology. *Polymer* 54: 589–595.

Chahoud, I., A. Gies, M. Paul, G. Schonfelder, and C. Talsness. 2001. Bisphenol A: Low dose effects—High dose effects. *Reprod. Toxicol.* 15: 587–599.

Chang, L., S. Chen, and X. Li. 2012. Synthesis and properties of core-shell magnetic molecular imprinted polymers. *Appl. Surf. Sci.* 258: 6660–6664.

Chang, L., Y. Li, J. Chu, J. Qi, and X. Li. 2010. Preparation of core-shell molecularly imprinted polymer via the combination of reversible addition-fragmentation chain transfer polymerization and click reaction. *Anal. Chim. Acta* 680: 65–71.

Coughlin, J.L., B. Winnik, and B. Buckley. 2011. Measurement of bisphenol A, bisphenol A β-D-glucuronide, genistein, and genistein 4'-β-D-glucuronide via SPE and HPLC-MS/MS. *Anal. Bioanal. Chem.* 401: 995–1002.

Delgado, L.F., P. Charles, K. Glucina, and C. Morlay. 2012. The removal of endocrine disrupting compounds, pharmaceutically activated compounds and cyanobacterial toxins during drinking water preparation using activated carbon—A review. *Sci. Total Environ.* 435–436: 509–525.

Dudziak, M. and M. Bodzek. 2009. Evaluation of xenoestrogens contents in water by extraction method (in Polish). *Ochrona Środowiska* 31: 9–14.

Fernandez-Alvarez, P., M. Le Noir, and B. Guieysse. 2009. Removal and destruction of endocrine disrupting contaminants by adsorption with molecularly imprinted polymers followed by simultaneous extraction and phototreatment. *J. Hazard. Mater.* 163: 1107–1112.

Guler, E., N. Kabay, M. Yuksel, N.O. Yigit, M. Kitis, and M. Bryjak. 2011. Integrated solution for boron removal from seawater using RO process and sorption-membrane filtration hybrid method. *J. Membr. Sci.* 375: 249–257.

Haginaka, J. and H. Sanbe. 2001. Uniformly sized molecularly imprinted polymer for (S)-naproxen retention and molecular recognition properties in aqueous mobile phase. *J. Chromatogr. A* 913: 141–146.

Haginaka, J., H. Sanbe, and H. Takehira. 1999b. Uniform-sized molecularly imprinted polymer for (S)-ibuprofen retention properties in aqueous mobile phase. *J. Chromatogr. A* 857: 117–125.

Haginaka, J., H. Takehira, K. Hosoya, and N. Tanaka. 1999a. Uniform-sized molecularly imprinted polymer for (S)-naproxen selectively modified with hydrophilic external layer. *J. Chromatogr. A* 849: 331–339.

Higuchi, A., B.-O. Yoon, T. Asano, K. Nakaegawa, S. Miki, M. Hara, Z. He, and I. Pinnau. 2002. Separation of endocrine disruptors from aqueous solutions by pervaporation. *J. Membr. Sci.* 198: 311–320.

Hosoya, K., K. Yoshizako, Y. Shirasu, K. Kimata, T. Araki, N. Tanaka, and J. Haginaka. 1996. Molecularly imprinted uniform-size polymer-based stationary phase for high-performance liquid chromatography structural contribution of cross-linked polymer network on specific molecular recognition. *J. Chromatogr. A* 728: 139–147.

Hua, Z., S. Zhou, and M. Zhao. 2009. Fabrication of a surface imprinted hydrogel shell over silica microspheres using bovine serum albumin as a model protein template. *Biosens. Bioelectron.* 25: 615–622.

Husain, Q. and S. Qayyun. 2013. Biological and enzymatic treatment of bisphenol A and other endocrine disrupting compounds: A review. *Crit. Rev. Biotechnol.* 33: 260–292.

Ifelebuegu, A.O. 2011. The fate and behavior of selected endocrine disrupting chemicals in full scale wastewater and sludge treatment unit processes. *Int. J. Environ. Sci. Technol.* 8: 245–254.

Joshi, V.P., R.N. Karmalkar, M.G. Kulkarni, and R.A. Mashelkar. 1999. Effect of solvents on selectivity in separation using molecularly imprinted adsorbents: Separation of phenol and bisphenol A. *Ind. Eng. Chem. Res.* 38: 4417–4423.

Kabay, N. and M. Bryjak. 2013. Hybrid processes combining sorption and membrane filtration. In *Encyclopedia of Membrane Science and Technology*, E. Hoek and V.V. Tarabara (eds.). New York: John Wiley, pp. 1–21.

Kabay, N., M. Bryjak, S. Schlosser, M. Kitis, S. Avlonitis, Z. Matejka, I. Al-Mutaz, and M. Yuksel. 2008. Adsorption–membrane filtration (AMF) hybrid process for boron removal from seawater: An overview. *Desalination* 223: 38–48.

Kawaguchi, M., Y. Hayatsu, H. Nakata, Y. Ishii, R. Ito, K. Saito, and H. Nakazawa. 2005. Molecularly imprinted solid phase extraction using stable isotope labeled compounds as template and liquid chromatography–mass spectrometry for trace analysis of bisphenol A in water sample. *Anal. Chim. Acta* 539: 83–89.

Kempe, H. and M. Kempe. 2004. Novel method for the synthesis of molecularly imprinted polymer bead libraries. *Macromol. Rapid Commun.* 25: 315–320.

Kempe, H. and M. Kempe. 2006. Development and evaluation of spherical molecularly imprinted polymer beads. *Anal. Chem.* 78: 3659–3666.

Kitahara, K.-I., I. Yoshihama, T. Hanada, H. Kokuba, and S. Arai. 2010. Synthesis of monodispersed molecularly imprinted polymer particles for high-performance liquid chromatographic separation of cholesterol using templating polymerization in porous silica gel bound with cholesterol molecules on its surface. *J. Chromatogr. A* 1217: 7249–7254.

Koltuniewicz, A., A. Witek, and K. Bezak. 2004. Efficiency of membrane-sorption integrated processes. *J. Membr. Sci.* 239: 129–141.

Kruithof, J.C. and B.J. Martijn. 2013. UV/H$_2$O$_2$ treatment: An essential process in a multi barrier approach against trace chemical contaminants. *Water Sci. Technol.* 13: 130–138.

Lee, H.-Y. and B.S. Kim. 2009. Grafting of molecularly imprinted polymers on iniferter-modified carbon nanotube. *Biosens. Bioelectron.* 25: 587–591.

Lee, J., B.C. Lee, J.S. Ra, J. Cho, I.S. Kim, N.I. Chang, H.K. Kim, and S.D. Kim. 2008. Comparison of the removal efficiency of endocrine disrupting compounds in pilot scale sewage treatment processes. *Chemosphere* 71: 1582–1592.

Margot, J., C. Kienle, A. Magne, M. Weil, L. Rossi, L.F. de Alemcastro, C. Abegglen et al. 2013. Treatment of micropollutants in municipal wastewater: Ozone or powdered activated carbon? *Sci. Total Environ.* 461–462: 480–498.

Mayes, A.G. and K. Mosbach. 1996. Molecularly imprinted polymer beads: Suspension polymerization using a liquid perfluorocarbon as the dispersing phase. *Anal. Chem.* 68: 3769–3774.

Omi, S., T. Taguchi, M. Nagai, and G.-H. Ma. 1997. Synthesis of 100 μm uniform porous spheres by SPG emulsification with subsequent swelling of the droplets. *J. Appl. Polym. Sci.* 63: 931–942.

Prieto, A., A. Vallejo, O. Zuloaga, A. Paschke, B. Sellergen, E. Schillinger, S. Schrader, and M. Möder. 2011. Selective determination of estrogenic compounds in water by microextraction by packed sorbents and a molecularly imprinted polymer coupled with large volume injection-in-port-derivatization gas chromatography–mass spectrometry. *Anal. Chim. Acta* 703: 41–51.

Sanbe, H. and J. Haginaka. 2002. Uniformly sized molecularly imprinted polymers for bisphenol A and β-estradiol: Retention and molecular recognition properties in hydro-organic mobile phases. *J. Pharm. Biomed.* 30: 1835–1844.

Sellergren, B. 2001. *Molecularly Imprinted Polymers*, Vol. 23: *Man-Made Mimics of Antibodies and Their Applications in Analytical Chemistry*. Amsterdam, the Netherlands: Elsevier.

Shin, H.-S., Ch.-H. Park, S.-J. Park, and H. Pyo. 2001. Sensitive determination of bisphenol A in environmental water by gas chromatography with nitrogen–phosphorus detection after cyanomethylation. *J. Chromatogr. A* 912: 119–125.

Staples, Ch.A., T.F. Parkerton, and D.R. Peterson. 2000. A risk assessment of selected phthalate esters in North American and Western European surface waters. *Chemosphere* 40: 885–891.

Sugiura, S., M. Nakajima, and M. Seki. 2002. Preparation of monodispersed polymeric microspheres over 50 μm employing microchannel emulsification. *Ind. Eng. Chem. Res.* 41: 4043–4047.

Walsh, R., Q. Osmani, H. Hughes, P. Duggan, and P. McLoughlin. 2011. Synthesis of imprinted beads by aqueous suspension polymerization for chiral recognition of antihistamines. *J. Chromatogr. B* 879: 3523–3530.

Wang, X., L. Chen, X. Xu, and Y. Li. 2011. Synthesis of molecularly imprinted polymers via ring-opening metathesis polymerization for solid-phase extraction of bisphenol A. *Anal. Bioanal. Chem.* 401: 1423–1432.

Witorsch, R.J. 2002. Endocrine disruptors: Can biological effects and environmental risks be predicted? *Reg. Toxicol. Pharmacol.* 36: 118–130.

Wolska, J. and M. Bryjak. 2012. Sorption of phthalates on molecularly imprinted polymers. *Sep. Sci. Technol.* 47: 1316–1321.

Wolska, J. and M. Bryjak. 2014. Removal of bisphenol A by means of hybrid membrane-sorption process. *Sep. Sci. Technol.*, in press.

Yan, H. and K.H. Row. 2006. Characteristic and synthetic approach of molecularly imprinted polymer. *Int. J. Mol. Sci.* 7: 155–178.

Yangali-Quitanilla, V., S.K. Maeng, T. Fujioka, M. Kennedy, Z. Li, and G. Amya. 2011. Nanofiltration vs. reverse osmosis for the removal of emerging organic contaminants in water reuse. *Des. Water Treat.* 34: 50–56.

Ye, L., P.A.G. Cormack, and K. Mosbach. 2001. Molecular imprinting on microgel spheres. *Anal. Chim. Acta* 435: 187–196.

Yoshimatsu, K., K. Reimhult, A. Krozer, K. Mosbach, K. Sode, and L. Ye. 2007. Uniform molecularly imprinted microspheres and nanoparticles prepared by precipitation polymerization: The control of particle size suitable for different analytical applications. *Anal. Chim. Acta* 584: 112–121.

Zhang, J.-H., M. Jiang, L. Zou, D. Shi, S.-R. Mei, Y.-X. Zhu, Y. Shi, K. Dai, and B. Lu. 2006. Selective solid-phase extraction of bisphenol A using molecularly imprinted polymers and its application to biological and environmental samples. *Anal. Bioanal. Chem.* 385: 780–786.

Zhao, Ch., Q. Wei, K. Yang, X. Liu, M. Nomizu, and N. Nishi. 2004. Preparation of porous polysulfone beads for selective removal of endocrine disruptors. *Sep. Purif. Technol.* 40: 297–302.

Zhu, G., J. Fan, Y. Gao, X. Gao, and J. Wang. 2011. Synthesis of surface molecularly imprinted polymer and the selective solid phase extraction of imidazole from its structural analogues. *Talanta* 84: 1124–1132.

9 Biopolymer-Based Sorbents for Metal Sorption

Eric Guibal, Thierry Vincent, and Ricardo Navarro

CONTENTS

9.1 INTRODUCTION

The regulations concerning the levels of concentration of metals in drinking water are becoming more and more stringent. As a consequence, the discharge levels for industrial wastewater are strictly controlled. Conventional processes such as precipitation, solvent extraction, membrane processes, and ion-exchange and chelating resins can be inappropriate for technical limitations (difficulty to reach discharge levels), environmental reasons (production of toxic sludge), or economic criteria. Biosorption has retained a great attention for the last decades as an alternative to conventional processes. This technique consists in using materials of biological origin for the sorption of metal ions, dyes, etc. The concept is based on the use of functional groups present at the surface of these materials for binding metal ions through adsorption, ion exchange, chelation, reductive precipitation, etc. A wide diversity of materials have been tested including agriculture subproducts, waste materials from food industry, and valorization of wastes from water compartments (e.g., invading algal biomass). Algal and fungal materials represent emblematic examples of biomass that were tested for metal sorption (Gonzalez Bermudez et al. 2012, Guibal et al. 1992, 1995, Kleinuebing et al. 2010, 2011, Svecova et al. 2006, Yipmantin et al. 2011). For example, algae are mainly constituted of cellulosic compounds, alginate, fucoidan biopolymers (Davis et al. 2003), and diatomaceous materials. Fungal material is characterized by the presence of proteins, carbohydrates, and more specifically chitin-based materials as main constituents of cell wall. However, the main commercial source of chitin-based products comes from the valorization of crustacean shells. The complexity of the structure of these raw materials, the possible interactions between their different components, and the possible variation in their composition (e.g., depending on the extraction process, on the growing conditions) may explain, in some cases, their limited efficiency, the variability in their sorption properties, and also the difficulty to elaborate some advanced materials based on these resources. This may also explain that a great attention has been paid to the extraction of their main active components (i.e., alginate for algae and chitin/chitosan for fungi and crustacean shells). Among the great diversity of biopolymers available in nature, the present work focuses on these two emblematic polysaccharides because of their wide availability and because they are complementary in terms of functional groups (i.e., carboxylic acid groups for alginate, basic amine groups for chitosan), stability range (acidic solutions for alginate, neutral/alkaline solutions for chitosan), and application (affinity for different metal ions due to the difference in their functional groups, with regard to hard and soft acid–base theory). Though these biopolymers can be readily modified by chemical grafting of specific reactive groups to improve their reactivity and their selectivity (Guibal et al. 2000b), the present study is limited to the use of raw materials (with the exception of chitosan grafting that is required for using the biopolymer in acidic solutions) (Alves and Mano 2008, Guibal et al. 2000b, 2002, Yang et al. 2011). These materials can be also used for the encapsulation and immobilization of liquid or solid compounds for preparing new sorbents (Guibal et al. 2009, Krys et al. 2013).

After describing the properties of these biopolymers, their ability to be conditioned under different forms will be discussed. The second part deals with sorption mechanism considering the different mechanisms involved in metal binding, the parameters that control sorption performance, the possibility to recycle the sorbent, and the different modes of application. The third section focuses on the encapsulation properties of these materials for immobilizing liquid compounds (ionic liquids, extractants) or solid compounds (mineral sorbents and ion exchangers). Finally, some examples are given regarding the possible use of composite materials (i.e., biopolymers combined with metal ions) for the synthesis of advanced materials. The aim of this chapter is not to draw an exhaustive compilation but to illustrate the wide diversity of possibilities opened by these materials of biological origin.

9.2 ALGINATE AND CHITOSAN: STRUCTURE AND PROPERTIES

9.2.1 ALGINATE

Alginate (or alginic acid) is an anionic polysaccharide, which is present in large quantities in the cell wall of brown algae. It is extracted from algal biomass by a three-step procedure consisting of (1) successive contact with formaldehyde solution and hydrochloric acid solution (to remove phenolic compounds), (2) treatment with sodium carbonate (to solubilize sodium alginate), and (3) reprecipitation of alginate in ethanol (Hernandez-Carmona et al. 1998, 1999, McHugh et al. 2001). Actually, alginic acid is a linear copolymer constituted of two blocks of 1,4-linked hexuronic acid residues: mannuronic acid residues (β (1 $\rightarrow$ 4)-linked-D-mannuronopyranosyl, M) and guluronic acid residues (α-L-guluronopyranosyl, G) (Yang et al. 2011). These residues are assembled as blocks of repeating G units (GG blocks), repeating M units (MM blocks), and blocks of mixed M and G residues (MG blocks) of different distributions (Figure 9.1). The properties of the biopolymer can thus be controlled by the

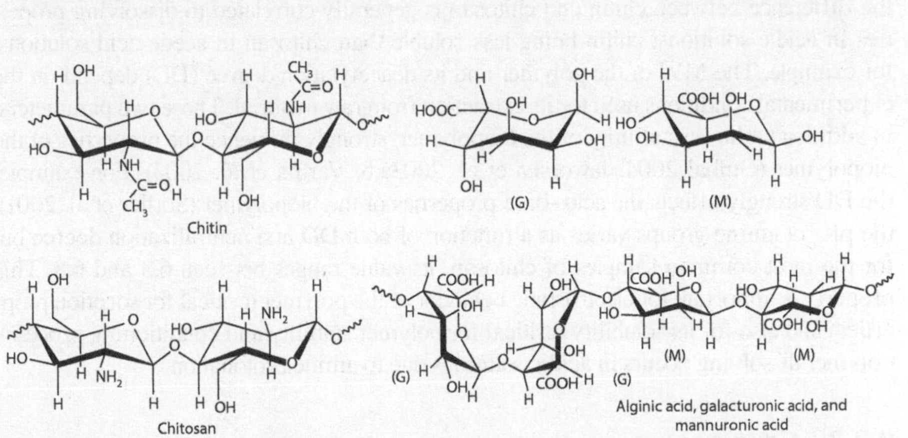

FIGURE 9.1 Structure of chitin, chitosan, and alginate (mannuronic [M] and guluronic [G] acid moieties).

distribution and the percentage of G and M residues. The molecular weight (MW) is another important parameter for characterizing the biopolymer.

The presence of carboxylic groups, as diuronate units, and their arrangement in the biopolymer chains may explain the strong interaction that may occur between alginate and metal cations and more specifically divalent cations (Agulhon et al. 2012a, Andersen et al. 2012). This is the basis of the affinity of the biopolymer for metal ions. This mechanism is frequently called "the egg-box" model (Christensen et al. 1990, Davis et al. 2003), though depending on the composition of alginate and the conditions selected for interacting with metal ions, other mechanisms could be involved (Agulhon et al. 2012b, Li et al. 2007, Sikorski et al. 2007). These interactions are responsible for the ionotropic gelation properties of alginate that can be used for preparing different stable conditionings of alginate under the form of beads, membranes, and scaffolds. Alginate is also gelling in HCl solutions. The acid–base properties of the biopolymer are conditioned by the presence of carboxylic groups, their G and M residues having pK_a of 3.65 and 3.38, respectively (Davis et al. 2003).

9.2.2 CHITOSAN

Chitin (2-acetamido-2-deoxy-β-D-glucose units bound by β(1 → 4) linkages) is the precursor of chitosan (partially or totally deacetylated form of chitin, Figure 9.1) (Roberts 1992). This biopolymer is present in the cell wall of fungi, the cuticle of insects, and the shell of crustacean. It is one of the most abundant polysaccharides in nature. It is commercially obtained from crustacean shells after grinding, acidic treatment (to remove carbonate-based materials), and alkaline treatment at ambient temperature (for protein and dye removal). The deacetylation of chitin proceeds with alkaline treatment at boiling temperature. Actually, it is quite difficult to elaborate a pure chitosan (i.e., totally deacetylated), and it results in the strong depolymerization of the biopolymer. Chitin and chitosan are thus, in most cases, heteropolymers with different levels of acetylation, and the difference between chitin and chitosan is generally correlated to dissolving properties in acidic solutions: chitin being less soluble than chitosan in acetic acid solutions, for example. The MW of the polymer and its deacetylation degree (DD) depend on the experimental conditions used for its extraction from raw material. These two parameters, in addition to the crystallinity of the biopolymer, strongly influence the properties of the biopolymer (Guibal 2004, Jaworska et al. 2003a,b, Varma et al. 2004). For example, the DD strongly affects the acid–base properties of the biopolymer (Sorlier et al. 2001): the pK_a of amine groups varies as a function of both DD and neutralization degree but for the most common samples of chitosan, its value ranges between 6.3 and 6.8. This property is important for electrostatic behavior of the polymer (critical for sorption properties) and also for its solubility (critical for polymer stability and conditioning process): polymer dissolving occurs in acidic solutions due to amine protonation.

9.2.3 CONDITIONING

One of the advantages of these materials (apart from their intrinsic sorption properties that will be explored in the next section) is the wide range of different conditionings that can be elaborated (Figure 9.2). The physical modification consists

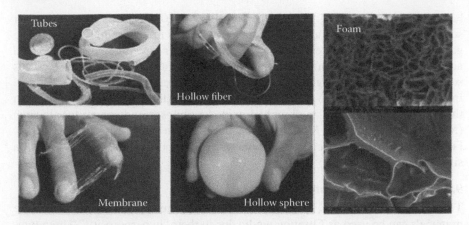

FIGURE 9.2 Examples of conditionings of chitosan.

in the dissolving of the biopolymer in a suitable solvent (water for alginate, acidic solutions for chitosan) followed by a shaping step and the gelling or neutralization step. Depending on the shape to elaborate, the sequence of treatment steps may vary.

9.2.3.1 Beads

The biopolymer solution is distributed through a thin nozzle into the neutralizing or gelling solution. Depending on the viscosity of the solution, the height of the nozzle must be adapted to produce spherical drops before they fall into the gelling/neutralization bath: the relaxation of the viscous solution influences the time necessary for restoring the spherical shape. Alternatively, some specific equipment can be used to produce well-calibrated (homogeneous in size) and well-dispersed (using electromagnetic charging) beads: Encapsulator B-390 (Büchi, Switzerland), for example. The coagulating bath may be, for example, sodium hydroxide or hexametaphosphate for chitosan (Sicupira et al. 2010, Wu et al. 2013) and hydrochloric acid or metal cations for alginate (Agulhon et al. 2012a,b, Rodrigues and Lagoa 2006).

9.2.3.2 Membranes

The synthesis of membranes may basically follow two procedures. The casting procedure consists of the spreading of the viscous biopolymer solution in a plan holder support followed by the solvent evaporation (Modrzejewska and Kaminski 1999). The evaporation can be complete or partial (in this case, it is necessary to reach a sufficient evaporation, greater than 50%, to produce a self-supported membrane that can be manipulated) (Rabelo et al. 2012). The membrane is then neutralized (or treated with phosphate compounds, see Section 9.2.3.1, for chitosan), or gelled (with hydrochloric acid or metal cations). The second process consists of the direct extrusion of the viscous solution into the neutralization/gelling bath through a tailored extrusion module.

9.2.3.3 Fibers, Hollow Fibers, and Tubes

Fibers, hollow fibers, and tubes have been prepared using the wet spinning method (Cuadros et al. 2012, Dresvyanina et al. 2013, Modrzejewska and Eckstein 2004, Vikhoreva 2012, Vincent and Guibal 2001). Dry- and gel-spinning processes have been also developed (Pillai et al. 2009, Tasselli et al. 2013): the viscous biopolymer solution (in acetic acid or acetic acid–alcohol solutions for chitosan) was spun using gaseous ammonia (Notin et al. 2006a,b) or in a dry air flow to remove alcohol and acetic acid solvent system) (Desorme et al. 2013). For hollow fibers, the viscous solution is extruded through a thin coaxial tube spinneret; a core liquid is pumped in the inner nozzle and the combined extruded hollow fiber is finally neutralized in the neutralizing bath (e.g., sodium hydroxide for chitosan and copper for alginate) (Mirmohseni et al. 2012, Modrzejewska and Eckstein 2004, Tasselli et al. 2013). For simple fibers, the viscous biopolymer solution is directly extruded through a simple thin nozzle.

9.2.3.4 Foams and Sponges

The synthesis of scaffolds, sponges, or foams is a promising alternative since these materials can be used as filtration media due to their "hyperporosity." Three typical procedures were developed for the synthesis of these materials based on the sequence of operation or temperature processing: (1) gelling followed by the conditioning/shaping (Hortiguela et al. 2011), (2) the conditioning/shaping followed by the gelling or coagulation/neutralization (Guibal et al. 2013, Li et al. 2011, Saravanan et al. 2011), or (3) the *in situ* generation of porosity during hydrogel formation (Andersen et al. 2012). For example, Hortiguela et al. (2011) reported the gelling of chitosan with $HAuCl_4$ through simultaneous formation of gold nanoparticles followed by the shaping of the hydrogels through a freezing step followed by a freeze-drying step. This process allows the shaping and the incorporation of metal ions in the same process. On the opposite hand, most of the processes that were developed consist of the synthesis of the biopolymer scaffold followed by metal binding (and possible *in situ* reduction). In most cases, the biopolymer solutions are frozen in a suitable mold (box, syringe, annular mold, etc.) before being gelled or freeze-dried: the temperature, the temperature ramp, and the biopolymer concentration are important criteria for setting the physical (porous microstructure: pore size and homogeneity of distribution of pores) and mechanical properties of the foams/sponges (Madihally and Matthew 1999). Andersen et al. (2012) reported the incorporation of (a) air bubbles by strong agitation and/or (b) bubbling agents (calcium carbonate associated with a slowly acidifying agent: glucono-δ-lactone) before the hydrogels were air-dried.

9.2.4 Post-Treatments

Chitosan being soluble in most acidic solutions (with the remarkable exception of sulfuric acid), the material can thus be poorly stable for further application, for metal binding in slightly acidic solutions, etc. It is thus necessary to improve the chemical resistance of these materials by cross-linking with chemical agents (such as glutaraldehyde and epichlorhydrin in the case of chitosan) (Kolodynska 2012, Park et al. 2013, Rabelo et al. 2012, Ruiz et al. 2001, Webster et al. 2007) or sulfate anions (Peirano Blondet et al. 2008) or by reacetylation (using acetic anhydride in the case of chitosan, to be converted to chitin) (Lavertu et al. 2012, Portero et al. 2002, Sorlier et al. 2001, Vincent and Guibal 2001).

9.3 ALGINATE AND CHITOSAN AS SORBENTS

The mechanisms occurring in metal biosorption are similar to those involved in metal binding on synthetic resins, based on the specific reactivity of chemical groups hold on these biopolymers: carboxylic groups for alginate and amine groups for chitosan. Their performances are also driven by the same operating parameters.

9.3.1 MECHANISMS

The mechanisms involved in metal uptake depend on the metal, its speciation (and the composition of the solution), the pH, and the functional groups of the biosorbent. Basically, metal binding operates through (1) ion-exchange/electrostatic attraction mechanisms, (2) complexation/chelation, or (3) binding of ternary compounds (though much less frequent than other mechanisms and concerning very specific systems). Additional mechanisms may partly contribute to remove metal ions from solutions such as *in situ* precipitation and/or reduction.

In the case of alginate, it is generally agreed that metal sorption occurs through the ion exchange of protons (from alginic acid) or sodium/calcium cations (from alginate) with the divalent metal cations to be bound (Chen et al. 2002, Deze et al. 2012, Jodra and Mijangos 2001, Karagunduz and Unal 2006, Papageorgiou et al. 2006, Pielesz and Bak 2008). The bond between carboxylic groups and Ca(II) is much stronger than that with Na(I) leading to easier ion exchange with Na(I) (Khotimchenko et al. 2008); however, sodium alginate being soluble in water is generally preferred using calcium alginate gels (especially when prepared under the form of alginate hydrogel beads). Comparing Pb(II) and Cu(II) binding on calcium alginate resins, Chen et al. (2002) reported some differences in the binding mechanisms for Cu(II) (combination of ion exchange and coordinative complexation) and for Pb(II) (pure ion exchange) based on XPS and FTIR analyses. Similar ion-exchange mechanisms were identified by Raman spectroscopy and wide-angle x-ray spectroscopy (WAXS) in the case of dye sorption by alginate hydrogels (Pielesz and Bak 2008). These interactions can be also associated with the gelling mechanisms occurring between carboxylic groups of alginate and metal ions through the so-called egg-box model (divalent metal cations interact with vicinal carboxylic groups from different chains to form compact gel structures). Recently, Agulhon et al. (2012a,b) utilized both small-angle x-ray spectroscopy (SAXS) and quantum chemical density functional theory (DFT methods) for clarifying the kind of interactions existing between different metal ions and mannuronic/guluronic acid units of alginate. They conclude that with transition metals, the complexation with the disaccharides proceeds through strong coordination covalent bonds, while alkaline-earth cations form ionic bonds with uronate groups. While the alginate affinity for transition metals is correlated to the strength of the chemical interaction with uronate, in the case of alkaline-earth metal cations, the binding energy is not a critical parameter. However, the interaction mode of alginate-based materials with metal ions remains debatable (Plazinski 2013).

In the case of chitosan-based sorbents, the reactivity of functional groups may completely change with the metal and the pH of the solution: the amine groups may have different modes of interactions with metal ions (Guibal 2004). The free electron doublet on nitrogen contributes to the binding of metal cation by complexation/chelation under near-neutral pH conditions. On the opposite hand, the acid–base properties of amine groups (weak base) may cause metal anion binding through ion exchange (with counteranions associated with protonated amine groups) and electrostatic attraction (to protonated amine groups) in acidic solutions. Metal binding on amine groups (and on chitosan derivatives; Varma et al. 2004) has been widely described and interpreted using different spectroscopic methods (Webster et al. 2007) and modeled using complexation concepts (Jeon and Holl 2004). The acid–base properties of amine groups of chitosan (pK_a in the range 6.3–6.8; Sorlier et al. 2001) may explain the strong competition of protons for the binding of metal cations in acidic solutions (Dzul Erosa et al. 2001) but also the possibility to bind metal anions by ion-exchange/electrostatic attraction mechanisms. Thus, metals that form anionic-hydrolyzed polynuclear species, such as molybdate (Guibal et al. 1999c, 2000a, Milot et al. 1998) or vanadate (Guzman et al. 2002) or metal chloro-anionic complexes in HCl solutions such as $PdCl_6^{2-}$, $PtCl_4^{2-}$, or $AuCl_4^{2-}$ (Arrascue et al. 2003, Guibal et al. 1999b, 2001, Ruiz et al. 2000) are bound on protonated amine groups in acidic solutions. A third mechanism of interaction of chitosan with metal ions has been identified. It consists of the formation of a ternary complex between the biopolymer, the metal (herein strontium), and a complexing agent (i.e., carbonate) (Piron and Domard 1998a). Under appropriate conditions (i.e., pH higher than 11 and carbonate concentration of 0.01 M) chitosan can bind the ion pair (Sr^{2+}, CO_3^{2-}) as a ternary complex. In the case of interactions of chitosan with polynuclear metal anions, ionotropic gelation mechanisms have been also identified (similar to those observed with alginate) (Dambies et al. 2001). These ionic interactions can be used for the conditioning and shaping of biopolymers and for the elaboration of new materials (similar to chitosan, Figure 9.2).

9.3.2 Parameters Influencing Sorption Performance

9.3.2.1 pH and Metal Speciation

The pH is a critical parameter for the binding of metal ions with alginate and chitosan. This parameter may control the interaction of the biopolymer with metal ions through different mechanisms: (1) competition of protons, (2) impact on chemical stability and reactivity of the sorbent, and (3) metal speciation. For example, chitosan (in its raw form) is soluble in acidic solutions by protonation of amine groups, while alginate is soluble in alkaline media (ion exchange of calcium, or other divalent metals, with sodium or potassium to give soluble forms of alginate). The stability (against dissolving) of the biopolymer is important for designing the optimum sorbent (selection of the acid to be used for pH control, cross-linking treatment, etc.). The acid–base properties of these biopolymers control the protonation of reactive groups, which, in turn, may influence the efficiency of the sorbent (competition effect, reactivity of functional groups) and up to the mechanisms involved in metal binding. The effect of pH on metal sorption on chitosan has been previously discussed in the section describing sorption mechanisms (see Section 9.3.1). Carboxylic groups on mannuronic and guluronic acid units of alginate have pK_a values in the range 3.4–3.7: the biopolymer will have different abilities to bind metal cations depending on the predominance of carboxylic acid (acid solutions) or carboxylate (weak acid/neutral solutions) groups. The increase in the pH improves metal-binding efficacy (for metal cations) (Plazinski 2012, 2013).

The speciation of metal ions is a critical parameter that is controlled by (1) the pH of the solution, (2) the concentration of the metal, and (3) the presence of ligands. Depending on the metal form, the affinity of the metal for the sorbent may change as well as the sorption mechanism. For example, the formation of polynuclear hydrolyzed species of uranium revealed favorable for metal binding on chitosan (Guibal et al. 1994). Similar conclusions were raised for molybdate and vanadate sorption on chitosan-based materials (Guibal et al. 1999c, 2000a, Guzman et al. 2002, Jansson-Charrier et al. 1996a): the sorption of target metal depends on the formation of polynuclear hydrolyzed species, which, in turn, was controlled by the pH of the solution and the concentration of the metal. Hence, for a given pH, the metal sorption began when the total concentration of the metal was sufficient (at this pH value) for forming polynuclear hydrolyzed species (Guibal et al. 2000a, Guzman et al. 2002) (Figure 9.3).

The presence of ligands may also interfere in the sorption of metal ions depending on the speciation of the metal (Guibal et al. 2000b, Guzman et al. 2003, Hernandez et al. 2007, Juang and Ju 1997, Juang et al. 1999, Nishad et al. 2012, Padala et al. 2011, Tseng et al. 1999, Wu et al. 1999). The concentration of chloride ions may displace the speciation of platinum or palladium to the formation of chloro-anionic species that can bind to protonated amine groups (Guibal et al. 2000b). To a certain extent, the presence of chloride ions may thus improve sorption performance, while an excess of chloride ions induces a strong competitor effect in the case of ion-exchange/electrostatic attraction mechanism. The presence of nitrilotriacetic acid changes the selectivity of chitosan for Co(II) and Cu(II), depending on the pH (Padala et al. 2011). While Cu(II) is bound to chitosan through chelation of amine

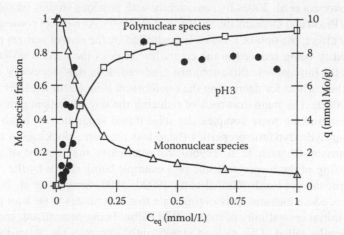

FIGURE 9.3 Example of correlation between the speciation of the metal ions (concentration of polynuclear species, □, and mononuclear species, Δ) and molybdate sorption capacity (sorption isotherms points: bold circles) using chitosan at pH 3. (From Guibal, E. et al., *Sep. Sci. Technol.*, 35, 1021, 2000a. With permission.)

groups under near-neutral pH conditions, the presence of citrate anions leads to the formation of anionic citrate–Cu anionic species that can bind to chitosan in acidic solutions: there is a change in the optimum pH range and in the binding mechanism (Guzman et al. 2003). Actually, a correlation was found between the predominance of copper–citrate complex anions over free anionic citrate species and the beginning of copper sorption by electrostatic attraction. In addition, the speciation diagram of Cu(II) in the presence of citrate was also correlated to the curve showing the maximum sorption capacity at different pHs.

9.3.2.2 Intrinsic Polymer Characteristics: Crystallinity and Structure

The crystallinity of materials may affect their hydration, which, in turn, may impact the accessibility of water and target metal ions to reactive groups in the core of sorbent particles (Piron and Domard 1998b, Piron et al. 1997). The sorption of uranyl ions on chitosan proceeds by coordination with amino groups (Piron and Domard 1997, 1998b), the stoichiometric ratio being close to two $-NH_2$ units present on the amorphous domain of the biopolymer with one uranyl ion. The impact of crystallinity properties of chitosan on molybdate and hexachloroplatinate sorption has been abundantly discussed (Guibal et al. 1999a, Jaworska et al. 2003a,b, Milot et al. 1998). It was found that the residual crystallinity of the biopolymer (and the origin of the material: squid vs. fungal or shrimp origin) strongly influences both the accessibility and the availability of reactive groups (protonated amine groups) for metal binding. For example, chitosan issued from β-chitin (squid chitosan) revealed much less efficient for Pt(IV) sorption than the material produced from α-chitin (shrimp, fungal biomass) (Jaworska et al. 2003a,b). X-ray diffraction patterns can be used for differentiating the different materials and the presence of specific peaks (shoulder at $2\theta \approx 22°$) can be correlated to worst sorption performance (equilibrium and kinetics) (Milot et al. 1998). A high crystallinity index can be correlated to less efficient sorption (Jaworska et al. 2003a,b), consistently with previous studies on sorption of uranyl ions (Piron and Domard 1998b, Piron et al. 1997). As a direct consequence of crystallinity effect, the uptake kinetics is controlled by the size of sorbent particles, the accessibility being reduced in the crystalline regions the mass transfer properties (especially intraparticle diffusion) are hindered. It is thus necessary reducing the size of the sorbent for decreasing the equilibrium time (Dzul Erosa et al. 2001, Ruiz et al. 2000). The main drawback of reducing the size of sorbent particles consists of either making more complex the solid/liquid separation in batch systems or decreasing hydrodynamic properties (head loss pressure, blocking) in fixed-bed columns (Jansson-Charrier et al. 1996b). An alternative may consist of changing the conditioning of the biopolymer, the best example being shown by the manufacturing of chitosan gel beads (Guibal et al. 1999d, 1998, Sicupira et al. 2010). The synthesis procedure includes a dissolving state that contributes to (at least partially) destroy the initial crystallinity of the material before being neutralized, coagulated, or ionotropically gelled. This method significantly improves the accessibility and availability of reactive groups by expansion of the structure of the material and decreases the crystallinity. However, at the end of the process, it is necessary to carefully control the drying process (water content being initially higher than 95%). Indeed, uncontrolled drying produces xerogels with collapse of the porous structure of the material increasing the resistance to intraparticle diffusion (Ruiz et al. 2002, Sicupira et al. 2010). Among the processes that can be used for controlled drying are freeze-drying (Ruiz et al. 2002), drying in the presence of spacer molecules (Ruiz et al. 2002), and drying under CO_2 supercritical conditions (Agulhon et al. 2012b, Deze et al. 2012, Quignard et al. 2010).

Another important criterion for metal-binding affinity is the structure of the material and the distribution of functional groups. Indeed, both chitosan and alginate are heteropolymers made of glucosamine/acetylglucosamine units and mannuronic/guluronic acid groups for chitosan and alginate, respectively. Acetylglucosamine being poorly reactive for metal binding, the degree of acetylation is a critical parameter for the efficiency of chitosan (Guibal et al. 1999a, Milot et al. 1998). Similar differences in the sorption performance are reported when varying the M/G ratio in alginate materials (Agulhon et al. 2012a,b, Papageorgiou et al. 2006). Depending on the repartition of the different groups and their organization in the polymer (blockwise, random, alternated), the spatial arrangement (which needs to be compatible with the steric organization of metal ions), and the type of metal ion, the affinity will be favored by the presence of mannuronic or guluronic acid groups (Agulhon et al. 2012a,b).

9.3.3 SELECTIVITY ISSUES

Due to the proper reactivity of functional groups, the sorbents are expected to have a selectivity for given metal ions (Vold et al. 2003), according, for example, to the principle of Pearson (hard and soft acid and base theory) (Ayers et al. 2006, Pearson 1966): "hard acids" (low polarizability and hard sphere type metals such as Mg(II), La(III), Fe(III), Cr(III), U(VI), V(IV)...) prefer to associate with "hard" bases ($N \gg P$, $O \gg S$, $F \gg Cl$), and "soft" acids (Ag(I), Cd(II), Hg(II) Au(III), and Sn(II)) prefer to associate with "soft" bases ($P \gg N$, $S \gg O$, $I \gg F$). Borderline metal cations (Fe(II), Pb(II), Zn(II), Co(II), Ni(II), V(V), Cu(II) ...) have intermediary binding properties and are generally classified according to the Irving–Williams series in terms of stability constants (Mn(II) < Fe(II) < Co(II) < Ni(II) < Cu(II) > Zn(II)). Hence, for alginate the following series was cited in terms of decreasing affinity: Pb(II) > Cu(II) > Zn(II) > Co(II) > Ni(II) > Mn(II) > Fe(II) (Chen et al. 2002, Karagunduz and Unal 2006). Selectivity for chitosan does not appear to depend on the ionic size and the hardness of metal cations (Rhazi et al. 2002). Alkaline and alkaline-earth metal ions are not bound to chitosan: transition metal ions can thus be efficiently recovered from brines and industrial effluents. The speciation of metal (formation of polynuclear species, chloro complexes) may affect the selectivity of the sorbent for target metals through speciation effects and change in the sorption mechanism (as discussed earlier).

The selectivity is generally difficult to reach for the effective separation of metals with similar chemistry. For example, in the case of the binding of precious metals (e.g., Pd(II), Pt(IV)) on chitosan, it is generally not possible to completely separate the two metals, despite the marked preference of the material for Pd(II), as shown by the Pd(II) overshoot observed on breakthrough curves from bicomponent solutions (in fixed-bed reactors, Figure 9.4) (Chassary et al. 2005). The desorption step can be used to increase the selectivity of the process but it remains difficult to reach the complete separation of the metals.

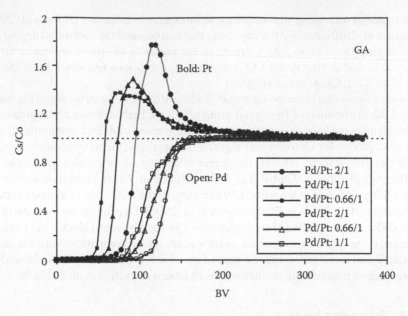

FIGURE 9.4 Example of breakthrough curves for the simultaneous sorption of Pd(II) and Pt(IV) in fixed-bed columns for glutaraldehyde cross-linked chitosan (different Pd/Pt ratio: the overshoot on Pt(IV) profiles shows the preference of the sorbent for Pd(II)). (Reprinted from *Hydrometallurgy*, 76, Chassary, P., Vincent, T., Marcano, J.S., Macaskie, L.E., and Guibal, E., Palladium and platinum recovery from bicomponent mixtures using chitosan derivatives, 131–147, Copyright 2005, with permission from Elsevier.)

9.3.4 RECYCLING

The recycling of the sorbent (and the possible valorization) of sorbed metals is a key parameter for the competitiveness of the sorption process. The possibility to desorb bound metal ions strongly depends on the type of metal, sorbent and mechanism involved in metal binding. A change in the pH is generally sufficient for achieving metal desorption, especially for metal cations. Metal cations being generally bound to chitosan at near-neutral pH can be readily desorbed when decreasing the pH: the protons contribute to the protonation of amine groups and compete with metal cations (Kamari and Ngah 2009, Kannamba et al. 2010, Osifo et al. 2009, Vieira et al. 2007). Since chitosan is soluble in acidic solutions (except in sulfuric acid solutions), this process is generally not applicable or requires the cross-linking of the material to prevent its dissolving during the desorption step. Glutaraldehyde can be used but at the expense of a loss of sorption efficiency in the case of metal cation binding by chelation (Dzul Erosa et al. 2001), because the amine groups bound to aldehyde groups do not react with metal cations. Metal binding can, in some cases, react with other reactive groups such as imino bonds (Vieira and Beppu 2006). Alternatively, metal ions can be desorbed using complexing agents such as EDTA (Chen et al. 2012, Kamari and Ngah 2009, Kannamba et al. 2010, Vieira and Beppu 2006). In the case of the desorption of metal anions loaded on chitosan, though some positive results were obtained changing the pH (using NaOH or NH$_4$OH, which can have a dual effect associating pH change and complexing effect) for molybdate, for example, the strength of the interaction is strong enough to maintain a significant amount of the metal (especially Pt(IV) and Pd(II)) on the sorbent (Chassary et al. 2005, Sicupira et al. 2010). Strong complexing agents are generally necessary for removing these metal ions such as thiourea (alone or associated with HCl), which can be efficiently used for removing precious metals from loaded sorbents (Chassary et al. 2005, Park et al. 2013, Santos Sopena et al. 2011, Sicupira et al. 2010, Zhou et al. 2010).

9.3.5 Mode of Operation

The mode of operation for these materials will strongly depend on their conditioning. Indeed, previous sections have identified the drastic effect of particle size of biopolymer powder on the mass transfer properties. The resistance to intraparticle diffusion requires selecting small particles that cannot be used at large scale for filling fixed-bed columns (due to pressure drop and clogging effects, hydrodynamic behavior) (Jansson-Charrier et al. 1996b). For short analytical-like columns, this parameter is less significant and microparticles can be used (Chassary et al. 2005, Ruiz et al. 2001, Singh et al. 2011). Small particles can then be only operated through batch reactor at the expense of the optimization of solid/liquid separation. The main parameters to be considered are the size of sorbent particle, the agitation speed, and the sorbent dosage (in relation with the metal concentration, and, more generally, the composition of the effluent). These reasons (combined with the analysis of diffusion properties) led to developing alternative conditionings of biopolymer to operate fixed-bed systems. Under the form of spherical particles with expanded structure (due to gelling procedure and decrease in material crystallinity) both the mass transfer performance and the hydrodynamic characteristics are favorable to the use in fixed-bed columns (Guibal et al. 1999d, Osifo et al. 2009). The main parameters to be optimized are thus the size of particles and the flow rate (superficial flow velocity and average contact time in the column).

An alternative application may consist of using chitosan hollow fibers: the concept is based on (1) the reactivity of the chitosan material (at the outer side of the fiber, shell side) to bind metal ions and (2) the morphology of the material that allows circulating the eluent (at the inner side of the fiber, core side). This makes possible the simultaneous sorption and desorption of the metal. For example, this concept was used for the recovery of chromate anions in acidic solutions (ion exchange/electrostatic attraction of chromate anions on protonated amine groups) and its elution/concentration using Aliquat 336 (a quaternary ammonium salt) (Vincent and Guibal 2000, 2001). The hollow fiber served as both a reactive support and a phase separation (aqueous/organic) barrier.

Ultrafiltration assisted by complexation (UFAC) is another way to use the biopolymer for metal binding, the concept being completely different. The principle used in the other systems consists of the transfer of the metal from the liquid phase to the solid phase. In UFAC, the biopolymer is dissolved and mixed with the metal solution: the biopolymer forms soluble complexes with metal ions and an ultrafiltration membrane is used for retaining the metal-bearing macromolecules (Fatin-Rouge et al. 2006, Rivas et al. 2005). The retentate concentrates the target metal ions, while the metal ions that do not react with the biopolymer can pass the membrane and are collected in the filtrate (Kuncoro et al. 2005). The main disadvantage of the process is related to the operating pH range that should conciliate both the reactivity of amine groups (in the case of chitosan) and the stability of the biopolymer (in this case, the biopolymer should remain in the dissolved state in the solution). To overcome this problem it is possible to modify the biopolymer to improve its dissolution in a wider range of pH. For example, the regioselective grafting of sulfonic groups on chitosan (grafting on C–OH groups) lets free amine groups for interacting with metal ions while maintaining the solubility of the biopolymer even in alkaline solutions (Carreon et al. 2010).

9.4 ALGINATE AND CHITOSAN AS ENCAPSULATING MATERIALS FOR THE ELABORATION OF NEW SORBENTS

Though this application is not directly connected to the affinity of the biopolymer for metal ions, the properties of these biopolymers for encapsulating microorganisms, enzymes, solid particles and liquids (oils, etc.) are well documented, and open the route for developing new sorbents. The potential of these materials to be elaborated under different forms (see Section 9.2.3) also justifies the interest for using these biopolymers for the design of new sorbing materials and new modes of application. The challenge in the synthesis of these materials is to manage and optimize both the accessibility and reactivity of the incorporated material (management of mass transfer properties) and the retention properties of the biopolymer matrix (confinement of micro- or nanoparticles, or retention of organic extractants).

9.4.1 ENCAPSULATION OF EXTRACTANTS AND IONIC LIQUIDS

Several recent studies have shown the possibility to incorporate extractants and/or ionic liquids into biopolymer capsules. The principle consists of the addition of the organic liquid phase into the biopolymer solution and the distribution of the viscous solution into neutralization, coagulation, or ionotropic gelation solution through a thin nozzle (Ngomsik et al. 2009, Wu et al. 2009). The main problem deals with the compatibility of the two phases: the organic (hydrophobic) and aqueous (hydrophilic) phases tend to segregate. The "compatibilization" of the two phases can be operated using an alkaline solution of gelatin (Campos et al. 2008a,b, Guibal and Vincent 2006, Guibal et al. 2008, 2009, 2010). In this case, the extractant is dispersed in the core of the alginate matrix under the form of small vesicles that are not necessarily interconnected (Figure 9.5).

The sorption properties in both terms of thermodynamics (equilibrium) and kinetics are controlled by the porosity of the matrix (drying method, characteristics of the biopolymer) and the properties of the extractant (in relation with the target metal). Though beads are the encapsulated materials most commonly prepared, it is also possible to elaborate composite materials under the form of foams using the same method for extractant immobilization (Guibal et al. 2010). The extractant is "compatibilized" with the biopolymer and poured into a mold before being frozen, reacted with the neutralizing or gelling agent and finally freeze-dried (see Section 9.2.3.4: the same procedure as for the preparation of biopolymer sponges and foams). These materials have been used for immobilizing ionic liquids for the binding of palladium, and they were finally used for supported catalysis applications (Jouannin et al. 2013, Vincent et al. 2013).

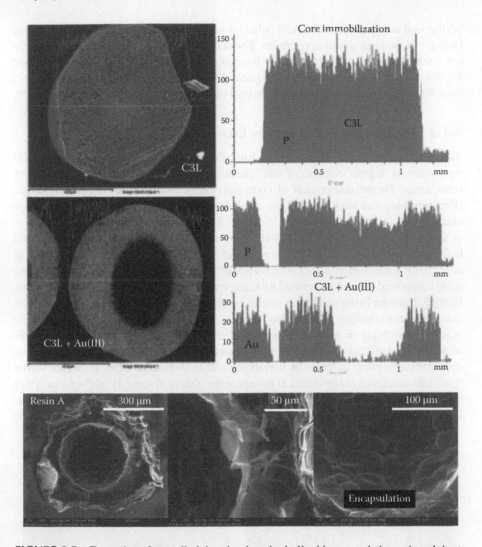

FIGURE 9.5 Examples of tetraalkylphosphonium ionic liquid encapsulation using alginate (core immobilization: dispersion in the whole mass of the biopolymer matrix [before and after Au(III) sorption]; encapsulation: liquid core and solid envelope, SEM photographs and SEM-EDX analysis). (Core immobilization: Reprinted from Campos, K. et al., *Solvent Extr. Ion Exch.*, 26, 270-601, 2008. With permission. Encapsulation: Reprinted from Kica, M. et al., *Solvent Extr. Ion Exch.*, 32, 543, 2014. With permission.)

Another concept has been recently tested for the immobilization of ionic liquid into biopolymer: the extractant phase is injected through a concentric nozzle into the coagulation/ionotropic gelation solution. The extractant forms the core of the spherical particle, while the biopolymer entraps the ionic liquid in a shell. In this case, the ionic liquid phase forms a unique internal vesicle and the challenge is the control of diffusion properties of the shell layer (Figure 9.5): this is especially important if the capsules need to be dried (Kica et al. 2014). The drying of the material has to be controlled, taking into account the mechanical behavior of the capsules, the retention of the ionic liquid, and the mass transfer properties. The stability of the material in terms of extractant retention depends on the loading of the composite material, the dry mass percentage for the encapsulated extractant being generally less than 30% (at the maximum less than 50%) to prevent disruption of the hybrid material, and the release of the extractant.

9.4.2 ENCAPSULATION OF SOLID ION EXCHANGERS

The same concepts can be used for the immobilization of solid particles into biopolymer matrices. A great diversity of solid particles having ion-exchange properties have been tested for the elaboration of composite adsorbents: ferrocyanide compounds (Prussian blue and analogues) (Lian et al. 2012, Lim and Kang 2013), mineral ion exchangers (ammonium molybdophosphate, zirconium phosphate, Figure 9.6) (Krys et al. 2013, Mimura et al. 2001, 2011, Ye et al. 2009), clays, and metal oxides and hydroxides (Boddu et al. 2008, Escudero et al. 2009, Lezehari et al. 2010, Sarkar et al. 2012, Yamani et al. 2012). The choice of the encapsulating agent depends on the solid compound to be incorporated and, more specifically, on its pH range of stability. Indeed, chitosan being soluble in acidic solutions and being gelled in alkaline solution, the ion exchanger needs to be resistant to these drastic conditions (to prevent material dissolving or acid–base degradation). On the other hand, alginate is stable in acidic solution and can dissolve in mild pH conditions. Chitosan and alginate are thus complementary in terms of dissolving and acid–base stability. It is noteworthy that chitosan can be reacetylated to reinforce the chemical stability of the biopolymer but at the expense of a procedure that could interfere with (or degrade) the encapsulated material. It is also possible in the case of the encapsulation of solid particles to

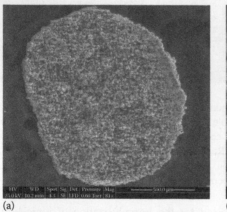
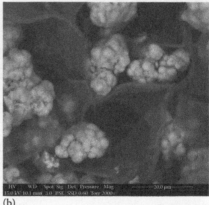

(a)   (b)

FIGURE 9.6 Encapsulation of zirconium phosphate (solid ion exchanger) in alginate matrix (SEM photographs): (a) general view of the cross-section of composite capsule; (b) focus on ion-exchanger particles in the alveoles of the bead. (Reprinted from *J. Colloid Interf. Sci.*, 409, Krys, P., Testa, F., Trochimczuk, A., Pin, C., Taulemesse, J.M., Vincent, T., and Guibal, E., Encapsulation of ammonium molybdophosphate and zirconium phosphate in alginate matrix for the sorption of rubidium(I), 141–150, Copyright 2013, with permission from Elsevier.)

elaborate alternative conditioning of hybrid supports: foams versus gel beads (Vincent et al. 2014). These systems can be used in fixed-bed columns using the foams as a reactive filtration unit.

9.5 IS THERE "ANOTHER LIFE" FOR METAL–BIOPOLYMER COMPOSITES AFTER METAL BINDING?

The interactions of the biopolymers with metal ions can serve not only for the recovery of metal ions but also for the elaboration of new hybrid materials and the design of new applications. Some examples are listed in the following as an illustration of the broad range of new materials that can be manufactured associating chitosan or alginate with metal ions.

9.5.1 New Materials for Environmental Applications

The affinity of a pollutant (metal ion, organic compound, etc.) for a given metal can be used for elaborating new sorbents making profit of the proper affinity of the metal ions for the biopolymers. For example, sulfur-bearing pesticides have affinity for silver (strong interaction between silver cation and sulfur groups) (Yoshizuka et al. 2000): binding silver ions on chitosan offers the possibility to recover methyl parathion from pH 5 to 6 solutions. The pesticide can be selectively desorbed from loaded supports using ammonium thiocyanide or ammonia, silver remaining bound to the biopolymer. Ammonia sorbents were prepared by immobilizing Cu(II) metal ions onto chitosan support (Li et al. 2012a). Chromate and vanadate were sorbed by hybrid Cu/chitosan membranes: Cu(II) is sorbed on chitosan membranes before being chemically reduced (using sodium borohydride). Copper nanoparticles bind metal anions by a combination of adsorption and reduction mechanisms (Condi de Godoi et al. 2013): chromate is mainly bound by reduction (copper being oxidized on the support), while the part of direct binding is more important for vanadate sorption. A number of studies have focused on the sorption of arsenic species (arsenate, arsenite, or organic species) using metal–chitosan hybrid materials. Molybdate anions are capable of reacting with arsenic to form a colored complex: the method is used for spectrophotometric determination of arsenic in solution. On the other side, molybdate is strongly bound to chitosan (with sorption capacities that can reach up to 800 mg Mo g^{-1}). Molybdate-impregnated chitosan beads are very efficient for removing As(III) and As(V) from slightly acidic solutions (Dambies et al. 2000, 2002). The reactivity and the selectivity are reinforced by the specific reaction of As(V) with molybdate and the sorbent can be regenerated and recycled using phosphoric acid (at fixed values of concentration and pH for preventing molybdate desorption). Iron-coated chitosan flakes and iron-doped chitosan granules were prepared by deposition of iron hydroxide at the surface of biopolymer particles or by alkaline coprecipitation (Gupta et al. 2009). These materials were efficiently tested for binding As(III) and As(V) from neutral solutions. The incorporation of zero-valent nanoparticles at the surface of chitosan fibers was obtained by impregnation of the biopolymer material with iron(III) chloride ions followed by their *in situ* reduction (using sodium borohydride) (Horzum et al. 2013): the composite material directly bound As(V) while As(III) was sorbed by a combined oxidation–sorption mechanism. As(V) was also recovered from aqueous solutions using chitosan complexed with different metal ions (Cu(II), Fe(III), La(III), Mo(VI) and Zr(IV)) (Shinde et al. 2013).

The interactions of chitosan with Ni(II) were used for the synthesis of affinity membranes (Ahmed et al. 2006). Histidine-tagged molecules are immobilized on a hybrid membrane (Ni(II) is sorbed on chitosan supported on polymeric support) for binding specific antibodies. This composite material allowed significant improvement in both the amount of bound protein and the affinity constant for the antibody. This enhancement is explained by the positive effect of Ni(II) that contributes to give to the protein a proper orientation (which increases its reactivity).

Another beneficial effect of the binding of metal ions on chitosan support is reported for the immobilization of trypsin (Wu et al. 2006). Chitosan was deposited on silica gel by impregnation and precipitation before binding metal cations by chelation at near-neutral pH (Cu(II), Zn(II), and Ni(II)). Trypsin is bound to the hybrid material at pH close to 7. The presence of the metal ions allows the reversible immobilization of the enzyme (by pH decrease); in addition, the immobilization is less inhibitory to enzyme activity compared to covalent immobilization. The stability of immobilized trypsin follows the sequence, Zn(II) > Ni(II) > Native > Cu(II).

9.5.2 NEW MATERIALS FOR SUPPORTED CATALYSIS

An abundant literature exists on the use of hybrid chitosan or alginate/metal materials for catalytic applications (Ai and Jiang 2013, Chtchigrovsky et al. 2012, Guibal 2005, Guibal et al. 2007, Macquarrie and Hardy 2005). In most cases, palladium or copper is used as the catalytic metal. Several procedures have been identified for the preparation of chitosan-supported Pd catalysts including (1) sorption followed by metal reduction (for hydrogenation reactions), (2) incorporation of metal precipitate (encapsulation), and (3) coprecipitation of metal and biopolymer (Kramareva et al. 2004, Leonhardt et al. 2010). These materials were used for hydrogenation (Guibal and Vincent 2004, Kramareva et al. 2003, Peirano Blondet et al. 2008, Schuessler et al. 2012, Vincent and Guibal 2002, 2003, 2004, Vincent et al. 2003), oxidation (Guibal et al. 2006, Kramareva et al. 2003, Kucherov et al. 2003), dehalogenation (Vincent et al. 2003), hydroamination (Corma et al. 2007), cycloaddition (Chtchigrovsky et al. 2009), and Suzuki cross-coupling reactions (Martina et al. 2011), depending on the catalytic metal and its oxidation state.

These materials have been manufactured under different forms: powder/flakes (Guibal et al. 2006, Vincent and Guibal 2002), gel beads (Guibal 2005), membranes, and hollow fibers (Guibal et al. 2005, Peirano et al. 2009) (Figure 9.7). It is noteworthy that the possibility to encapsulate ionic liquids (or incorporate these ionic liquids in the opened porosity of the support) for the binding of some catalytic metals (especially palladium) opens the route for elaborating new catalysts (called supported ionic liquid catalysts [SILC] or supported ionic liquid-phase catalysts [SILP]) under the form of gel beads or sponges/foams (Czulak et al. 2012, Jouannin et al. 2011, 2013, Moucel et al. 2010, Vincent et al. 2013).

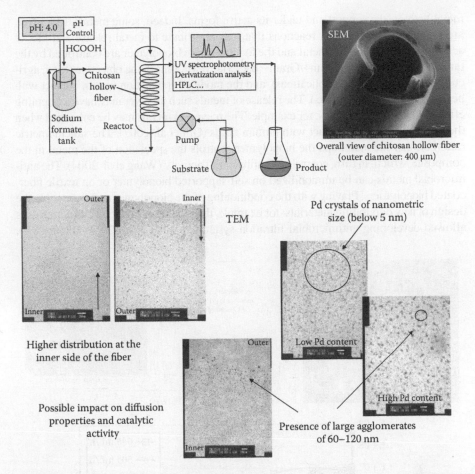

FIGURE 9.7 Experimental setup for the testing of Pd supported catalyst on chitosan hollow fiber—SEM image of the fiber and TEM analysis of catalytic hollow fibers (impact of metal loading on the size and distribution of the metal nanoparticles). (Reprinted from *J. Membr. Sci.*, 329, Peirano, F., Vincent, T., Quignard, F., Robitzer, M., and Guibal, E., Palladium supported on chitosan hollow fiber for nitrotoluene hydrogenation, 30–45, Copyright 2009; *Int. J. Biol. Macromol.*, 43, Peirano Blondet, F., Vincent, T., and Guibal, E., Hydrogenation of nitrotoluene using palladium supported on chitosan hollow fiber: Catalyst characterization and influence of operative parameters studied by experimental design methodology, 69–78, Copyright 2008, with permission from Elsevier.)

9.5.3 New Materials for Antimicrobial/Antiviral Effects

The affinity of biopolymers for metal ions that have a bactericidal effect can be used for elaborating antibacterial materials or antiviral materials (Mori et al. 2013). Chitosan, by itself, has a weak bacteriostatic effect that can be improved by binding metal ions such as Ag(I) (Hoang Vinh et al. 2010, Jena et al. 2012, Saravanan et al. 2011, Thomas et al. 2009), Zn(II) (Patale and Patravale 2011, Wang et al. 2004), and Cu(II) (Patale and Patravale 2011, Qin et al. 2007). The challenge is to maintain a good stability of the metal both on the support and under its active forms. Indeed, some metals, like silver, are readily subject to redox reactions that may contribute to metal release and loss of activity. The choice of the metal and the loading of the biopolymer are controlled by the target microorganisms (Gram+/Gram− bacteria, yeast, fungi), the objective (food, agriculture products, health applications), and the mode of application (water, contact with healing parts, burn parts, etc.). The release of metals such as silver may have a crippling effect for health applications, for example. The release of metals may be enhanced when the material comes in contact with human fluids (Qin et al. 2007). The stoichiometric ratio between the metal and the biopolymer controls the speciation of the metal in the composite, which, in turn, controls its antibacterial activity (Wang et al. 2004). The antimicrobial metals can be immobilized on self-supported biopolymer or on textile fiber-coated biopolymers. Playing with the conditioning of the biopolymer makes possible the design of new depurating materials: for example, the binding of silver on chitosan foams allowed developing antimicrobial filtration systems (Guibal et al. 2013) (Figure 9.8),

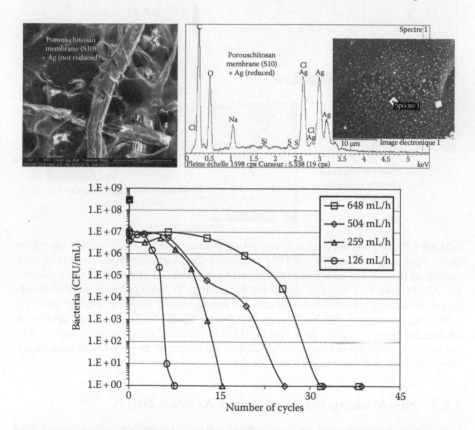

FIGURE 9.8 SEM photograph of silver/cellulose/chitosan foam, SEM-EDX analysis of silver micro-crystals on composite material, and effect of flow rate (with recirculation) on the abatement of *Escherichia coli* population through composite antibacterial foams. (Reprinted from *J. Colloid Interf. Sci.,* 393, Guibal, E., Cambe, S., Bayle, S., Taulemesse, J.-M., and Vincent, T., Silver/chitosan/cellulose fibers foam composites: From synthesis to antibacterial properties, 411–420, Copyright 2013, with permission from Elsevier.)

or manufacturing highly efficient wound dressings (Vimala et al. 2010). Combined with hydroxyapatite, silver/chitosan scaffolds were tested for controlling implant during reconstructive surgery of bones (with complementary antibacterial effects) (Saravanan et al. 2011). The association of chitosan with copper showed a very interesting synergistic effect in the protection of cucumber seedlings against *Botrytis cinerea* (Ben-Shalom and Fallik 2003): the presence of copper prevents chitosan degradation and enhances the elicitor effect of the biopolymer.

9.5.4 NEW MATERIALS FOR SENSORS

Chitosan has retained a great attention for the last decades in the field of the elaboration of sensors. Indeed, the sensors are requiring in many cases gold (or other metals) nanoparticles at the surface of electrodes. Gold nanoparticles have a strong tendency to agglomerate/aggregate, which, in turn, reduces the activity of these nanoparticles. Citrate anions are frequently used for stabilizing these nanoparticles, but chitosan is also a promising substrate for achieving gold stabilization (Lim and Kang 2013, Sugunan et al. 2005). Chitosan has different modes of action in the synthesis of these materials: encapsulating the metal nanoparticles (with stabilization), binding of metals (with subsequent reduction), and encapsulating other supports that immobilize metal nanoparticles (Khun et al. 2012, Lian et al. 2012, Lim and Kang 2013, Mathew et al. 2012). These materials can be used for the detection of metal ions (Mathew et al. 2012, Sugunan et al. 2005), glucose (Khun et al. 2012, Li et al. 2012b), organic compounds (Lian et al. 2012, Lim and Kang 2013), and humidity (Condi de Godoi et al. 2011). A similar stabilizing effect is reported on the synthesis and immobilization of silver nanoparticles for hydrogen peroxide detection (Tian et al. 2012, Zhang et al. 2012), and trichloroacetic acid analysis (Liu et al. 2012).

Chitosan can be also used in sensor manufacturing through electrodeposition mechanism for immobilization of analytical nanoparticles and nanocomposites (Liu et al. 2012, Payne 2007, Yang et al. 2013). When applying a cathodic potential, protons in the acidic solution are consumed at the surface of the electrode. This causes a localized strong increase in the pH; as a result, the chitosan chains become insoluble at the surface of the electrode with entrapment of materials in suspension (in the chitosan solution: enzymes, cells, nanoparticles, etc.). Alginate can be also deposited on chips (Cheng et al. 2011): in this case, the sol-gel transition is driven by anodic electrodeposition.

9.6 CONCLUSION

Chitosan and alginate are promising renewable resources for the binding of metal ions. Sorption mechanisms and performances are mainly controlled by the type of functional groups, the pH and the composition of the solution (metal speciation, protonation of reactive groups, etc.), and the conditioning of the biopolymers (size of particles, morphology, arrangement, etc.). Sorption may proceed through chelation/complexation, ion exchange/electrostatic attraction, and formation of ternary complex. The choice of the biopolymer depends on the target metal and the metal speciation. The great versatility of this material (both in terms of chemical modifications, not discussed in this chapter, and physical modifications) is a great interest for developing materials with improved diffusion properties, enhanced hydrodynamic behavior, and innovative application modes. Depending on the type of metal, the sorbent can be recycled (taking into account the strength of the interaction but also the cost of the metal compared to the cost of the sorbent: thermal degradation makes sense in the case of precious metals). In addition, these biopolymers can be used for encapsulating reactive compounds (ionic liquids, extractants, ion exchangers) in order to improve the reactivity, the selectivity, or the sorption efficiency of these materials, making also profit of the possibility to condition these composite sorbents under different forms (beads, membranes, foams, etc.).

The hybrid materials (i.e., metal-loaded biopolymers) can be also used to design new materials and new applications. Some examples have been reported to show how a second life can be offered to the biopolymers after metal binding. These composite materials are specially designed for high-added-value applications.

Despite the great potential of these biopolymers, the transfer to "true life" remains difficult. This is probably due to a lack of standardization in the characterization and production of these materials. Their variability in composition (degree of acetylation for chitosan, percentage and distribution of guluronic and mannuronic acid groups, MW, crystallinity, etc.) may explain that the final industrial users could be reluctant to use these materials. A great effort is thus necessary to convert these promising properties to successful applications.

ACKNOWLEDGMENTS

E.G. acknowledges the support of all PhD and master-level students and academic partners that collaborated with the research group over the last two decades for the development of biopolymer-based materials.

REFERENCES

Agulhon, P., V. Markova, M. Robitzer, F. Quignard, and T. Mineva. 2012a. Structure of algi-
    nate gels: Interaction of diuronate units with divalent cations from density functional
    calculations. *Biomacromolecules* 13: 1899–1907.
Agulhon, P., M. Robitzer, L. David, and F. Quignard. 2012b. Structural regime identifica-
    tion in ionotropic alginate gels: Influence of the cation nature and alginate structure.
    Biomacromolecules 13: 215–220.
Ahmed, S.R., A.B. Kelly, and T.A. Barbari. 2006. Controlling the orientation of immobilized
    proteins on an affinity membrane through chelation of a histidine tag to a chitosan-Ni[++]
    surface. *J. Membr. Sci.* 280: 553–559.
Ai, L. and J. Jiang. 2013. Catalytic reduction of 4-nitrophenol by silver nanoparticles stabilized
    on environmentally benign macroscopic biopolymer hydrogel. *Bioresour. Technol.* 132:
    374–377.
Alves, N.M. and J.F. Mano. 2008. Chitosan derivatives obtained by chemical modifications for
    biomedical and environmental applications. *Int. J. Biol. Macromol.* 43: 401–414.

Andersen, T., J.E. Melvik, O. Gasered, E. Alsberg, and B.E. Christensen. 2012. Ionically gelled alginate foams: Physical properties controlled by operational and macromolecular parameters. *Biomacromolecules* 13: 3703–3710.

Arrascue, M.L., H.M. Garcia, O. Horna, and E. Guibal. 2003. Gold sorption on chitosan derivatives. *Hydrometallurgy* 71: 191–200.

Ayers, P.W., R.G. Parr, and R.G. Pearson. 2006. Elucidating the hard/soft acid/base principle: A perspective based on half-reactions. *J. Chem. Phys.* 124: 194107.

Ben-Shalom, N. and E. Fallik. 2003. Further suppression of *Botrytis cinerea* disease in cucumber seedlings by chitosan-copper complex as compared with chitosan alone. *Phytoparasitica* 31: 99–102.

Boddu, V.M., K. Abburi, J.L. Talbott, E.D. Smith, and R. Haasch. 2008. Removal of arsenic(III) and arsenic(V) from aqueous medium using chitosan-coated biosorbent. *Water Res.* 42: 633–642.

Campos, K., R. Domingo, T. Vincent, M. Ruiz, A.M. Sastre, and E. Guibal. 2008a. Bismuth recovery from acidic solutions using Cyphos IL-101 immobilized in a composite biopolymer matrix. *Water Res.* 42: 4019–4031.

Campos, K., T. Vincent, P. Bunio, A. Trochimczuk, and E. Guibal. 2008b. Gold recovery from HCl solutions using Cyphos IL-101 (a quaternary phosphonium ionic liquid) immobilized in biopolymer capsules. *Solv. Extr. Ion Exch.* 26: 570–601.

Carreon, J., I. Saucedo, R. Navarro, M. Maldonado, R. Guerra, and E. Guibal. 2010. Mercury recovery from aqueous solutions by polymer-enhanced ultrafiltration using a sulfate derivative of chitosan. *Membr. Water Treat.* 1: 231–251.

Chassary, P., T. Vincent, J.S. Marcano, L.E. Macaskie, and E. Guibal. 2005. Palladium and platinum recovery from bicomponent mixtures using chitosan derivatives. *Hydrometallurgy* 76: 131–147.

Chen, J.P., L.A. Hong, S.N. Wu, and L. Wang. 2002. Elucidation of interactions between metal ions and Ca alginate-based ion-exchange resin by spectroscopic analysis and modeling simulation. *Langmuir* 18: 9413–9421.

Chen, Y., J. Hu, and J. Wang. 2012. Kinetics and thermodynamics of Cu(II) biosorption on to a novel magnetic chitosan composite bead. *Environ. Technol.* 33: 2345–2351.

Cheng, Y., X. Luo, J. Betz, G.F. Payne, W.E. Bentley, and G.W. Rubloff. 2011. Mechanism of anodic electrodeposition of calcium alginate. *Soft Matter* 7: 5677–5684.

Christensen, B.E., M. Indergaard, and O. Smidsrod. 1990. Polysaccharide research in Trondheim. *Carbohydr. Polym.* 13: 239–255.

Chtchigrovsky, M., Y. Lin, K. Ouchaou, M. Chaumontet, M. Robitzer, F. Quignard, and F. Taran. 2012. Dramatic effect of the gelling cation on the catalytic performances of alginate-supported palladium nanoparticles for the Suzuki–Miyaura reaction. *Chem. Mater.* 24: 1505–1510.

Chtchigrovsky, M., A. Primo, P. Gonzalez, K. Molvinger, M. Robitzer, F. Quignard, and F. Taran. 2009. Functionalized chitosan as a green, recyclable, biopolymer-supported catalyst for the 3+2 Huisgen cycloaddition. *Angew. Chem. Int. Ed.* 48: 5916–5920.

Condi de Godoi, F., R.B. Rabelo, F.d.C. Vasconcellos, and M.M. Beppu. 2011. Preparation of copper nanoparticles in chitosan membranes and their application as irreversible humidity indicators, in Pierucci, S. (ed.), *Icheap-10: 10th International Conference on Chemical and Process Engineering*. Italian Association of Chemical Engineering, Milano, Italy. Pts 1–3, pp. 217–222.

Condi de Godoi, F., E. Rodriguez-Castellon, E. Guibal, and M.M. Beppu. 2013. An XPS study of chromate and vanadate sorption mechanism by chitosan membrane containing copper nanoparticles. *Chem. Eng. J.* 234: 423–429.

Corma, A., P. Concepcion, I. Dominguez, V. Fornes, and M.J. Sabater. 2007. Gold supported on a biopolymer (chitosan) catalyzes the regioselective hydroamination of alkynes. *J. Catal.* 251: 39–47.

Cuadros, T.R., O. Skurtys, and J.M. Aguilera. 2012. Mechanical properties of calcium alginate fibers produced with a microfluidic device. *Carbohydr. Polym.* 89: 1198–1206.

Czulak, J., C. Jouannin, T. Vincent, I. Dez, A.C. Gaumont, and E. Guibal. 2012. Nitrophenol hydrogenation using Pd immobilized on ionic liquid alginate spherical resins. *Sep. Sci. Technol.* 47: 2166–2176.

Dambies, L., E. Guibal, and A. Roze. 2000. Arsenic(V) sorption on molybdate-impregnated chitosan beads. *Colloids Surf. A* 170: 19–31.

Dambies, L., T. Vincent, A. Domard, and E. Guibal. 2001. Preparation of chitosan gel beads by ionotropic molybdate gelation. *Biomacromolecules* 2: 1198–1205.

Dambies, L., T. Vincent, and E. Guibal. 2002. Treatment of arsenic-containing solutions using chitosan derivatives: Uptake mechanism and sorption performances. *Water Res.* 36: 3699–3710.

Davis, T.A., B. Volesky, and A. Mucci. 2003. A review of the biochemistry of heavy metal biosorption by brown algae. *Water Res.* 37: 4311–4330.

Desorme, M., A. Montembault, J.-M. Lucas, C. Rochas, T. Bouet, and L. David. 2013. Spinning of hydroalcoholic chitosan solutions. *Carbohydr. Polym.* 98: 50–63.

Deze, E.G., S.K. Papageorgiou, E.P. Favvas, and F.K. Katsaros. 2012. Porous alginate aerogel beads for effective and rapid heavy metal sorption from aqueous solutions: Effect of porosity in Cu^{2+} and Cd^{2+} ion sorption. *Chem. Eng. J.* 209: 537–546.

Dresvyanina, E.N., I.P. Dobrovol'skaya, P.V. Popryadukhin, V.E. Yudin, E.M. Ivan'kova, V.Y. Elokhovskii, and A.Y. Khomenko. 2013. Influence of spinning conditions on properties of chitosan fibers. *Fibre Chem.* 44: 280–283.

Dzul Erosa, M.S., T.I. Saucedo Medina, R. Navarro Mendoza, M. Avila Rodriguez, and E. Guibal. 2001. Cadmium sorption on chitosan sorbents: Kinetic and equilibrium studies. *Hydrometallurgy* 61: 157–167.

Escudero, C., N. Fiol, I. Villaescusa, and J.-C. Bollinger. 2009. Arsenic removal by a waste metal (hydr)oxide entrapped into calcium alginate beads. *J. Hazard. Mater.* 164: 533–541.

Fatin-Rouge, N., A. Dupont, A. Vidonne, J. Dejeu, P. Fievet, and A. Foissy. 2006. Removal of some divalent cations from water by membrane-filtration assisted with alginate. *Water Res.* 40: 1303–1309.

Gonzalez Bermudez, Y., I.L. Rodriguez Rico, E. Guibal, M. Calero de Hoces, and M. Angeles Martin-Lara. 2012. Biosorption of hexavalent chromium from aqueous solution by *Sargassum muticum* brown alga. Application of statistical design for process optimization. *Chem. Eng. J.* 183: 68–76.

Guibal, E. 2004. Interactions of metal ions with chitosan-based sorbents: A review. *Sep. Purif. Technol.* 38: 43–74.

Guibal, E. 2005. Heterogeneous catalysis on chitosan-based materials: A review. *Prog. Polym. Sci.* 30: 71–109.

Guibal, E., S. Cambe, S. Bayle, J.-M. Taulemesse, and T. Vincent. 2013. Silver/chitosan/cellulose fibers foam composites: From synthesis to antibacterial properties. *J. Colloid Interface Sci.* 393: 411–420.

Guibal, E., L. Dambies, C. Milot, and J. Roussy. 1999a. Influence of polymer structural parameters and experimental conditions on metal anion sorption by chitosan. *Polym. Int.* 48: 671–680.

Guibal, E., A. Figuerola Pinol, M. Ruiz, T. Vincent, C. Jouannin, and A.M. Sastre. 2010. Immobilization of Cyphos ionic liquids in alginate capsules for Cd(II) sorption. *Sep. Sci. Technol.* 45: 1935–1949.

Guibal, E., K.C. Gavilan, P. Bunio, T. Vincent, and A. Trochimczuk. 2008. Cyphos IL 101 (tetradecyl(trihexyl)phosphonium chloride) immobilized in biopolymer capsules for Hg(II) recovery from HCl solutions. *Sep. Sci. Technol.* 43: 2406–2433.

Guibal, E., A. Larkin, T. Vincent, and J.M. Tobin. 1999b. Chitosan sorbents for platinum sorption from dilute solutions. *Ind. Eng. Chem. Res.* 38: 4011–4022.

Guibal, E., C. Milot, O. Eterradossi, C. Gauffier, and A. Domard. 1999c. Study of molyb-date ion sorption on chitosan gel beads by different spectrometric analyses. *Int. J. Biol. Macromol.* 24: 49–59.

Guibal, E., C. Milot, and J. Roussy. 1999d. Molybdate sorption by cross-linked chitosan beads: Dynamic studies. *Water Environ. Res.* 71: 10–17.

Guibal, E., C. Milot, and J. Roussy. 2000a. Influence of hydrolysis mechanisms on molybdate sorption isotherms using chitosan. *Sep. Sci. Technol.* 35: 1021–1038.

Guibal, E., C. Milot, and J.M. Tobin. 1998. Metal-anion sorption by chitosan beads: Equilibrium and kinetic studies. *Ind. Eng. Chem. Res.* 37: 1454–1463.

Guibal, E., C. Roulph, and P. Lecloirec. 1992. Uranium biosorption by a filamentous fungus *Mucor miehei*—pH effect on mechanisms and performances of uptake. *Water Res.* 26: 1139–1145.

Guibal, E., C. Roulph, and P. Lecloirec. 1995. Infrared spectroscopic study of uranyl biosorp-tion by fungal biomass and materials of biological origin. *Environ. Sci. Technol.* 29: 2496–2503.

Guibal, E., M. Ruiz, T. Vincent, A. Sastre, and R. Navarro-Mendoza. 2001. Platinum and pal-ladium sorption on chitosan derivatives. *Sep. Sci. Technol.* 36: 1017–1040.

Guibal, E., I. Saucedo, J. Roussy and P. Lecloirec. 1994. Uptake of uranyl ions by new sorbing polymers—Discussion of adsorption isotherms and pH effect. *React. Polym.* 23: 147–156.

Guibal, E., N.V. Sweeney, T. Vincent, and J.M. Tobin. 2002. Sulfur derivatives of chitosan for palladium sorption. *React. Funct. Polym.* 50: 149–163.

Guibal, E. and T. Vincent. 2004. Chitosan-supported palladium catalyst. IV. Influence of temperature on nitrophenol degradation and thermodynamic parameters. *J. Environ. Manage.* 71: 15–23.

Guibal, E. and T. Vincent. 2006. Palladium recovery from dilute effluents using biopolymer-immobilized extractant. *Sep. Sci. Technol.* 41: 2533–2553.

Guibal, E., T. Vincent, and F. Peirano Blondet. 2007. Biopolymers as supports for heterogeneous catalysis: Focus on chitosan, a promising aminopolysaccharide, in Sengupta, A.K. (ed.), *Ion Exchange and Solvent Extraction.* CRC Press, Boca Raton, FL. Vol. 18, pp. 151–292.

Guibal, E., T. Vincent, and C. Jouannin. 2009. Immobilization of extractants in biopolymer capsules for the synthesis of new resins: A focus on the encapsulation of tetraalkyl phos-phonium ionic liquids. *J. Mater. Chem.* 19: 8515–8527.

Guibal, E., T. Vincent, and R.N. Mendoza. 2000b. Synthesis and characterization of a thiourea derivative of chitosan for platinum recovery. *J. Appl. Polym. Sci.* 75: 119–134.

Guibal, E., T. Vincent, and S. Spinelli. 2005. Environmental application of chitosan-supported catalysts: Catalytic hollow fibers for the degradation of phenolic derivatives. *Sep. Sci. Technol.* 40: 633–657.

Guibal, E., T. Vincent, E. Touraud, S. Colombo, and A. Ferguson. 2006. Oxidation of hydro-quinone to *p*-benzoquinone catalyzed by Cu(II) supported on chitosan flakes. *J. Appl. Polym. Sci.* 100: 3034–3043.

Gupta, A., V.S. Chauhan, and N. Sankararamakrishnan. 2009. Preparation and evaluation of iron-chitosan composites for removal of As(III) and As(V) from arsenic contaminated real life groundwater. *Water Res.* 43: 3862–3870.

Guzman, J., I. Saucedo, R. Navarro, J. Revilla, and E. Guibal. 2002. Vanadium interactions with chitosan: Influence of polymer protonation and metal speciation. *Langmuir* 18: 1567–1573.

Guzman, J., I. Saucedo, J. Revilla, R. Navarro, and E. Guibal. 2003. Copper sorption by chito-san in the presence of citrate ions: Influence of metal speciation on sorption mechanism and uptake capacities. *Int. J. Biol. Macromol.* 33: 57–65.

Hernandez-Carmona, G., D.J. McHugh, D.L. Arvizu-Higuera, and Y.E. Rodriguez-Montesinos. 1998. Pilot plant scale extraction of alginate from *Macrocystis pyrifera*. 1. Effect of pre-extraction treatments on yield and quality of alginate. *J. Appl. Phycol.* 10: 507–513.

Hernandez-Carmona, G., D.J. McHugh, and F. Lopez-Gutierrez. 1999. Pilot plant scale extraction of alginates from *Macrocystis pyrifera*. 2. Studies on extraction conditions and methods of separating the alkaline-insoluble residue. *J. Appl. Phycol.* 11: 493–502.

Hernandez, R.B., O.R. Yola, and A.L.R. Merce. 2007. Chemical equilibrium in the complexation of first transition series divalent cations Cu^{2+}, Mn^{2+} and Zn^{2+} with chitosan. *J. Braz. Chem. Soc.* 18: 1388–1396.

Hoang Vinh, T., T. Lam Dai, B. Cham Thi, V. Hoang Dinh, N. Thinh Ngoc, P. Dien Gia, and N. Phuc Xuan. 2010. Synthesis, characterization, antibacterial and antiproliferative activities of monodisperse chitosan-based silver nanoparticles. *Colloids Surf. A* 360: 32–40.

Hortigueela, M.J., I. Aranaz, M.C. Gutierrez, M. Luisa Ferrer, and F. del Monte. 2011. Chitosan gelation induced by the in situ formation of gold nanoparticles and its processing into macroporous scaffolds. *Biomacromolecules* 12: 179–186.

Horzum, N., M.M. Demir, M. Nairat, and T. Shahwan. 2013. Chitosan fiber-supported zerovalent iron nanoparticles as a novel sorbent for sequestration of inorganic arsenic. *RSC Adv.* 3: 7828–7837.

Jansson-Charrier, M., E. Guibal, J. Roussy, B. Delanghe, and P. LeCloirec. 1996a. Vanadium (IV) sorption by chitosan: Kinetics and equilibrium. *Water Res.* 30: 465–475.

Jansson-Charrier, M., E. Guibal, J. Roussy, R. Surjous, and P. LeCloirec. 1996b. Dynamic removal of uranium by chitosan: Influence of operating parameters. *Water Sci. Technol.* 34: 169–177.

Jaworska, M., K. Kula, P. Chassary, and E. Guibal. 2003a. Influence of chitosan characteristics on polymer properties: II. Platinum sorption properties. *Polym. Int.* 52: 206–212.

Jaworska, M., K. Sakurai, P. Gaudon, and E. Guibal. 2003b. Influence of chitosan characteristics on polymer properties. I: Crystallographic properties. *Polym. Int.* 52: 198–205.

Jena, P., S. Mohanty, R. Mallick, B. Jacob, and A. Sonawane. 2012. Toxicity and antibacterial assessment of chitosan-coated silver nanoparticles on human pathogens and macrophage cells. *Int. J. Nanomed.* 7: 1805–1818.

Jeon, C. and W.H. Holl. 2004. Application of the surface complexation model to heavy metal sorption equilibria onto aminated chitosan. *Hydrometallurgy* 71: 421–428.

Jodra, Y. and F. Mijangos. 2001. Ion exchange selectivities of calcium alginate gels for heavy metals. *Water Sci. Technol.* 43: 237–244.

Jouannin, C., I. Dez, A.C. Gaumont, J.M. Taulemesse, T. Vincent, and E. Guibal. 2011. Palladium supported on alginate/ionic liquid highly porous monoliths: Application to 4-nitroaniline hydrogenation. *Appl. Catal. B* 103: 444–452.

Jouannin, C., C. Vincent, I. Dez, A.-C. Gaumont, T. Vincent, and E. Guibal. 2013. Highly porous catalytic materials with Pd and ionic liquid supported on chitosan. *J. Appl. Polym. Sci.* 128: 3122–3130.

Juang, R.S. and C.Y. Ju. 1997. Equilibrium sorption of copper(II)-ethylenediaminetetraacetic acid chelates onto cross-linked, polyaminated chitosan beads. *Ind. Eng. Chem. Res.* 36: 5403–5409.

Juang, R.S., F.C. Wu, and R.L. Tseng. 1999. Adsorption removal of copper(II) using chitosan from simulated rinse solutions containing chelating agents. *Water Res.* 33: 2403–2409.

Kamari, A. and W.S.W. Ngah. 2009. Isotherm, kinetic and thermodynamic studies of lead and copper uptake by H_2SO_4 modified chitosan. *Colloids Surf. B* 73: 257–266.

Kannamba, B., K.L. Reddy, and B.V. AppaRao. 2010. Removal of Cu(II) from aqueous solutions using chemically modified chitosan. *J. Hazard. Mater.* 175: 939–948.

Karagunduz, A. and D. Unal. 2006. New method for evaluation of heavy metal binding to alginate beads using pH and conductivity data. *Adsorption* 12: 175–184.

Khotimchenko, M., V. Kovalev, and Y. Khotimchenko. 2008. Comparative equilibrium studies of sorption of Pb(II) ions by sodium and calcium alginate. *J. Environ. Sci.* 20: 827–831.

Khun, K., Z.H. Ibupoto, J. Lu, M.S. AlSalhi, M. Atif, A.A. Ansari, and M. Willander. 2012. Potentiometric glucose sensor based on the glucose oxidase immobilized iron ferrite magnetic particle/chitosan composite modified gold coated glass electrode. *Sens. Actuators B* 173: 698–703.

Kica, M., T. Vincent, A. Trochimczuk, R. Navarro, and E. Guibal. 2014. Tetraalkylphosphonium ionic liquid encapsulation in alginate beads for Cd(II) sorption from HCl solutions. *Solv. Extr. Ion Exch.*, in press.

Kleinuebing, S.J., E.A. da Silva, M.G.C. da Silva, and E. Guibal. 2011. Equilibrium of Cu(II) and Ni(II) biosorption by marine alga *Sargassum filipendula* in a dynamic system: Competitiveness and selectivity. *Bioresour. Technol.* 102: 4610–4617.

Kleinuebing, S.J., R.S. Vieira, M.M. Beppu, E. Guibal, and M.G. Carlos da Silva. 2010. Characterization and evaluation of copper and nickel biosorption on acidic algae *Sargassum filipendula*. *Mater. Res.* 13: 541–550.

Kolodynska, D. 2012. Adsorption characteristics of chitosan modified by chelating agents of a new generation. *Chem. Eng. J.* 179: 33–43.

Kramareva, N.V., E.D. Finashina, A.V. Kucherov, and L.M. Kustov. 2003. Copper complexes stabilized by chitosans: Peculiarities of the structure, redox, and catalytic properties. *Kinet. Catal.* 44: 793–800.

Kramareva, N.V., A.Y. Stakheev, O.P. Tkachenko, K.V. Klementiev, W. Grunert, E.D. Finashina, and L.M. Kustov. 2004. Heterogenized palladium chitosan complexes as potential catalysts in oxidation reactions: Study of the structure. *J. Mol. Catal. A* 209: 97–106.

Krys, P., F. Testa, A. Trochimczuk, C. Pin, J.M. Taulemesse, T. Vincent, and E. Guibal. 2013. Encapsulation of ammonium molybdophosphate and zirconium phosphate in alginate matrix for the sorption of rubidium(I). *J. Colloid Interface Sci.* 409: 141–150.

Kucherov, A., E. Finashina, N. Kramareva, V. Rogacheva, A. Zezin, E. Said-Galiyev, and L. Kustov. 2003. Comparative study of Cu(II) catalytic sites immobilized onto different polymeric supports. *Macromol. Symp.* 204: 175–189.

Kuncoro, E.P., J. Roussy, and E. Guibal. 2005. Mercury recovery by polymer-enhanced ultrafiltration: Comparison of chitosan and poly(ethylenimine) used as macroligand. *Sep. Sci. Technol.* 40: 659–684.

Lavertu, M., V. Darras, and M.D. Buschmann. 2012. Kinetics and efficiency of chitosan reacetylation. *Carbohydr. Polym.* 87: 1192–1198.

Leonhardt, S.E.S., A. Stolle, B. Ondruschka, G. Cravotto, C. De Leo, K.D. Jandt, and T.F. Keller. 2010. Chitosan as a support for heterogeneous Pd catalysts in liquid phase catalysis. *Appl. Catal. A* 379: 30–37.

Lezehari, M., J.-P. Basly, M. Baudu, and O. Bouras. 2010. Alginate encapsulated pillared clays: Removal of a neutral/anionic biocide (pentachlorophenol) and a cationic dye (safranine) from aqueous solutions. *Colloids Surf. A* 366: 88–94.

Li, C., W. Li, and L. Wei. 2012a. Removal of ammonia from aqueous solution using copper-incorporated chitosan. *Energy Educ. Sci. Technol. A* 30: 223–230.

Li, D., J. Diao, J. Zhang, and J. Liu. 2011. Fabrication of new chitosan-based composite sponge containing silver nanoparticles and its antibacterial properties for wound dressing. *J. Nanosci. Nanotechnol.* 11: 4733–4738.

Li, L., Y. Fang, R. Vreeker, and I. Appelqvist. 2007. Reexamining the egg-box model in calcium-alginate gels with X-ray diffraction. *Biomacromolecules* 8: 464–468.

Li, Y., F. Wang, F. Huang, Y. Li, and S. Feng. 2012b. Direct electrochemistry of glucose oxidase and its biosensing to glucose based on the Chit-MWCNTs-AuNRs modified gold electrode. *J. Electroanal. Chem.* 685: 86–90.

Lian, W., S. Liu, J. Yu, X. Xing, J. Li, M. Cui, and J. Huang. 2012. Electrochemical sensor based on gold nanoparticles fabricated molecularly imprinted polymer film at chitosan-platinum nanoparticles/graphene-gold nanoparticles double nanocomposites modified electrode for detection of erythromycin. *Biosens. Bioelectron.* 38: 163–169.

Lim, J.-W. and I.-J. Kang. 2013. Chitosan-gold nano composite for dopamine analysis using Raman scattering. *Bull. Korean Chem. Soc.* 34: 237–242.

Liu, B., Y. Deng, X. Hu, Z. Gao, and C. Sun. 2012. Electrochemical sensing of trichloroacetic acid based on silver nanoparticles doped chitosan hydrogel film prepared with controllable electrodeposition. *Electrochim. Acta* 76: 410–415.

Macquarrie, D.J. and J.J.E. Hardy. 2005. Applications of functionalized chitosan in catalysis. *Ind. Eng. Chem. Res.* 44: 8499–8520.

Madihally, S.V. and H.W.T. Matthew. 1999. Porous chitosan scaffolds for tissue engineering. *Biomaterials* 20: 1133–1142.

Martina, K., S.E.S. Leonhardt, B. Ondruschka, M. Curini, A. Binello, and G. Cravotto. 2011. In situ cross-linked chitosan Cu(I) or Pd(II) complexes as a versatile, eco-friendly recyclable solid catalyst. *J. Mol. Catal. A* 334: 60–64.

Mathew, M., S. Sureshkumar, and N. Sandhyarani. 2012. Synthesis and characterization of gold-chitosan nanocomposite and application of resultant nanocomposite in sensors. *Colloids Surf. B* 93: 143–147.

McHugh, D.J., G. Hernandez-Carmona, D. Luz Arvizu-Higuera, and Y.E. Rodriguez-Montesinos. 2001. Pilot plant scale extraction of alginates from *Macrocystis pyrifera*—3. Precipitation, bleaching and conversion of calcium alginate to alginic acid. *J. Appl. Phycol.* 13: 471–479.

Milot, C., J. McBrien, S. Allen, and E. Guibal. 1998. Influence of physicochemical and structural characteristics of chitosan flakes on molybdate sorption. *J. Appl. Polym. Sci.* 68: 571–580.

Mimura, H., M. Saito, K. Akiba, and Y. Onodera. 2001. Selective uptake of cesium by ammonium molybdophosphate (AMP)-calcium alginate composites. *J. Nucl. Sci. Technol.* 38: 872–878.

Mimura, H., W. Yan, Y. Wang, Y. Niibori, I. Yamagishi, M. Ozawa, T. Ohnishi, and S. Koyama. 2011. Selective separation and recovery of cesium by ammonium tungstophosphate-alginate microcapsules. *Nucl. Eng. Des.* 241: 4750–4757.

Mirmohseni, A., M.S.S. Dorraji, A. Figoli, and F. Tasselli. 2012. Chitosan hollow fibers as effective biosorbent toward dye: Preparation and modeling. *Bioresour. Technol.* 121: 212–220.

Modrzejewska, Z. and W. Eckstein. 2004. Chitosan hollow fiber membranes. *Biopolymers* 73: 61–68.

Modrzejewska, Z. and W. Kaminski. 1999. Separation of Cr(VI) on chitosan membranes. *Ind. Eng. Chem. Res.* 38: 4946–4950.

Mori, Y., T. Ono, Y. Miyahira, N. Vinh Quang, T. Matsui, and M. Ishihara. 2013. Antiviral activity of silver nanoparticle/chitosan composites against H1N1 influenza A virus. *Nanoscale Res. Lett.* 8: 93.

Moucel, R., K. Perrigaud, J.-M. Goupil, P.-J. Madec, S. Marinel, E. Guibal, A.-C. Gaumont, and I. Dez. 2010. Importance of the conditioning of the chitosan support in a catalyst-containing ionic liquid phase immobilised on chitosan: The palladium-catalysed allylation reaction case. *Adv. Synth. Catal.* 352: 433–439.

Ngomsik, A.-F., A. Bee, J.-M. Siaugue, D. Talbot, V. Cabuil, and G. Cote. 2009. Co(II) removal by magnetic alginate beads containing Cyanex 272. *J. Hazard. Mater.* 166: 1043–1049.

Nishad, P.A., A. Bhaskarapillai, S. Velmurugan, and S.V. Narasimhan 2012. Cobalt (II) imprinted chitosan for selective removal of cobalt during nuclear reactor decontamination. *Carbohydr. Polym.* 87: 2690–2696.

Notin, L., C. Viton, L. David, P. Alcouffe, C. Rochas, and A. Domard. 2006a. Morphology and mechanical properties of chitosan fibers obtained by gel-spinning: Influence of the dry-jet-stretching step and ageing. *Acta Biomater.* 2: 387–402.

Notin, L., C. Viton, J.-M. Lucas, and A. Domard. 2006b. Pseudo-dry-spinning of chitosan. *Acta Biomater.* 2: 297–311.

Osifo, P.O., H.W.J.P. Neomagus, R.C. Everson, A. Webster, and M.A.v. Gun. 2009. The adsorption of copper in a packed-bed of chitosan beads: Modeling, multiple adsorption and regeneration. *J. Hazard. Mater.* 167: 1242–1245.

Padala, A.N., A. Bhaskarapillai, S. Velmurugan, and S.V. Narasimhan. 2011. Sorption behaviour of Co(II) and Cu(II) on chitosan in presence of nitrilotriacetic acid. *J. Hazard. Mater.* 191: 110–117.

Papageorgiou, S.K., F.K. Katsaros, E.P. Kouvelos, J.W. Nolan, H. Le Deit, and N.K. Kanellopoulos. 2006. Heavy metal sorption by calcium alginate beads from *Laminaria digitata. J. Hazard. Mater.* 137: 1765–1772.

Park, S.-I., I.S. Kwak, S.W. Won, and Y.-S. Yun. 2013. Glutaraldehyde-crosslinked chitosan beads for sorptive separation of Au(III) and Pd(II): Opening a way to design reduction-coupled selectivity-tunable sorbents for separation of precious metals. *J. Hazard. Mater.* 248: 211–218.

Patale, R.L. and V.B. Patravale. 2011. *O,N*-carboxymethyl chitosan-zinc complex: A novel chitosan complex with enhanced antimicrobial activity. *Carbohydr. Polym.* 85: 105–110.

Payne, G.F. 2007. Biopolymer-based materials: The nanoscale components and their hierarchical assembly. *Curr. Opin. Chem. Biol.* 11: 214–219.

Pearson, R.G. 1966. Acids and bases. *Science (New York, N.Y.)* 151(3707): 172–177.

Peirano Blondet, F., T. Vincent, and E. Guibal. 2008. Hydrogenation of toluene using palladium supported on chitosan hollow fiber: Catalyst characterization and influence of operative parameters studied by experimental design methodology. *Int. J. Biol. Macromol.* 43: 69–78.

Pielesz, A. and M.K.K. Bak. 2008. Raman spectroscopy and WAXS method as a tool for analysing ion-exchange properties of alginate hydrogels. *Int. J. Biol. Macromol.* 43: 438–443.

Pillai, C.K.S., W. Paul, and C.P. Sharma. 2009. Chitin and chitosan polymers: Chemistry, solubility and fiber formation. *Prog. Polym. Sci.* 34: 641–678.

Piron, E., M. Accominotti, and A. Domard. 1997. Interaction between chitosan and uranyl ions. Role of physical and physicochemical parameters on the kinetics of sorption. *Langmuir* 13: 1653–1658.

Piron, E. and A. Domard. 1997. Interaction between chitosan and uranyl ions—Part 1. Role of physicochemical parameters. *Int. J. Biol. Macromol.* 21: 327–335.

Piron, E. and A. Domard. 1998a. Formation of a ternary complex between chitosan and ion pairs of strontium carbonate. *Int. J. Biol. Macromol.* 23: 113–120.

Piron, E. and A. Domard. 1998b. Interaction between chitosan and uranyl ions. Part 2. Mechanism of interaction. *Int. J. Biol. Macromol.* 22: 33–40.

Plazinski, W. 2012. Sorption of lead, copper, and cadmium by calcium alginate. Metal binding stoichiometry and the pH effect. *Environ. Sci. Pollut. Res.* 19: 3516–3524.

Plazinski, W. 2013. Binding of heavy metals by algal biosorbents. Theoretical models of kinetics, equilibria and thermodynamics. *Adv. Colloid Interface Sci.* 197–198: 58–67.

Portero, A., C. Remunan-Lopez, M.T. Criado, and M.J. Alonso. 2002. Reacetylated chitosan microspheres for controlled delivery of anti-microbial agents to the gastric mucosa. *J. Microencapsul.* 19: 797–809.

Qin, Y., C. Zhu, J. Chen, D. Liang, and G. Wo. 2007. Absorption and release of zinc and copper ions by chitosan fibers. *J. Appl. Polym. Sci.* 105: 527–532.

Quignard, F., F. Di Renzo, and E. Guibal. 2010. From natural polysaccharides to materials for catalysis, adsorption, and remediation, in Rauter, A.P., Vogel, P., and Queneau, Y. (eds.), *Carbohydrates in Sustainable Development I: Renewable Resources for Chemistry and Biotechnology.* Springer, pp. 165–197.

Rabelo, R.B., R.S. Vieira, F.M.T. Luna, E. Guibal, and M.M. Beppu. 2012. Adsorption of copper(II) and mercury(II) ions onto chemically-modified chitosan membranes: Equilibrium and kinetic properties. *Adsorpt. Sci. Technol.* 30: 1–21.

Rhazi, M., J. Desbrieres, A. Tolaimate, M. Rinaudo, P. Vottero, A. Alagui, and M. El Meray. 2002. Influence of the nature of the metal ions on the complexation with chitosan. Application to the treatment of liquid waste. *Eur. Polym. J.* 38: 1523–1530.

Rivas, B.L., E. Pereira, R. Cid, and K.E. Geckeler. 2005. Polyelectrolyte-assisted removal of metal ions with ultrafiltration. *J. Appl. Polym. Sci.* 95: 1091–1099.

Roberts, G.A.F. 1992. *Chitin Chemistry*. The Macmillan Press Limited, London, U.K.

Rodrigues, J.R. and R. Lagoa. 2006. Copper ions binding in Cu-alginate gelation. *J. Carbohydr. Chem.* 25: 219–232.

Ruiz, M., A.M. Sastre, and E. Guibal. 2000. Palladium sorption on glutaraldehyde-crosslinked chitosan. *React. Funct. Polym.* 45: 155–173.

Ruiz, M., A.M. Sastre, and E. Guibal. 2002. Pd and Pt recovery using chitosan gel beads. I. Influence of the drying process on diffusion properties. *Sep. Sci. Technol.* 37: 2143–2166.

Ruiz, M., A.M. Sastre, M.C. Zikan, and E. Guibal. 2001. Palladium sorption on glutaraldehyde-crosslinked chitosan in fixed-bed systems. *J. Appl. Polym. Sci.* 81: 153–165.

Santos Sopena, L.A., M. Ruiz, A.V. Pestov, A.M. Sastre, Y. Yatluk, and E. Guibal. 2011. *N*-(2-(2-Pyridyl)ethyl)chitosan (PEC) for Pd(II) and Pt(IV) sorption from HCl solutions. *Cellulose* 18: 309–325.

Saravanan, S., S. Nethala, S. Pattnaik, A. Tripathi, A. Moorthi, and N. Selvamurugan. 2011. Preparation, characterization and antimicrobial activity of a bio-composite scaffold containing chitosan/nano-hydroxyapatite/nano-silver for bone tissue engineering. *Int. J. Biol. Macromol.* 49: 188–193.

Sarkar, S., E. Guibal, F. Quignard, and A.K. SenGupta. 2012. Polymer-supported metals and metal oxide nanoparticles: Synthesis, characterization, and applications. *J. Nanopart. Res.* 14: 715.

Schuessler, S., N. Blaubach, A. Stolle, G. Cravotto, and B. Ondruschka. 2012. Application of a cross-linked Pd-chitosan catalyst in liquid-phase-hydrogenation using molecular hydrogen. *Appl. Catal. A* 445: 231–238.

Shinde, R.N., A.K. Pandey, R. Acharya, R. Guin, S.K. Das, N.S. Rajurkar, and P.K. Pujari. 2013. Chitosan-transition metal ions complexes for selective arsenic(V) preconcentration. *Water Res.* 47: 3497–3506.

Sicupira, D., K. Campos, T. Vincent, V. Leao, and E. Guibal. 2010. Palladium and platinum sorption using chitosan-based hydrogels. *Adsorption* 16: 127–139.

Sikorski, P., F. Mo, G. Skjak-Braek, and B.T. Stokke. 2007. Evidence for egg-box-compatible interactions in calcium-alginate gels from fiber X-ray diffraction. *Biomacromolecules* 8: 2098–2103.

Singh, P., J. Bajpai, A.K. Bajpai, and R.B. Shrivastava. 2011. Fixed-bed studies on removal of arsenic from simulated aqueous solutions using chitosan nanoparticles. *Bioremed. J.* 15: 148–156.

Sorlier, P., A. Denuziere, C. Viton, and A. Domard. 2001. Relation between the degree of acetylation and the electrostatic properties of chitin and chitosan. *Biomacromolecules* 2: 765–772.

Sugunan, A., C. Thanachayanont, J. Dutta, and J.G. Hilborn. 2005. Heavy-metal ion sensors using chitosan-capped gold nanoparticles. *Sci. Technol. Adv. Mater.* 6: 335–340.

Svecova, L., M. Spanelova, M. Kubal, and E. Guibal. 2006. Cadmium, lead and mercury biosorption on waste fungal biomass issued from fermentation industry. 1. Equilibrium studies. *Sep. Purif. Technol.* 52: 142–153.

Tasselli, F., A. Mirmohseni, M.S. Seyed Dorraji, and A. Figoli. 2013. Mechanical, swelling and adsorptive properties of dry–wet spun chitosan hollow fibers crosslinked with glutaraldehyde. *React. Funct. Polym.* 73: 218–223.

Thomas, V., M.M. Yallapu, B. Sreedhar, and S.K. Bajpai. 2009. Fabrication, characterization of chitosan/nanosilver film and its potential antibacterial application. *J. Biomater. Sci. Polym. Ed.* 20: 2129–2144.

Tian, L., Y. Feng, Y. Qi, B. Wang, Y. Chen, and X. Fu. 2012. Non-enzymatic amperometric sensor for hydrogen peroxide based on a biocomposite made from chitosan, hemoglobin, and silver nanoparticles. *Microchim. Acta* 177: 39–45.

Tseng, R.L., F.C. Wu, and R.S. Juang. 1999. Effect of complexing agents on liquid-phase adsorption and desorption of copper(II) using chitosan. *J. Chem. Technol. Biotechnol.* 74: 533–538.

Varma, A.J., S.V. Deshpande, and J.F. Kennedy. 2004 Metal complexation by chitosan and its derivatives: A review. *Carbohydr. Polym.* 55: 77–93.

Vieira, R.S. and M.M. Beppu. 2006. Interaction of natural and crosslinked chitosan membranes with Hg(II) ions. *Colloids Surf. A* 279: 196–207.

Vieira, R.S., E. Guibal, E.A. Silva, and M.M. Beppu. 2007. Adsorption and desorption of binary mixtures of copper and mercury ions on natural and crosslinked chitosan membranes. *Adsorption* 13: 603–611.

Vikhoreva, G.A. 2012. Processing of chitosan biopolymer into granules, films, and fibers. *Fibre Chem.* 44: 210–216.

Vimala, K., Y.M. Mohan, K.S. Sivudu, K. Varaprasad, S. Ravindra, N.N. Reddy, Y. Padma, B. Sreedhar, and K. MohanaRaju. 2010. Fabrication of porous chitosan films impregnated with silver nanoparticles: A facile approach for superior antibacterial application. *Colloids Surf. B* 76: 248–258.

Vincent, C., A. Hertz, T. Vincent, Y. Barré, and E. Guibal. 2014. Immobilization of inorganic ion-exchanger into biopolymer foams—Application to cesium sorption. *Chem. Eng. J.* 236: 202–211.

Vincent, T. and E. Guibal. 2000. Non-dispersive liquid extraction of Cr(VI) by TBP/Aliquat 336 using chitosan-made hollow fiber. *Solv. Extr. Ion Exch.* 18: 1241–1260.

Vincent, T. and E. Guibal. 2001. Cr(VI) extraction using Aliquat 336 in a hollow fiber module made of chitosan. *Ind. Eng. Chem. Res.* 40: 1406–1411.

Vincent, T. and E. Guibal. 2002. Chitosan-supported palladium catalyst. 1. Synthesis procedure. *Ind. Eng. Chem. Res.* 41: 5158–5164.

Vincent, T. and E. Guibal. 2003. Chitosan-supported palladium catalyst. 3. Influence of experimental parameters on nitrophenol degradation. *Langmuir* 19: 8475–8483.

Vincent, T. and E. Guibal. 2004. Chitosan-supported palladium catalyst. 5. Nitrophenol degradation using palladium supported on hollow chitosan fibers. *Environ. Sci. Technol.* 38: 4233–4240.

Vincent, T., P. Krys, C. Jouannin, A.C. Gaumont, I. Dez, and E. Guibal. 2013. Hybrid macroporous Pd catalytic discs for 4-nitroaniline hydrogenation: Contribution of the alginate-tetraalkylphosphonium ionic liquid support. *J. Organomet. Chem.* 723: 90–97.

Vincent, T., S. Spinelli, and E. Guibal. 2003. Chitosan-supported palladium catalyst. II. Chlorophenol dehalogenation. *Ind. Eng. Chem. Res.* 42: 5968–5976.

Vold, I.M.N., K.M. Varum, E. Guibal, and O. Smidsrod. 2003. Binding of ions to chitosan—Selectivity studies. *Carbohydr. Polym.* 54: 471–477.

Wang, X., Y.M. Du, and H. Liu. 2004. Preparation, characterization and antimicrobial activity of chitosan—Zn complex. *Carbohydr. Polym.* 56: 21–26.

Webster, A., M.D. Halling, and D.M. Grant. 2007. Metal complexation of chitosan and its glutaraldehyde cross-linked derivative. *Carbohydr. Res.* 342: 1189–1201.

Wu, F.C., R.L. Tseng, and R.S. Juang. 1999. Role of pH in metal adsorption from aqueous solutions containing chelating agents on chitosan. *Ind. Eng. Chem. Res.* 38: 270–275.

Wu, J., M. Luan, and J. Zhao. 2006. Trypsin immobilization by direct adsorption on metal ion chelated macroporous chitosan-silica gel beads. *Int. J. Biol. Macromol.* 39: 185–191.

Wu, S.-J., T.-H. Liou, C.-H. Yeh, F.-L. Mi, and T.-K. Lin. 2013. Preparation and characterization of porous chitosan-tripolyphosphate beads for copper(II) ion adsorption. *J. Appl. Polym. Sci.* 127: 4573–4580.

Wu, Y., H. Mimura, and Y. Niibori. 2009. Selective uptake of plutonium (IV) on calcium alginate gel polymer and TBP microcapsule. *J. Radioanal. Nucl. Chem.* 281: 513–520.

Yamani, J.S., S.M. Miller, M.L. Spaulding, and J.B. Zimmerman. 2012. Enhanced arsenic removal using mixed metal oxide impregnated chitosan beads. *Water Res.* 46: 4427–4434.

Yang, J., J.-H. Yu, J.R. Strickler, W.-J. Chang, and S. Gunasekaran. 2013. Nickel nanoparticle-chitosan-reduced graphene oxide-modified screen-printed electrodes for enzyme-free glucose sensing in portable microfluidic devices. *Biosens. Bioelectron.* 47: 530–538.

Yang, J.-S., Y.-J. Xie, and W. He. 2011. Research progress on chemical modification of alginate: A review. *Carbohydr. Polym.* 84: 33–39.

Ye, X.S., Z.J. Wu, W. Li, H.N. Liu, Q. Li, B.J. Qing, M. Guo, and F. Go. 2009. Rubidium and cesium ion adsorption by an ammonium molybdophosphate-calcium alginate composite adsorbent. *Colloids Surf. A* 342: 76–83.

Yipmantin, A., H.J. Maldonado, M. Ly, J.M. Taulemesse, and E. Guibal. 2011. Pb(II) and Cd(II) biosorption on *Chondracanthus chamissoi* (a red alga). *J. Hazard. Mater.* 185: 922–929.

Yoshizuka, K., Z.R. Lou, and K. Inoue. 2000. Silver-complexed chitosan microparticles for pesticide removal. *React. Funct. Polym.* 44: 47–54.

Zhang, M., Y. Song, L. Wang, L. Wan, X. Xiao, and S. Ye. 2012. Novel hydrogen peroxide sensor based on chitosan-Ag nanoparticles electrodeposited on glassy carbon electrode. *Asian J. Chem.* 24: 18–22.

Zhou, L., J. Xu, X. Liang, and Z. Liu. 2010. Adsorption of platinum(IV) and palladium(II) from aqueous solution by magnetic cross-linking chitosan nanoparticles modified with ethylenediamine. *J. Hazard. Mater.* 182: 518–524.

10 Mixed-Mode Sorbents in Solid-Phase Extraction

N. Fontanals, F. Borrull, and Rosa Maria Marcé

CONTENTS

10.1 INTRODUCTION

Over recent decades, solid-phase extraction (SPE) has emerged as the most commonly used and successful sample extraction technique, since it can enrich different types of analytes and extract them from their liquid matrices efficiently. One of the main advantages of SPE is its versatility, mainly due to the availability of different materials that cover the interactions with various analytes. Thus, one of the main aims of SPE research focuses on the development of novel materials that improve the properties of the already existing materials and, therefore, the results of SPE (Ramos 2012).

First, chromatographic materials were adapted and used as SPE sorbents. Originally, the first SPE materials were silica-based and modified with groups such as C_{18}, C_8, phenyl, CH, CN, and NH_2. However, silica-based materials present several disadvantages, such as instability at extreme pHs, low recovery in the extraction of polar analytes, and the presence of some residual silanol groups. Carbon-based sorbents were another type of materials used, including graphitized carbon blacks (GCBs) and porous graphitic carbon (PGC). However, the disadvantage of these sorbents is that they involve some difficulty when eluting certain compounds, and some of them even remain irreversibly adsorbed. Polymer-based sorbents were the latest development in materials. However, the conventional hydrophobic macroporous polystyrene–divinylbenzene (PS–DVB) still displays poor capacity and selectivity, as it has a specific surface area up to 800 m²/g and interacts with the analytes through the hydrophobic interaction that occurs (van der Waals force and π–π interactions of the aromatic rings that make up the sorbent structure).

To improve capacity, hypercross-linked sorbents, with higher specific surface areas (up to 2000 m²/g), provide enhanced interaction with the analytes and, therefore, higher retention (Tsyurupa and Davankov 2002). The hydrophobic structure of the original porous polymers has also been improved by generating both hydrophilic macroporous and hypercross-linked sorbents, which can also display hydrophilic interactions (hydrogen bonding and dipole–dipole interactions). The hydrophilicity of the sorbents can be introduced either by a hydrophilic precursor monomer or by chemically modifying the PS–DVB polymer skeleton (Fontanals et al. 2007).

With respect to selectivity, some tailor-made sorbents have been designed to interact selectively with the target compound(s) but remove all other analytes, including interferences. The first sorbents to be considered exclusively selective were immunosorbents (ISs), which have an immobilized antibody, which presents specific and selective interactions for the target compound (antigen). However, ISs have several drawbacks, such as being time-consuming to prepare, irreproducibility between batches, instability, and limited use in aqueous media. To overcome these drawbacks, the 1990s saw the emergence of molecularly imprinted materials (MIPs), which are synthetic polymers that have specific cavities designed for a template molecule. During SPE, and because of their molecular recognition retention, MIPs are selective with respect to the target analyte or structurally related compounds (Martín-Esteban 2013).

In recent years, research into SPE sorbents has focused on improving capacity and selectivity within a single material, leading to the emergence of what are known as mixed-mode polymeric sorbents. These sorbents combine a polymeric skeleton with ionic groups, with two types of interactions available: reverse phase (RP) from the skeleton and ionic exchange from the ionic groups. Mixed-mode sorbents are classified depending on whether the ionic group attached to the resin is cationic or anionic and, at the same time, whether it is strong or weak. The most common of these are sulfonic and carboxylic acid for strong (SCX) and weak cation exchange (WCX) sorbents, respectively, and amines, quaternary for strong (SAX) and tertiary and secondary for weak anion exchange (WAX) sorbents. The benefit of mixed-mode sorbents is that the ion exchange interaction between the sorbent and the analytes and/or interferences is turned on and off by the careful selection of washing and elution solvent pH, resulting in the selective protonation or deprotonation of the analytes or interferences, and even the sorbent (in the case of weak ion exchange sorbents). Thus, the interferences and analytes can be eluted separately during the washing and elution steps, respectively, thanks to the careful selection of pH and the solvent in each SPE step (Fontanals et al. 2010a).

At present, mixed-mode sorbents are one of the main focuses of research for manufacturers and companies. One of the reasons for this is, generally, the need for cleaner extracts from SPE and, in particular, preventing ion suppression/enhancement when these extracts are injected into liquid chromatography–mass spectrometry (LC–MS) systems. Therefore, despite being relatively new, they have been applied in various fields to extract different types of analytes in a selective manner from the matrix interference usually present in complex samples, such as those of biological, foodstuff, and environmental origin (Fontanals et al. 2010a).

In view of the wide acceptance of these mixed-mode materials in SPE, in the present chapter, all of the aspects related with them are covered. The chapter is divided into three main sections that include a description of the different approaches for their preparation, a discussion of the interactions displayed and the SPE protocol responsible for these interactions, and a selection of applications in different field analyses.

10.2 FUNDAMENTALS

As stated earlier, depending on the ionic group that functionalizes the mixed-mode polymeric sorbent, they are classified and intended for selectively retaining certain types of compounds. So, on one side, SCX sorbents have a polymeric skeleton modified with a SAX group, with the sulfonic group being most commonly used. The anionic group promotes cation exchange interactions with the cationic compounds in the sample, while most of the analytes (whatever their ionic state) and the interferences in the sample can also establish RP interactions with the skeleton. These features make them suitable for the selective extraction of (weak) basic compounds from complex matrices. In contrast, SAX sorbents have a polymeric skeleton modified with a SCX group, such as quaternary amine. The cationic moiety promotes anion exchange interactions with the anionic compounds in the sample, while, in a similar way to the SCX sorbents, most of the analytes (whatever their ionic state) and the interferences in the sample can also establish RP interactions with the skeleton. These features of SAX sorbents make them suitable for the selective extraction of (weak) acidic compounds from complex matrices.

With respect to weak ionic exchange sorbents, the difference in comparison to their strong counterparts is that, in weak ion exchange technology, the weak ionic group in the sorbent may be charged or not depending on the pH, whereas, in strong ion exchange, the ionic moiety always remains charged. Specifically, WCX sorbents are modified with WAX groups, with carboxylic acid being one of the most common. These groups promote cation exchange interactions with the anionic compounds in the sample, and, like other mixed-mode sorbents, the rest of compounds can establish RP interactions. WCX sorbents are designed to extract (strong) basic compounds selectively from complex samples. WAX sorbents are modified with WCX groups, such as tertiary or secondary amines. These protonated amines promote anion exchange interactions with the anionic compounds in the samples, while the rest of the compounds can also be interacted through RP interactions. This type of mixed-mode sorbents are designed to extract (strong) acidic compounds selectively from complex samples.

10.3 PREPARATION OF THE MIXED-MODE SORBENTS

This section, which is divided into commercially available and in-house sorbents, describes the preparation of the different mixed-mode polymeric sorbents or the moiety that modifies them. It also includes, when available, information regarding the morphology of the different mixed-mode sorbents as well as their ion exchange capacity (IEC). This information is really useful for predicting the retention mechanism involved in each type of sorbent.

10.3.1 COMMERCIAL SORBENTS

To the best of our knowledge, all of the commercially available mixed-mode sorbents are based on a previously designed and commercialized polymeric sorbent that displays hydrophobic and/or hydrophilic interactions with the skeleton itself and then further modified with ion exchange groups. Table 10.1 summarizes the commercially available mixed-mode ion exchange sorbents together with their main features.

One such sorbent is Oasis HLB (Waters Corporation, Milford, MA), which is based on polyvinylpyrrolidone–divinylbenzene (PVP–DVB) and has a specific surface area of $800\ m^2/g$. It has been further modified with each of the four different types of ions in order to be converted into each type of mixed-mode sorbent: SCX, SAX, WCX, and WAX, the resulting versions being known as Oasis MCX, Oasis MAX, Oasis WCX, and Oasis WAX, respectively. In particular, Oasis MCX is modified with sulfonic groups ($pK_a < 1$) after reacting the precursor resin (Oasis HLB) with concentrated sulfonic acid (Brousmiche et al. 2008). Figure 10.1 outlines the synthetic routes to obtaining all of the Oasis-based mixed-mode sorbents. Oasis MAX is modified with a dimethylbutylamine (DMBA) moiety, which is the amine that worked best in comparison to the rest of the amines tested: triethylamine, diethylethylamine, dimethylmethylamine, and DMBA. Oasis MAX is obtained from an intermediate chlorinated resin from the precursor (Oasis HLB). In addition, the reaction conditions have been also optimized, and research revealed that the amination reaction was complete after 2 h at reflux (93°C), since no significant increase in the IEC took place beyond this point (Brousmiche et al. 2008). In a similar way, Oasis WAX is obtained by further modifying the chlorinated Oasis HLB skeleton with the piperazine group, which displays a higher IEC (0.86 meq/g) compared to the other amines tested: dibutylamine (0.35 meq/g), diisopropylamine (0.15 meq/g), and morpholine (0.59 meq/g). In all cases, however, the reaction time rose to 18 h compared to the 2 h reaction time for Oasis MAX. Oasis WCX is also obtained from the intermediate chlorinated resin, which is further oxidized into carboxylic groups in the presence of hydrogen peroxide and various catalysts (Brousmiche et al. 2008).

It should be mentioned that the preparation procedure of the Oasis-based mixed-mode sorbents is reported in a published study (Brousmiche et al. 2008), and this is why it is possible to describe the preparation procedure in detail. This is not the case for the rest of the mixed-mode sorbents detailed in Table 10.1, for which, in most cases, the maximum information available is the moiety that modifies the resin. However, not even this information is available in other cases. Nevertheless, from the properties and the interactions displayed, significant differences in their preparation are not expected in comparison to those already reported for Oasis-based mixed-mode sorbents.

TABLE 10.1
Structure and Properties of Mixed-Mode Polymeric Sorbents Commercially Available

Supplier	Sorbent	Type of Mixed Mode	Sorbent Structure — Polymer Based	Ionic Group
Waters	Oasis MCX	SCX	Oasis HLB PVP–DVB (~800 m²/g)	—SO₃H
	Oasis MAX	SAX		(amine)
	Oasis WCX	WCX		COOH
	Oasis WAX	WAX		NH (piperazine)
Phenomenex	Strata-X-C	SCX	Strata-X PS–DVB chemically modif. Pyrrolidone (~800 m²/g)	—SO₃H
	Strata-X-A	SAX		R_1—N^+—R_3 / R_2
	Strata-X-WC	WCX		COOH
	Strata-X-AW	WAX		NH—NH₂

(continued)

TABLE 10.1 (continued)
Structure and Properties of Mixed-Mode Polymeric Sorbents Commercially Available

Supplier	Sorbent	Type of Mixed Mode	Sorbent Structure	
			Polymer Based	Ionic Group
Agilent Technol.	Bond Elut Plexa PCX	SCX	Bond Elut Plexa PS–DVB modif. hydroxyl groups (~550 m²/g)	n.d.
	Bond Elut Plexa PAX	SAX		n.d.
	Abselut NEXUS WCX	WCX	Abselut NEXUS (MAA–DVB) (~575 m²/g)	MAA
	Not available	WAX		

Biotage	Evolute CX	SCX	—SO₃H
	Evolute AX	SAX	
	Evolute ABN PS–DVB hydroxylated		
	Evolute WCX	WCX	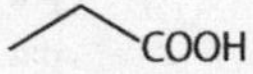
	Evolute WAX	WAX	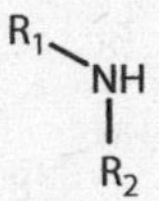
	Speed H₂O–phobic SC–DVB	SCX	Speed H₂O–phobic DVB PS–DVB
J.T. Baker			

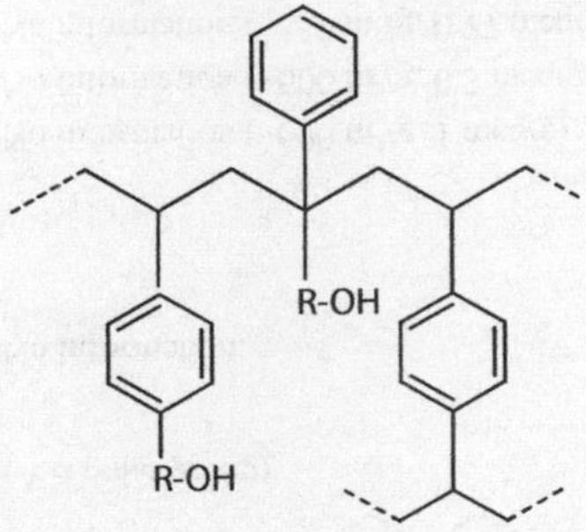

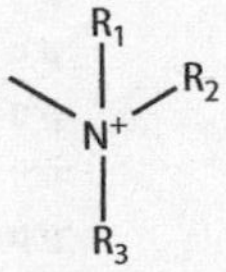

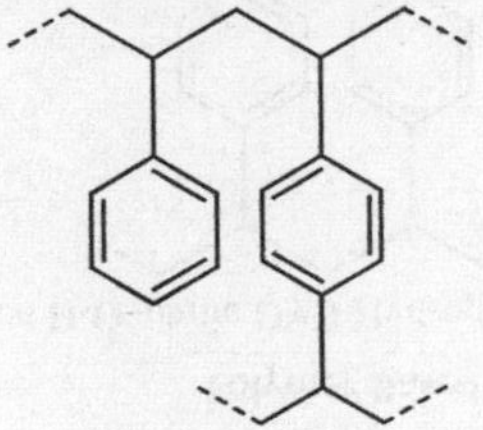

—SO$_3$H

R_1—$\overset{\displaystyle R_2}{\underset{\displaystyle R_3}{N^+}}$—COOH

R_1—NH—R_2

—SO$_3$H

R–OH

R–OH

(continued)

TABLE 10.1 (continued)
Structure and Properties of Mixed-Mode Polymeric Sorbents Commercially Available

Supplier	Sorbent	Type of Mixed Mode	Sorbent Structure — Polymer Based	Ionic Group
	Speed H$_2$O–philic SC–DVB	SAX	Speed H$_2$O–philic DVB Hydrophilic PS–DVB	$R_1{-}\overset{+}{N}(R_2){-}R_3$
	Not available	WCX		
	Speed H$_2$O–phobic WA–DVB	WAX	(A = polar group)	$R_1{-}NH{-}R_2$
Bonna-Agela Technol.	Cleanert PCX	SCX	No information	n.d.
	Cleanert PAX	SAX		n.d.
	Cleanert PWCX	WCX		n.d.
	Cleanert PWAX	WAX		n.d.
Polyintell	AttractSPE SCX	SCX	No information (~600 m^2/g, 1 meq/g)	n.d.
	AttractSPE SSX	SAX	No information (~600 m^2/g, 0.3 meq/g)	n.d.
	AttractSPE WCX	WCX	No information (~850 m^2/g, 0.77 meq/g)	n.d.
	AttractSPE WAX	WAX	No information (~650 m^2/g, 0.5 meq/g)	n.d.

Manufacturer	Product	Mode	Structure	
Scharlau	Extrabond ECX	SCX	Extrabond PS–DVB (~700 m²/g)	n.d.
	Extrabond EAX	SAX		n.d.
Macherey-Nagel	CHROMABOND HR-XC	SCX	CHROMABOND HR-X PS–DVB	
	CHROMABOND HR-XA	SAX		
	CHROMABOND HR-XCW	WCX		
	CHROMABOND HR-XAW	WAX		

SO_3^-

R_1—$\overset{R_2}{\underset{}{N^+}}$—$R_3$

COOH

R_1—NH—R_2

(continued)

TABLE 10.1 (continued)
Structure and Properties of Mixed-Mode Polymeric Sorbents Commercially Available

Supplier	Sorbent	Type of Mixed Mode	Sorbent Structure		Ionic Group
			Polymer Based		
UTC	Styrene screen BXC	SCX	Styrene screen PS–DVB		$-Si-(CH_2)_2-C_6H_6-$ SO_3H
	Styrene screen QAX	SAX			$-Si-(CH_2)_3N(CH_3)_3$
	Styrene screen CCX	WCX			$Si-CH_2COOH$
	Clean screen THC	WAX	Clean screen Hybrid (polymer + silica)		
	Clean screen DAU	SCX			

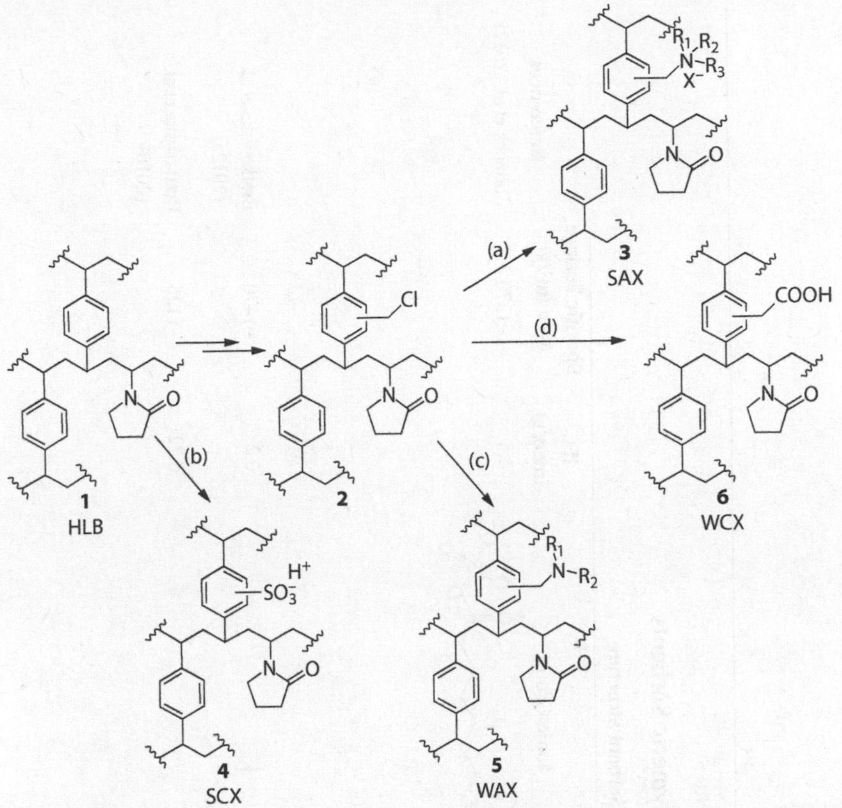

FIGURE 10.1 Examples of the synthetic route for a commercially available mixed-mode polymeric sorbent, (a) tertiary amine, Δ; (b) H_2SO_4, r.t./Δ; (c) secondary amine, Δ; and (d) 30% H_2O_2, Δ. (Reprinted from *J. Chromatogr. A*, 1191, Brousmiche, D.W., O'Gara, J.E., Walsh, D.P., Lee, P.J., Iraneta, P.C., Trammell, B.C., Xu, Y., and Mallet, C.R., Functionalization of divinylbenzene/N-vinylpyrrolidone copolymer particles: Ion exchangers for solid phase extraction, 108–117, Copyright 2008, with permission from Elsevier.)

10.3.2 IN-HOUSE SORBENTS

Our research group was the first to pioneer the preparation of in-house mixed-mode polymeric sorbents. Resins were synthesized for each type of mixed-mode sorbent (i.e., SCX, SAX, WCX, and WAX). Table 10.2 lists all of the in-house mixed-mode sorbents described in this section. The unique feature compared to commercially available sorbents comes from the polymeric skeleton. While commercially available sorbents are prepared from macroporous structures, our in-house mixed-mode sorbents are prepared from hypercross-linked structures, which enhance RP interactions through their greater specific surface area (1000–1500 m^2/g). Specifically, the hypercross-linked resins are prepared from vinylbenzylchloride (VBC)–DVB precursors obtained by precipitation polymerization (PP), which produces low micron-size particles that are more suitable for SPE. The particles obtained, named as PP, were further hypercross-linked

TABLE 10.2
Structure and Properties of "In-House" Mixed-Mode Polymeric Sorbents

Sorbent	Type of Mixed Mode	Sorbent Structure		IEC (meq/g)	Specific Surface Area (m²/g)	References
		Polymer Based	Ionic Group			
HXLPP–SCX	SCX	HXLPP VBC–DVB		2.5	~1370	Cormack et al. (2012)
HXLPP–SAX	SAX		MAA	0.2	~1470	Bratkowska et al. (2012a)
HXLPP–WCX	WCX	HXLPP MAA–VBC–DVB		0.72	~1125	Bratkowska et al. (2010)

				Reference
HXLPP–WAX–EDA	WAX	HXLPP VBC–DVB	0.75	~1000	Fontanals et al. (2008)
HXLPP–WAX–piperazine NVIm–DVB	WAX	NVIm–DVB	0.90	~1000	Fontanals et al. (2008)
	SAX		n.d.	~625	Fontanals et al. (2006)
SILPs	SAX	Imidazolium	2.7	<5	Bratkowska et al. (2012b), Fontanals et al. (2009)

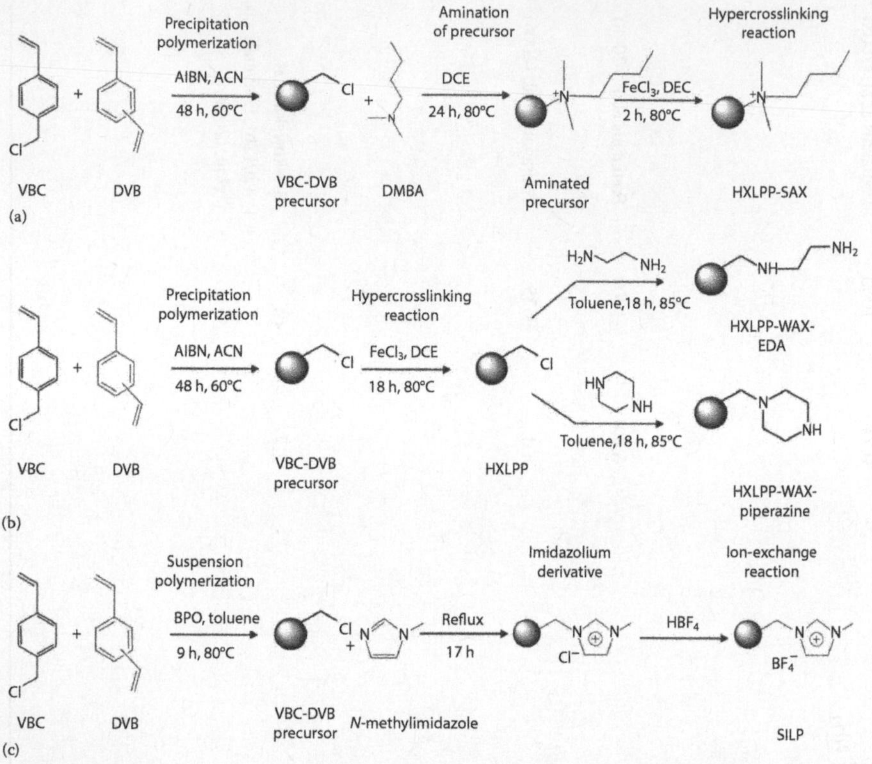

FIGURE 10.2 Synthetic approaches to prepare different "in-house" mixed-mode polymeric sorbents: (a) HXLPP–SAX, (b) HXLPP–WAX–EDA and HXLPP–WAX–piperazine, and (c) SILPs. See the text for the synthetic conditions and sorbent properties.

(hypercross-linked precipitation polymerization [HXLPP]) by means of the Friedel–Crafts reaction, where the chlorine moiety in the VBC monomer acts as an internal electrophile. Figure 10.2 shows the different approaches adopted to obtain the various in-house mixed-mode sorbents.

In particular, for the synthesis of SCX, the HXLPP resins were post-functionalized with either acetyl sulfate or lauroyl sulfate, with the latter proving to be the most effective reagent for the sulfonation. After optimizing different variables affecting the synthesis, three HXLPP–SCX resins were prepared with different contents in terms of the percentage of the sulfate reagent (i.e., 15%, 20%, and 50%). Finally, the HXLPP–SCX sorbent modified with 50% sulfate reagent displayed the highest IEC (2.5 meq/g) and the largest specific surface area (1370 m²/g) that might be indicative of sulfone bridge formation (Cormack et al. 2012).

The procedures to obtain HXLPP–SAX (modified with 5% and 10% DMBA) resins were slightly different. In this case, the DMBA in the appropriate ratio was firstly reacted with the VBC–DVB precursor (named PP) that had been already obtained via the usual PP procedure, to obtain the PP–SAX particles. The quaternized precursors (PP–SAX) were then hypercross-linked following the usual procedure, and HXLPP–SAX resins were obtained. The authors discovered that amination was less efficient after hypercross-linking due to the bulky nature of the tertiary amine (DMBA) restricting the amine's access to the free chloromethyl groups; therefore, amination was conducted before hypercross-linking (Bratkowska et al. 2012a). Figure 10.2a outlines the synthetic route for obtaining the HXLPP–SAX sorbents as an example of the pre-hypercross-linked chemical modification.

In contrast, post-hypercross-linked chemical modification was used in order to prepare HXLPP–WAX (modified with piperazine, HXLPP–WAX–piperazine, and ethylenediamine [HXLPP–WAX–EDA]) (Figure 10.2b). In these cases, the amine moieties were introduced by modification through the remaining chlorine groups already present in the HXLPP resin. Specifically, firstly, HXLPP resin was swollen in dried toluene, and then a five-molar excess of the secondary amine (either EDA or piperazine) relative to chlorine was reacted for 18 h at 85°C (Fontanals et al. 2008).

Alternatively, in the synthesis of the HXLPP–WCX sorbent, the carboxyl moiety was introduced in the precursor monomer (methacrylic acid [MAA]) used in the hypercross-linking process. In other words, the terpolymer MAA–VBC–DVB in a 10/50/40, w/w% ratio was the precursor polymer in this hypercross-linking reaction (Bratkowska et al. 2010). In this way, the carboxylic moieties were already present from the beginning of the reaction, and as their incorporation into the HXLPP–WCX was satisfactory (equivalent to 0.72 meq/g as IEC), the authors did not evaluate the possibility of introducing the carboxylic group in a post-hypercross-linked modification. Moreover, it should be mentioned that the IEC obtained for the HXLPP–WCX (0.72 meq/g) was similar to that reported for the commercially available sorbents, such as Oasis WCX (0.75 meq/g) and Strata-X-CW (0.74 meq/g). However, the specific surface area was larger (1125 m²/g for HXLPP–WCX compared to ~800 m²/g).

In an initial study, N-vinylimidazole–DVB (NVIm–DVB), which was designed as hydrophilic sorbent, prepared by conventional suspension polymerization, was subsequently classified as a SAX sorbent because the imidazole group that it contains may or may not be protonated depending on the pH and thus may interact ionically with the analytes (Fontanals et al. 2006). The main difference between NVIm–DVB as the SAX sorbent and others is that the positive charge in the imidazole group of NVIm–DVB is delocalized across the imidazole ring, whereas, with the other SAX sorbents, the quaternary ammonium group bonds to the polymer network through an aliphatic carbon and, as such, cannot be delocalized because the charge center is not in conjugation with the polymer.

Another group of materials that have been also classified as SAX mixed-mode polymeric sorbents are supported ionic liquid phases (SILPs), as their nature is mostly based on imidazolium-based functional groups immobilized onto silica or polymer supports, giving rise to SAX interactions through the nitrogen moiety cation. To date, the SILPs have been prepared onto a polymeric support that is based on VBC–DVB combining N-methylimidazole cations with trifluoroacetate (Fontanals et al. 2009), tetrafluoroborate and trifluoromethanesulfonate (Bratkowska et al. 2012b), and N-butylimidazole with chloride (Zhu et al. 2011). Figure 10.2c shows the synthetic route for obtaining one of these SILPs modified with tetrafluoroborate. One feature of these SILPs is that they are prepared from a polymer support that contains 2% of cross-linker (DVB), which has a very low specific surface area. Therefore, their main retention mechanisms are through the cationic groups attached to this polymer. It should be noted that, to date, there are no commercially available SILPs. In addition, there are more SILPs that were prepared and evaluated under SAX interaction mechanisms (Kirchner 2009, Vidal et al. 2012). However, they are silica-based and are not therefore considered in this chapter, which is more dedicated to mixed-mode polymeric sorbent.

Along similar lines, other research groups have developed mixed-mode materials, but without a polymeric skeleton. An interesting example of this is the hybrid organic–inorganic silica monolith initially modified with mercapto moieties, which after oxidation were converted into sulfonic acid groups that provide SAX interactions in the monolith (Zheng et al. 2009). Multiwalled carbon nanotubes have been noncovalently functionalized with poly(diallyldimethylammonium chloride) to create SAX sorbents (Kanaujia et al. 2011). The mixed-mode properties of the admicelles (based on sodium dodecyl sulfate–tetrabutylammonium) have also been reported as an alternative material for retaining ionic compounds (Luque and Rubio 2012).

10.4 SOLID-PHASE EXTRACTION CONDITIONS

As mentioned earlier, one of the critical aspects of mixed-mode sorbent technology is the selection of a suitable SPE protocol which, as is well known, includes four steps, namely, (1) conditioning, (2) sample loading, (3) washing, and (4) elution. In this section, the SPE protocols recommended for each type of mixed-mode sorbent are described. It should be mentioned that the volumes in each step are not indicated, as they depend on different factors, such as the amount of sorbent, the type and complexity of sample, type of application, and the strength or weakness of the interaction, among others. Schematically, Figure 10.3 also summarizes all four protocols and the general properties of the analytes most likely to ionically interact with each type of sorbent. It should remembered that the conditioning step is the same in all four protocols and consists of organic solvent followed by aqueous solution, under conditions very similar to those applied for the sample.

It should be also pointed out that, in some studies, the mixed-mode polymeric sorbents are evaluated or applied using an SPE protocol that differs from the recommended protocol, insofar as the pH and solvent are not optimized in each SPE step. This might lead to drawing erroneous conclusions from the performance of mixed-mode polymeric sorbents. This issue will be addressed after presenting each of the recommended protocols.

10.4.1 SCX Sorbents

When working with SCX sorbents, the aim of the various SPE steps is to switch the chargeability of the analytes (or interferences) because of the SAX properties ($pK_a < 1$) of the modifying group in the sorbent, which remains deprotonated under all of the conditions of the SPE steps. In this respect, weak basic compounds are the most suitable analytes for extraction with SCX sorbents (as they are cations that can switch chargeability).

After conditioning, the sample is adjusted to a low pH (usually about pH 3) in order to protonate the analytes and establish ionic interactions with the SAX moieties in the sorbent, and then loaded. The washing step can then be split into two parts.

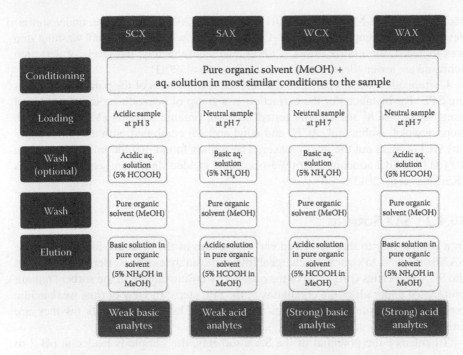

SCX	SAX	WCX	WAX
Conditioning	Pure organic solvent (MeOH) + aq. solution in most similar conditions to the sample		
Loading	Acidic sample at pH 3	Neutral sample at pH 7	Neutral sample at pH 7	Neutral sample at pH 7
Wash (optional)	Acidic aq. solution (5% HCOOH)	Basic aq. solution (5% NH₄OH)	Basic aq. solution (5% NH₄OH)	Acidic aq. solution (5% HCOOH)
Wash	Pure organic solvent (MeOH)	Pure organic solvent (MeOH)	Pure organic solvent (MeOH)	Pure organic solvent (MeOH)
Elution	Basic solution in pure organic solvent (5% NH₄OH in MeOH)	Acidic solution in pure organic solvent (5% HCOOH in MeOH)	Acidic solution in pure organic solvent (5% HCOOH in MeOH)	Basic solution in pure organic solvent (5% NH₄OH in MeOH)
	Weak basic analytes	Weak acid analytes	(Strong) basic analytes	(Strong) acid analytes

FIGURE 10.3 SPE protocols and type of analytes recommended for each type of mixed-mode ion exchange polymeric sorbent. In brackets, an example of most usual conditions is shown.

First, the acidic aqueous solution is compromised with the aim of removing the water-soluble compounds poorly retained by RP interactions as well as further activating the ionic interactions between the analytes and the sorbent. This step is not included in all protocols. The second part of the washing always forms part of the protocol and involves adding pure organic solvent (usually pure methanol [MeOH] and acetonitrile [ACN]) that disrupts the RP interactions between the neutral and acidic compounds or interferences in the sample and they elute. Finally, the basic analytes elute with basic solution (generally from 2% to 10% NH₄OH) in organic solvent (MeOH or ACN). The basic solution ensures the protonation of the analytes (becoming neutral), and the elution strength of the solvent elutes the analytes, while the sorbent remains deprotonated.

Most of the examples in the literature use these guidelines as the optimal SPE protocol that ensures the efficient use of the sorbent. However, there are some studies that vary slightly. For instance, in the extraction of illicit drugs from wastewaters using Oasis MCX in all instances, some authors (Bijlsma et al. 2009, Pedrouzo et al. 2011) performed the washing step only with a basic aqueous solution, while the recommended solution should be acidic. In any case, this basic solution might further deprotonate the acidic analytes and release them from the cartridge, if these analytes are water soluble. Other studies (González-Mariño et al. 2011, Sousa et al. 2011), however, performed the washing step with the solutions recommended in the protocol (i.e., acidic aqueous solution followed by pure MeOH), and the only observation is that the acidic illicit drugs (e.g., THC and its metabolite THC–COOH) eluted with the pure

MeOH (González-Mariño et al. 2011, Sousa et al. 2011). In any case, under similar detection instrumentation (i.e., ESI–QqQ–MS/MS), when the methanol washing step was included, the matrix effect is lower, and, consequently, the levels of detection achieved are lower (Pedrouzo et al. 2011, Sousa et al. 2011).

Certain other studies performed a detailed optimization of the parameters affecting SPE. For instance, for the extraction of a group of benzodiazepines from blood using Oasis MCX, the authors tested different organic solvents (MeOH, ethanol, isopropanolol, isobutanol, ACN, and ethyl acetate) combined with an initial aqueous HCl to wash out the interferences. The authors found that the mixture based on 40% 0.15 M HCl aqueous and 60% isopropanol yielded significantly cleaner extracts (Karlonas et al. 2013).

10.4.2 SAX Sorbents

In a similar way to that described earlier, the aim of the different SPE steps in the SAX sorbents is to switch the chargeability of the analytes (or interferences) because the SCX properties ($pK_a > 18$) of the quaternary amine moiety in the sorbent remain protonated under all of the conditions in the SPE steps. In view of this, weak acidic compounds are suitable candidates for extraction by SAX sorbents (as they are anions that can switch chargeability).

To enhance the potential of the SAX sorbents, the sample is loaded at pH 7 to ensure that the acidic analytes become deprotonated and can establish ionic interactions with the quaternary amines in the sorbent. The subsequent washing step optionally includes basic aqueous solution for rinsing the water-soluble compounds and further activating the ionic interactions. The second part of the washing step is based on pure organic solvent (MeOH or ACN), which washes the basic and neutral analytes or interferences that only interact with the sorbent by RP interactions. The elution step is performed with acidic solution (2%–10% HCOOH) in MeOH or ACN. This acidic solution ensures that the acids are protonated (become neutral) and the ionic interactions with the sorbent (that remains charged) are disrupted, while the elution strength of the organic solvent elutes these acidic analytes.

There are some examples, however, that modified the recommended protocol. Some of them refer to loading the sample in organic solvent that has been basified. This practice is quite usual when the sample has already been treated in another extraction step, such as the protein precipitation with ACN (Xia et al. 2009) or with MeOH (Landberg et al. 2009). The analytes now dissolved in the organic solvent can be loaded directly into the SAX cartridge, thereby avoiding the time-consuming step of evaporating and redissolving the sample in an aqueous-based solvent.

In another example, a group of benzotriazoles in aqueous matrices were extracted using Oasis MAX and eluting the analytes merely with MeOH. These basic analytes were not able to interact ionically with Oasis MAX sorbent. However, the authors found that MAX cartridges provided cleaner extracts with lower complexity compared to Oasis HLB, which might be attributed to the proper interaction of the acidic interferences with the amine moieties of the sorbent (Emotte et al. 2012). The authors, however, did not test any cationic exchange sorbent, which might produce cleaner extracts and selective extraction of the target compounds.

It should be pointed out that when SILPs were tested as SAX sorbents, the volume and percentage of the acidic additive were larger, which might be necessary to disrupt stronger ionic interactions between the analytes and the sorbent (Bratkowska et al. 2012b). For instance, for the elution of a group of acidic pharmaceuticals from SILPs (200 mg packed in a cartridge), it was necessary to use 15 mL of 10% HCOOH in MeOH, whereas the most common elution conditions involve 5 mL of 5% HCOOH in MeOH (Bratkowska et al. 2012b). Meanwhile, it should be also noted that SILPs also tolerate washing with larger volumes of MeOH (e.g., 20 mL [Fontanals et al. 2009] and 10 mL [Bratkowska et al. 2012b]) without losses, while the usual washing volume is not higher than 5 mL in other SAX sorbents.

10.4.3 WCX SORBENTS

In weak ion exchange technology, the aim of the different SPE steps is to switch the chargeability of the analytes or that of the sorbents, since the ionic group that modifies the sorbent has more tunable pK_a, which enables more successful changes depending on the pH in the different SPE steps. Specifically, in WCX, the carboxylic acid (which is the most usual anionic group for modifying the resins) can be protonated (neutral) or deprotonated (anionic) depending on the pH. Under these circumstances, the most suitable compounds to be extracted with WCX sorbents are (strong) basic compounds that might be in their cationic form and therefore able to establish cationic interactions.

Following the protocol's recommendations, the sample is loaded under neutral pH conditions so that the carboxylic groups in the sorbent are deprotonated and the basic compounds protonated, as they are able to interact ionically. Apart from the ionic interactions, all of the compounds can be bonded to the sorbent through RP interactions. The subsequent washing step optionally begins with aqueous ammonium solution in order to further enhance the ionic interactions. However, at the same time, all of the water-soluble compounds that are weakly bonded are released. In addition, the compulsory second part of the washing step involves pure organic solvent that elutes (or washes out) all of the nonprotonated (neutral and anionic acidic) analytes merely bonded through RP interactions in the sorbent. Finally, in the elution step, an acidic (2%–10% HCOOH) solution in organic solvent ensures the protonation of the carboxylic groups in the sorbent, and, thus, the ionic interactions with the basic (protonated) analytes are disrupted. The organic solvent elutes these basic analytes from the sorbent.

Some studies further optimize the SPE protocol. For instance, Bratkowska et al. (2010) evaluated the performance of the extraction of an in-house HXLPP–WCX sorbent for a group of basic pharmaceuticals. One part of the study focused on investigating the elution efficiency of different acidic solvent compositions that included 5 mL aliquots of 2% HCOOH in MeOH, 2% trifluoroacetic (TFA) in MeOH, and 2% TFA in MeOH/ACN (1/4). Eventually, 2% TFA in MeOH was selected for use in the elution step as it delivered the best results (higher recoveries than with 2% HCOOH in MeOH) and did not cause any significant disturbance in the LC separation of the analytes. TFA acted as the acidic additive in a solution composed of ACN/H_2O (80/20, v/v) that was also used as an elution solvent to elute a group of peptides from Oasis WCX (Shi et al. 2009).

Some other studies evaluated WCX sorbents without using the recommended protocol (Allanson et al. 2007, Batt et al. 2008, Tylová et al. 2011, Weigel et al. 2004). For instance, in the determination of a group of pharmaceutical compounds from environmental waters, Oasis WCX, Oasis HLB, and Oasis MCX were evaluated and compared using the recommended SPE protocol for SCX sorbents in all of the sorbents, which led to confusing results. Finally, Oasis MCX was the sorbent of choice, probably because it was the only option that was evaluated under its optimal conditions (Batt et al. 2008).

10.4.4 WAX SORBENTS

Similarly to the WCX sorbents, the aim of WAX technology is to switch either the chargeability of the weak cation moieties (i.e., secondary or tertiary amines) in the sorbent or the chargeability of the analytes/or interferences during the SPE steps. Indeed, the suitable candidates for WAX sorbents are (strong) acid compounds, which might be in their anionic form all or most of the time, and therefore able to establish anionic interactions.

The recommended protocol started with the loading of the sample into the cartridge at neutral pH, with the aim of deprotonating (anionic) the acidic compounds, while the amines in the sorbent remain protonated (cationic) and are therefore able to establish ionic interactions. Similarly to the other mixed-mode sorbents, the rest of analytes and interferences can also display RP interactions with the skeleton of the sorbent. As in the other mixed-mode protocols, the washing step can be divided into two parts. First, an aqueous acidic solution (normally using HCOOH as the acidic additive) is responsible for the solubilization, and so washing out, of water-soluble compounds, as well as for reinforcing the ionic interaction points. The second part of the washing step consists of pure methanol solvent (most commonly) that disrupts the RP interactions between the sorbent and analytes/interferences (neutrals and cationic basics) and washes them from the cartridge. Finally, the elution step is performed with a basic solution (2%–10% NH_4OH) in the organic solvent. The basicity neutralizes the sorbent and disrupts the ionic interaction, to elute the acidic analytes thanks to the elution strength of the organic solvent.

In the following examples, the WAX protocol used was the same as the one recommended. However, the authors take advantage of the dual properties of the sorbent to elute two groups of analytes separately and selectively, one group in the washing step and the other in the elution step. It should be noted that this strategy is also feasible with the other mixed-mode sorbents. In this line, a group of pharmaceuticals that includes basic and acidic compounds were loaded together in the in-house HXLPP–WAX sorbents. Subsequently, 4 mL of MeOH eluted the basic pharmaceuticals, as well as washing out all the interferences. Finally, 2 mL of 2% NH_4OH in MeOH/ACN (1/4, v/v) eluted the acidic pharmaceuticals of the group (Fontanals et al. 2008). A similar strategy was used for the selective elution of the glucuronide and sulfate steroid conjugates. The glucuronide form was neutralized and eluted with 4 mL of 10% HCOOH in MeOH/ H_2O (95/5, v/v) (which also served as the washing solution), whereas the sulfate species remained charged and retained in the cartridge. Then, 4 mL of 5% NH_4OH MeOH/H_2O (90/10, v/v) was passed through, which neutralized the secondary amine in the cartridge and activated the elution of sulfate conjugates as the interactions between the polymer and the sulfate moieties were disrupted (Strahm et al. 2008b).

10.5 APPLICATION OF THE MIXED-MODE SORBENTS

10.5.1 SCX SORBENTS

Mixed-mode SCX polymeric sorbents are the type of mixed-mode sorbent that have been applied most often. One reason might be due to the fact that they are more widely available in different commercial forms or because their features make them more suitable for selective application to extract a group of compounds from their complex matrices. Table 10.3 shows selected examples of different applications of SCX sorbents to extract different types of compounds from samples with different complexities and the techniques used to determine the compounds. It should be noted that this table (as well as other tables of applications for each type of mixed mode in this chapter) aims to illustrate the disparity of applications, rather than duplicates of the same application.

In most cases, the SCX sorbents are used strictly in off-line mode in SPE cartridges, as detailed in Table 10.3. In other cases, online SPE–LC (Li et al. 2011, Zhou et al. 2009) and 96-well plate (Cunliffe et al. 2009, Fountain et al. 2009, Harris et al. 2004, Xu et al. 2005) approaches have been also been successful in their applications. In particular, in the online SPE–LC setup using Oasis MCX for the determination of a group of steroidal alkaloids, the challenge was to find a suitable solution that acts at the same time as acting as an eluting solution for the analytes ionically trapped on the SCX sorbent and also a mobile phase to separate them in the LC. The different solutions tested included 0.1% NH_4OH aqueous solution or 0.05% diethylamine aqueous solution. The latter gave the best results and peak performance and was selected for the study (Zhou et al. 2009).

Other examples included, for instance, Bond Elut Plexa PCX being used in a flow-injection column followed by flame atomic absorption spectrometry to preconcentrate and determine trace amounts of metals from urine (Anthemidis et al. 2012). Another example is the hyphenated technique LC–SPE–NMR for analyzing alkaloids of complex mixtures, in which the role of the SPE with SCX sorbent was to retain these basic alkaloids selectively, while excluding the remaining interferences from the LC (Johansen et al. 2012).

Other studies focused on the comparison between different SCX sorbents (Klinke and Linnet 2007, López et al. 2011, Msagati and Nindi 2006). In particular, five different sorbents that include three different mixed-mode SCX polymeric sorbents (Strata-X-C, Oasis MCX, and Bond Elut Plexa PCX) along with Bond Elut Certify (SCX silica-based sorbent) and Lichrolut EN (hypercross-linked polymeric sorbent) were compared for the extraction of piperazine derivatives in wine. Initial experiments pointed to all of the mixed-mode SCX polymeric sorbents tested as the best, as the results did not show significant differences between them. However, when the volume of wine was increased, only Bond Elut Plexa PCX was able to be loaded with volumes up to 200 mL of wine, while the breakthrough volumes of Oasis MCX and Strata-X-C were lower than 40 and 120 mL, respectively (López et al. 2011).

When the duality of the mixed-mode polymeric sorbent is not sufficient to extract all of the target analytes selectively and quantitatively, the tandem approach can be adopted. A number of studies have tested how the performance of the whole SPE method

TABLE 10.3
Applications of the Mixed-Mode SCX Polymeric Sorbents

Compounds	Application	Sorbent	SPE Mode	Analysis	References
Therapeutic drugs	Environment	Strata-X-C	Off-line	LC-MS/MS (QqQ)	Van De Steene et al. (2006)
		Strata-X-C	LLE + off-line	GC-MS	Tauxe-Wuersch et al. (2006)
	Biological fluids	Oasis MCX	Off-line	CE-UV	Lehtonen et al., (2004)
		Oasis MCX	PPT + off-line	LC-MS/MS (QqQ)	Xue et al. (2006)
		Oasis MCX	Online	LC-MS/MS (QqQ)	Li et al. (2011)
		Oasis MCX	96-well plate	LC-MS/MS (QqQ)	Cunliffe et al. (2009)
		Bond Elute Plexa PCX	96-well plate	LC-Orbitrap	Musenga and Cowan (2013)
	Food	Cleanert PCX	Off-line	GC-MS	Wu et al. (2009)
		Oasis MCX	Off-line	LC-UV	Nochetto et al. (2009)
		Bond Elute Plexa PCX	Off-line	HILIC-MS/MS (QqQ)	Wu et al. (2012)
		Strata-X-C	Off-line	LC-MS/MS	Huq et al. (2006)
Illicit drugs	Environment	Oasis MCX	Off-line	UHPLC-MS/MS (QqQ)	Bijlsma et al. (2009), Gheorghe et al. (2008), González-Mariño et al. (2011), Pedrouzo et al. (2011)
	Biological fluids	Oasis MCX	Off-line	UHPLC-MS/MS (QqQ)	Berg et al. (2009), Fountain et al. (2009)
		Cleanert PCX	Off-line	LC-MS/MS (QTRAP)	Dowling and Regan (2011)
Biological comp.	Biological fluids	Oasis MCX	Off-line	CE-MS	Sentellas et al. (2004)
		Strata-X-C	Off-line	LC-MS/MS (QqQ)	Coles and Kharasch (2007), Kollroser and Schober (2002), Lin et al. (2005), Wang et al. (2007), Zhao et al. (2004)
	Food	Oasis MCX	96-well plate	HILIC-MS/MS (QqQ)	Heinig and Wirz (2009)
		Oasis MCX	96-well plate	HILIC-MS/MS (QqQ)	Xu et al. (2005)
		Oasis MCX	Off-line	LC-FL-MS	Malakova et al. (2007)
Herbicides	Environment	Bond Elute Plexa PCX	Off-line	LC-MS/MS (QqQ)	Zheng et al. (2010)
		Oasis MCX	Off-line	LC-MS/MS (QqQ)	Nanita et al. (2008)
Additives	Food	Oasis MCX	PLE + off-line	CE-DAD or LC-DAD	Rodríguez-Gonzalo et al. (2009)
	Environment	Strata-X-C	Off-line	LC-MS/MS (QqQ)	Bermudo et al. (2006)
	Biological fluids	Evolute CX	Off-line	LC-MS/MS (QqQ)	Williams and Caulfield (2009)

improves by connecting one mixed-mode SCX sorbent in series with a C18 sorbent (Ivanov Dobrev and Kaminek 2002, Park et al. 2007, Tauxe-Wuersch et al. 2006) or with a polymeric sorbent (Díaz-Cruz et al. 2008, Gros et al. 2009, Izumi et al. 2009, Roberts and Bersuder 2006). In these studies, the main aim of the tandem approach is both to clean up the matrix of interferences (usually achieved by the mixed-mode sorbent) and to recover the target analytes. For example, Izumi et al. (2009) compared Oasis HLB with the tandem Oasis HLB and MCX for the extraction of hormones from plants. The purification procedure was more efficient when using the tandem approach and it was selected for further analysis. It should be mentioned, however, that some studies rejected the tandem option, since the SPE protocol resulted more complex and the results were not significantly better compared to those with a single cartridge (Díaz-Cruz et al. 2008, Gros et al. 2009, Roberts and Bersuder 2006).

10.5.2 SAX SORBENTS

Mixed-mode SAX sorbents, and in particular Oasis MAX due to its long use in the field, have also been extensively applied for the extraction of different types of acidic analytes in different fields. Table 10.4 summarizes the details for the application of SAX sorbents most of which, as mentioned earlier, correspond to the use of Oasis MAX that has been used in a variety of applications.

For instance, closantel, an antiparasitic drug administrated to animals, was analyzed in animal tissue, using Oasis MAX for the SPE in all cases (Lai et al. 2011, Sun et al. 2007, Yeung et al. 2010). Initially, it was adapted to the off-line SPE mode (Sun et al. 2007, Yeung et al. 2010), but later, authors easily adapted the SPE protocol to the online mode (Lai et al. 2011), as the acidic organic mobile phase used in LC also worked as an elution solvent in the SAX protocol. Comparing both SPE modes, the authors found that the online mode provides higher sensitivity and a less painstaking and time-consuming procedure. Interestingly, Hewitt et al. (Hewitt et al. 2011) used a 20 mm × 2.1 mm pre-column packed with Oasis MAX as the stationary phase in LC to separate a group of nonionic surfactants commonly used in the formulation of protein pharmaceuticals, monitoring its hydrolysis in different formulations from different vendors.

In other studies, its performance has been compared to other sorbents. For instance, Kojima et al. (2004) compared Oasis MAX with the silica-based SAX sorbent (Bond Elut SAX) for the extraction of alkylphenols from river water samples. The authors ruled out Bond Elut SAX because it provided lower recoveries, which were attributed to the decomposition of the sorbent due to the strong alkaline solutions (1 M NaOH in aqueous sample). The Cela research group compared the performance of Oasis MAX and Oasis HLB in two separated studies (Carpinteiro et al. 2010, Montes et al. 2010) to extract neutral compounds (trans-resveratrol—a phytochemical present in vegetables [Carpinteiro et al. 2010, Montes et al. 2010], and a group of fungicides [Carpinteiro et al. 2010, Montes et al. 2010]) in wine samples. Despite extracting neutral compounds, in both studies, Oasis MAX was the sorbent of choice, as it provided cleaner extracts. The authors justified these results because, after loading the sample, the basic washing step enabled acids and phenols (present in wine and acting as interferences) to become ionized and retained through ionic interaction with the amine moieties in the MAX cartridges. Subsequently, in the

TABLE 10.4
Applications of the Mixed-Mode SAX Polymeric Sorbents

Compounds	Application	Sorbent	SPE Mode	Analysis	References
Therapeutic drugs	Environment	Oasis MAX	Off-line	LC-DAD	Benito-Peña et al. (2006), Gil-Garcia et al. (2008), Tansupo et al. (2010)
		HXLPP-SAX	Off-line	LC-UV	Bratkowska et al. (2012a)
	Biological fluids	Oasis MAX	Off-line	LC-MS/MS (QqQ)	Abdel-Hamid et al. (2006)
	Biological fluids	Oasis MAX	96-well plate	LC-MS/MS (QqQ)	Li et al. (2006)
Biological comp.	Tissue	Oasis MAX	Off-line	LC-MS/MS (QqQ)	Kakimoto et al. (2008)
	Hair	Oasis MAX	Off-line	GC-MS	Kharbouche et al. (2009)
	Food	Oasis MAX	Off-line	CE-MS	Ge et al. (2007)
Fungicides	Food	Oasis MAX	Off-line	GC-MS	Montes et al. (2010)
Additives	Environment	Oasis MAX	Off-line	LC-MS/MS (QqQ)	García-López et al. (2010)
	Biological fluids	Oasis MAX	Off-line	GC-MS	Landberg et al. (2009)
Organic acids	Environment	CHROMABOND HR-AX	Passive sampler	LC-MS	Fauvelle et al. (2012)

elution step with 1 mL of MeOH, just neutral and weak bases (as the target analytes) were recovered, free from acidic and phenolic interferences.

Other interesting studies compare different classes of mixed-mode sorbents, particularly those that compared SCX sorbents to SAX ones (Ge et al. 2006, Josefsson and Sabanovic 2006, Lara et al. 2006, Lavén et al. 2009, Siwek et al. 2008). For example, Oasis HLB, Oasis MCX, and Oasis MAX were compared for the extraction of a group of quinolones from raw milk. The authors found that Oasis MAX provided better recoveries (%R ~ 90%) compared to Oasis MCX (0%) or Oasis HLB (%R ~ 50%). However, the elution conditions in Oasis MAX were not suitable for subsequent electrophoretic analysis. For this reason, a two-step procedure was adopted. This involved a first step with Oasis MAX to eliminate the proteins and remaining fat through the washing step and a second step in which the cleaner extract was loaded into Oasis HLB, which offered an elution solution that was more suitable for capillary electrophoresis (CE) (Lara et al. 2006). Another study compared Oasis MAX and Oasis MCX for the extraction of a group of beta-agonists and beta-antagonists from blood. With Oasis MCX, most of the analytes eluted in the elution step after an effective washing step with methanol solution. In contrast, with Oasis MAX, the target analytes were fractionated and some eluted in the washing step. Therefore, Oasis MCX was selected for this study (Josefsson and Sabanovic 2006).

In multiresidue analysis, a development of a generic analytical method remains difficult when a high number of compounds have to be considered simultaneously. As a result, testing different sorbents with different properties is a common practice in multiresidue analysis, in which the main difficulty is selecting the best SPE sorbent and conditions to give acceptable recoveries for all compounds with different physicochemical properties (Batt et al. 2008, Culleré et al. 2010, Fauvelle et al. 2012, Kasprzyk-Hordern et al. 2007, Marchi et al. 2009, Musenga and Cowan 2013, Tylová et al. 2011, Weigel et al. 2004). For instance, Culleré et al. (2010) tested the retention ability of 13 different sorbents (7 RP sorbents, 4 mixed-mode SCX, and 2 mixed-mode SAX) and 18 different aroma compounds (covering neutral compounds, organic acids, and organic bases) at different pHs. Results showed that two of the RP polymeric sorbents presented the highest retention for most of the compounds. These results partly surprised the authors, who expected that basic compounds at acid pHs and acid compounds at basic pHs would have been more efficiently extracted by mixed-mode SCX and SAX, respectively. In any case, this suggests that ion exchange interactions for these volatile compounds might be weak and the RP interactions may be the most important mechanisms in this extraction. It should be noted that Tables 10.3 through 10.6 (which show the applications for the different mixed-mode sorbents) only provided those applications where the mixed-mode sorbents were selected after a comparative evaluation with other sorbents.

10.5.3 WCX SORBENTS

In a similar way, WCX sorbents have been also applied for the extraction of basic compounds, as summarized in Table 10.5.

As can be seen in this table, they mainly focus on the determination of therapeutic and illicit drugs in environmental samples and fluids. Just two of these studies focus

TABLE 10.5
Applications of the Mixed-Mode WCX Polymeric Sorbents

Compounds	Application	Sorbent	SPE Mode	Analysis	References
Therapeutic drugs	Environment	Oasis WCX	Off-line	LC-FL	Lee et al. (2007)
		HXLPP-WCX	Off-line	LC-UV	Bratkowska et al. (2010)
		Oasis WCX	Off-line	LC-MS/MS (QqQ)	Lee et al. (2007), Zorita et al. (2008)
		Oasis WCX	Off-line	UHPLC-MS/MS (QqQ)	Batt et al. (2008)
		Oasis WCX	Off-line	HILIC-MS/MS (QqQ)	Peru et al. (2006)
	Biological fluids	Strata-X-WC	Off-line	LC-MS	Allanson et al. (2007)
	Food	Strata-X-WC	Off-line	LC-MS/MS (QqQ)	Huq et al. (2006)
Illicit drugs	Environment	Oasis WCX	Online	LC-MS	Fontanals et al. (2013)
	Biological fluids	Strata-X-WC	Online	LC-DAD	Schonberg et al. (2006)
		Strata-X-WC	Online	LC-MS/MS (QqQ)	Chiuminatto et al. (2010)
Biological comp.	Biological fluids	Oasis WCX	96-well plate	LC-MS/MS(QqQ)	Shi et al. (2009)
		Strata-X-WC	96-well plate	LC-MS/MS (QqQ)	Clark and Frank (2011)

on the determination of biological compounds, using the 96-well plate mode. In the rest of the studies, the WCX sorbent is used in the off-line SPE approach, with the exception of five studies that applied the online SPE mode. The online SPE with WCX sorbent is perfectly feasible since the acidic elution solvent (described in the WCX protocol) is matched with a typical LC mobile phase. These circumstances have been exploited by different authors that applied the online SPE (WCX)–LC approaches (Chiuminatto et al. 2010, de Jong et al. 2007, Fontanals et al. 2013). In addition, in the determination of a group of therapeutic and illicit drugs from human urine using Strata-X-CW in online SPE mode, the authors claimed that changing of the pH value from 7 (SPE loading) to 2.5 (chromatographic mobile phase) enhances the separation in terms of both resolution and analysis time since it increases the elution of two of the drugs from the SPE column and decreases their retention time, as well as avoiding the peak tailing (Chiuminatto et al. 2010). Another study (Fontanals et al. 2013) that also determined a group of illicit drugs from complex wastewater samples using Oasis WCX in online SPE–LC–MS reported the reduction of the matrix effect when a mixed-mode sorbent involved an effective washing step based on methanol. In this way, signal suppression and ion enhancement up to 95% and 256% were reported, respectively, when the washing step was not included. Meanwhile, the matrix effect was considered negligible (10%–20% at highest) when the washing step was included. Indeed, the washing step permitted the proper quantification of the target analytes without the use of expensive deuterated internal standards.

With respect to the sorbent comparison, in one study, Oasis WCX was the sorbent of choice from among different types of silica and polymer-based sorbents (no specific details were given), for the quantitative determination of six potential breast cancer biomarker peptides in human serum. The negatively charged WCX sorbent appeared to be only sorbent able to retain one of the peptides (bradykinin), which exhibited a high pK_a value. Therefore, it displayed good recoveries for all of the peptides studied (Shi et al. 2009). Huq et al. (2006) also compared how Strata-X, Strata-X-C, and Strata-X-CW performed in the extraction of a group of tetracyclines from honey samples. In this comparison, they demonstrated that the neutral sorbent, Strata-X, was not efficient at extracting compounds like tetracyclines and also that the elution conditions used to extract tetracyclines from a SCX sorbent such as Strata X-C were not suitable for the MS detector. Therefore, the WCX sorbent, Strata X-CW, was a suitable alternative for efficiently extracting tetracyclines under appropriate conditions as well as for obtaining cleaner extracts. A similar conclusion was drawn in other studies in which a group of fluoroquinolones were extracted from wastewaters using Oasis WCX as sorbent in both cases, which was selected after comparing it with Oasis HLB and Oasis MCX in terms of recoveries and cleanness of the extract (Lee et al. 2007, Zorita et al. 2008). The previous results might indicate that both WCX sorbents have similar features. In the results from the comparison, one should bear in mind that a balanced option should be adopted. For instance, when Oasis HLB, Oasis MCX, and Oasis WCX were compared for the extraction of tetracyclines in wastewaters, on one hand, the neutral sorbent provided good recoveries for all the compounds, but the extracts contained a lot of interferences, which affect the ion suppression in the tandem mass spectrometry (MS/MS) detection.

On the other hand, Oasis MCX strongly retained the target analytes, and it was not possible to quantify them properly. Therefore, the alternative was Oasis WCX, which provided better recoveries and cleaner extracts than its neutral analogue (Oasis HLB) (Lee et al. 2007, Zorita et al. 2008).

10.5.4 WAX SORBENTS

WAX sorbents have been used to some extent for the extraction of compounds with weak acidic or strong acidic; the latter, in principle, are more suitable for extraction with this type of sorbent. Table 10.6 presents all of the detailed information that, in principle, covers all the examples in the literature that uses WAX sorbents.

For the extraction of domoic acid (a neurotoxin that causes amnesic shellfish poisoning) that carboxylic acid groups have (weak anion moieties), Regueiro et al. (2011) selected WAX sorbent instead of the SAX sorbent previously reported in literature for the extraction of domoic acid from shellfish. The authors justified their selection because SAX sorbents are always charged at any pH, which leads to irreversible retention of strong acids present in the sample matrix, and so its extraction efficiency is reduced after every extraction. In contrast, WAX sorbents can be easily reactivated by neutralizing their charge with a basic solution, which results in much longer life span.

The performance of Oasis HLB, Oasis MAX, and Oasis WAX were compared for the determination of estrogens and their conjugates in river sediments. The results obtained showed that, when the targets were extracted from ultrapure matrices (e.g., Milli-Q water), both Oasis HLB and Oasis WAX displayed a similar performance. However, when dealing with the sediments, better results were achieved with Oasis WAX, which allowed a washing step to clean up the matrix (Matejicek et al. 2007). Similar arguments (cleaner extracts obtained in WAX sorbent) were also used for the selection of Oasis WAX sorbent instead of Oasis HLB, for the extraction of fluorescent whitening agents from environmental waters (Chen et al. 2006) or perfluorinated acids in water and biota (Taniyasu et al. 2005).

Other studies, however, compared the performance of the four types of mixed-mode sorbents. For instance, eight different sorbents, Oasis HLB, MAX, MCX, WAX, and WCX, CHROMABOND C_{18}, Isolute ENV+, and Isolute HCX (a silica-based SCX sorbent), were compared for the determination of 28 therapeutic and illicit drugs from surface waters. It was found that Oasis MCX provided the best extraction since the extracts were cleaner (suitable for injection into the UHPLC–MS/MS) and higher recoveries were achieved for all the compounds studied (Kasprzyk-Hordern et al. 2007). The Oasis MCX option was also considered among the other mixed-mode Oasis-based sorbents (i.e., MAX, WCX, and WAX) for the extraction of a group of 34 pharmaceuticals from urine. In this case, MAX and WAX sorbents were ruled out because they were unable to retain basic analytes, whereas WCX was unable to retain acidic compounds during the loading (Marchi et al. 2009).

As well as their comparison, to further exploit the complementary features of the different mixed-mode sorbents, they have been combined in a highly feasible tandem approach that consists of the combination of two (Lavén et al. 2009) or three (Yang et al. 2011) mixed-mode polymeric sorbents in series. For instance, Yang et al. (2011)

TABLE 10.6
Applications of the Mixed-Mode WAX Polymeric Sorbents

Compounds	Application	Sorbent	SPE Mode	Analysis	References
Therapeutic drugs	Environment	Oasis WAX	Off-line	LC-MS/MS (QqQ)	Matejicek et al. (2007)
		HXLPP-WAX		LC-UV	Fontanals et al. (2008)
		HXLPP-WAX	Online		Fontanals et al. (2010b)
	Biological fluids	Oasis WAX	Off-line	GC-MS	Strahm et al. (2008b)
		Oasis WAX	Off-line	LC-MS/MS (QqQ)	Carli et al. (2009), Strahm et al. (2008a,b)
Biological comp.	Food	Biobasic AX	Online	LC-MS/MS (QqQ)	Regueiro et al. (2011)
Additives	Environment	Oasis WAX	Off-line	LC-MS/MS (QqQ)	Chen et al. (2006), Taniyasu et al. (2005)
		Strata-X-AW	Passive sampler	LC-MS	Kaserzon et al. (2012)
Phenolic comp.	Paper and clothes	Oasis WAX	Off-line	LC-MS/MS (QqQ)	Chen and Ding (2006)
	Environment	Oasis WAX	96-well plate	LC-MS/MS (QqQ)	Zedda et al. (2010)
Natural organic matter		Strata-X-AW	Off-line	^{13}C-NMR	Ratpukdi et al. (2009)

developed an SPE procedure that consisted of three mixed-mode polymeric sorbents, (1) Oasis MCX, (2) Oasis MAX, and (3) Oasis WAX, to recover 51 out of 56 metabolites involved in biological pathways (including carboxylic acids, sugar phosphates, amino acids, acyl derivatives, and nucleotides) because, individually, Oasis MCX was unable to retain carboxylic acids, which were suitably retained in Oasis MAX, and Oasis WAX was the only option that enabled the retention of phosphate sugars. Therefore, the tandem approach might be suitable when analytes with disparate physicochemical properties need to be extracted simultaneously.

On other occasions, the mixed-mode sorbents do not provide the best performance (Culleré et al. 2010, Emotte et al. 2012, Fauvelle et al. 2012, Gilart et al. 2012, Gros et al. 2006, Kusch et al. 2006, Seitz et al. 2006, Van De Steene et al. 2006, Weigel et al. 2004). For example, a group of artificial sweeteners in environmental water were determined by SPE–LC–MS/MS. At first, seven different sorbents (Oasis HLB, Isolute ENV+, Bond Elut Plexa, Strata-X as hydrophilic polymeric sorbents, Oasis WAX, Oasis MAX, and Bond Elut Plexa PAX as mixed-mode sorbents) were compared. After preliminary experiments, the authors discarded these tested mixed-mode sorbents due to either excessive or poor retention of the analytes. Moreover, the recovery results obtained with the RP sorbents were unaffected by the matrix effect when dealing with more complex samples (Seitz et al. 2006). In another study (Gilart et al. 2012), Oasis HLB, Oasis MAX, Oasis WAX, and Affinilute MIP–NSAIDs (a commercially available MIP selective for nonsteroidal anti-inflammatory drugs) were compared for the determination of a group of acidic pharmaceuticals in terms of selectivity and capacity. In preliminary results involving SPE–LC–UV, Oasis MAX and Oasis WAX were discarded because, in both cases, a broad band appeared at the beginning of the chromatogram. This was explained due to the elution steps in basic and acidic media, which promoted the removal of acidic interferences from the matrix. Finally, the MIP was selected since it provided a very effective reduction of matrix interferences and selective extraction of target analytes. In any case, it should be highlighted that MIPs are not always available or designed for broader classes of compounds. In such cases, mixed-mode sorbents can play an important role since they have a certain degree of selectivity and greater capacity than MIPs.

In general, the selection of one type of sorbent depends on parameters such as the type of analyte and sample, the extraction conditions, and the compatibility between solutions. Moreover, there are other variables, such as the degree of selectivity or sensitivity achieved in each particular study. In view of this diversity of factors, selecting a suitable sorbent is not a simple or trivial matter, and some experimental tests are recommended to be performed before selecting one type of mixed-mode polymeric sorbent.

10.6 CONCLUSIONS

The development of mixed-mode polymeric sorbents that combine selectivity and capacity is a growing topic of research in SPE. This is widely demonstrated by the range of commercially available materials as well as numerous applications in the field. The applications of mixed-mode polymeric sorbents mainly focus on enhancing sensitivity while reducing the matrix effect usually encountered in complex samples analyzed in MS/MS detectors. The optimization of the parameters involved in the extraction should be carefully performed and should take into account the type of analytes and matrix, the type of mixed-mode sorbent, and the SPE protocol.

Although the application in mixed-mode polymeric sorbents and its suitable protocols are starting to become well established, further applications should be expected involving a broader range of extraction techniques (e.g., SPME, SBSE) or by preparing other materials with more sophisticated features.

ABBREVIATIONS

CE	Capillary electrophoresis
DAD	Diode-array detector
FL	Fluorescence
GC	Gas chromatography
HILIC	Hydrophilic interaction liquid chromatography
HPLC	High-pressure liquid chromatography
IEC	Ion exchange capacity
MS	Mass spectrometry
MS/MS	Tandem mass spectrometry
QqQ	Triple quadrupole
SAX	Strong anion exchange
SCX	Strong cation exchange
SPE	Solid-phase extraction
UHPLC	Ultrahigh-pressure liquid chromatography
UV	Ultraviolet
WAX	Weak anion exchange
WCX	Weak cation exchange

ACKNOWLEDGMENTS

The authors thank the Ministry of Science and Innovation for the Project CTQ 2011-24179 and the Department of Innovation, Universities and Enterprise (Project 2009 SGR 223) for the financial support.

REFERENCES

Abdel-Hamid, M., L. Sharaf, S. Kombian, and F. Diejomaoh. 2006. Determination of dydrogesterone in human plasma by tandem mass spectrometry: Application to therapeutic drug monitoring of dydrogesterone in gynecological disorders. *Chromatographia* 64: 287–292.

Allanson, A.L., M.M. Cotton, J.N.A. Tettey, and A.C. Boyter. 2007. Determination of rifampicin in human plasma and blood spots by high performance liquid chromatography with UV detection: A potential method for therapeutic drug monitoring. *J. Pharm. Biomed. Anal.* 44: 963–969.

Anthemidis, A.N., S. Xidia, and G. Giakisikli. 2012. Study of bond Elut® Plexa™ PCX cation exchange resin in flow injection column preconcentration system for metal determination by flame atomic absorption spectrometry. *Talanta* 97: 181–186.

Batt, A.L., M.S. Kostich, and J.M. Lazorchak. 2008. Analysis of ecologically relevant pharmaceuticals in wastewater and surface water using selective solid-phase extraction and UPLC&MS/MS. *Anal. Chem.* 80: 5021–5030.

Benito-Peña, E., A.I. Partal-Rodera, M.E. León-González, and M.C. Moreno-Bondi. 2006. Evaluation of mixed mode solid phase extraction cartridges for the preconcentration of beta-lactam antibiotics in wastewater using liquid chromatography with UV-DAD detection. *Anal. Chim. Acta* 556: 415–422.

Berg, T., E. Lundanes, A.S. Christophersen, and D.H. Strand. 2009. Determination of opiates and cocaine in urine by high pH mobile phase reversed phase UPLC-MS/MS. *J. Chromatogr. B* 877: 421–432.

Bermudo, E., E. Moyano, L. Puignou, and M.T. Galceran. 2006. Determination of acrylamide in foodstuffs by liquid chromatography ion-trap tandem mass-spectrometry using an improved clean-up procedure. *Anal. Chim. Acta* 559: 207–214.

Bijlsma, L., J.V. Sancho, E. Pitarch, M. Ibáñez, and F. Hernández. 2009. Simultaneous ultra-high-pressure liquid chromatography-tandem mass spectrometry determination of amphetamine and amphetamine-like stimulants, cocaine and its metabolites, and a cannabis metabolite in surface water and urban wastewater. *J. Chromatogr. A* 1216: 3078–3089.

Bratkowska, D., A. Davies, N. Fontanals, P.A.G. Cormack, F. Borrull, D.C. Sherrington, and R.M. Marcé. 2012a. Hypercrosslinked strong anion-exchange resin for extraction of acidic pharmaceuticals from environmental water. *J. Sep. Sci.* 35: 2621–2628.

Bratkowska, D., N. Fontanals, S. Ronka, F. Borrull, A.W. Trochimczuk, and R.M. Marcé. 2012b. Comparison of different imidazolium supported ionic liquid polymeric phases with strong anion-exchange character for the extraction of acidic pharmaceuticals from complex environmental samples. *J. Sep. Sci.* 35: 1953–1958.

Bratkowska, D., R.M. Marcé, P.A.G. Cormack, D.C. Sherrington, F. Borrull, and N. Fontanals. 2010. Synthesis and application of hypercrosslinked polymers with weak cation-exchange character for the selective extraction of basic pharmaceuticals from complex environmental water samples. *J. Chromatogr. A* 1217: 1575–1582.

Brousmiche, D.W., J.E. O'Gara, D.P. Walsh, P.J. Lee, P.C. Iraneta, B.C. Trammell, Y. Xu, and C.R. Mallet. 2008. Functionalization of divinylbenzene/N-vinylpyrrolidone copolymer particles: Ion exchangers for solid phase extraction. *J. Chromatogr. A* 1191: 108–117.

Carli, D., M. Honorat, S. Cohen, M. Megherbi, B. Vignal, C. Dumontet, L. Payen, and J. Guitton. 2009. Simultaneous quantification of 5-FU, 5-FUrd, 5-FdUrd, 5-FdUMP, dUMP and TMP in cultured cell models by LC-MS/MS. *J. Chromatogr. B* 877: 2937–2944.

Carpinteiro, I., M. Ramil, I. Rodríguez, and R. Cela. 2010. Determination of fungicides in wine by mixed-mode solid phase extraction and liquid chromatography coupled to tandem mass spectrometry. *J. Chromatogr. A* 1217: 7484–7492.

Chen, H.-C. and W.-H. Ding. 2006. Hot-water and solid-phase extraction of fluorescent whitening agents in paper materials and infant clothes followed by unequivocal determination with ion-pair chromatography-tandem mass spectrometry. *J. Chromatogr. A* 1108: 202–207.

Chen, H.-C., S.-P. Wang, and W.-H. Ding. 2006. Determination of fluorescent whitening agents in environmental waters by solid-phase extraction and ion pair liquid chromatography-tandem mass spectrometry. *J. Chromatogr. A* 1102: 135–142.

Chiuminatto, U., F. Gosetti, P. Dossetto, E. Mazzucco, D. Zampieri, E. Robotti, M.C. Gennaro, and E. Marengo. 2010. Automated online solid phase extraction ultra high performance liquid chromatography method coupled with tandem mass spectrometry for determination of forty-two therapeutic drugs and drugs of abuse in human urine. *Anal. Chem.* 82: 5636–5645.

Clark, Z.D. and E.L. Frank. 2011. Urinary metanephrines by liquid chromatography tandem mass spectrometry: Using multiple quantification methods to minimize interferences in a high throughput method. *J. Chromatogr. B* 879: 3673–3680.

Coles, R. and E.D. Kharasch. 2007. Stereoselective analysis of bupropion and hydroxybupropion in human plasma and urine by LC/MS/MS. *J. Chromatogr. B* 857: 67–75.

Cormack, P.A.G., A. Davies, and N. Fontanals. 2012. Synthesis and characterisation of microporous polymers microspheres with strong-cation exchange character. *React. Funct. Polym.* 72: 939–946.

Culleré, L., M. Bueno, J. Cacho, and V. Ferreira. 2010. Selectivity and efficiency of different reversed-phase and mixed-mode sorbents to preconcentrate and isolate aroma molecules. *J. Chromatogr. A* 1217: 1557–1566.

Cunliffe, J.M., C.F. Noren, R.N. Hayes, R.P. Clement, and J.X. Shen. 2009. A high-throughput LC-MS/MS method for the quantitation of posaconazole in human plasma: Implementing fused core silica liquid chromatography. *J. Pharm. Biomed. Anal.* 50: 46–52.

de Jong, W.H.A., K.S. Graham, J.C. van der Molen, T.P. Links, M.R. Morris, H.A. Ross, E.G.E. de Vries, and I.P. Kema. 2007. Plasma free metanephrine measurement using automated online solid-phase extraction HPLC-tandem mass spectrometry. *Clin. Chem.* 53: 1684–1693.

Díaz-Cruz, M.S., M.J. García-Galán, and D. Barceló. 2008. Highly sensitive simultaneous determination of sulfonamide antibiotics and one metabolite in environmental waters by liquid chromatography-quadrupole linear ion trap-mass spectrometry. *J. Chromatogr. A* 1193: 50–59.

Dowling, G. and L. Regan. 2011. A new mixed-mode solid phase extraction strategy for opioids, cocaines, amphetamines and adulterants in human blood with hybrid liquid chromatography tandem mass spectrometry detection. *J. Pharm. Biomed. Anal.* 54: 1136–1145.

Emotte, C., O. Heudi, F. Deglave, A. Bonvie, L. Masson, F. Picard, A. Chaturvedi et al. 2012. Validation of an on-line solid-phase extraction method coupled to liquid chromatography-tandem mass spectrometry detection for the determination of Indacaterol in human serum. *J. Chromatogr. B* 895–896: 1–9.

Fauvelle, V., N. Mazzella, F. Delmas, K. Madarassou, M. Eon, and H. Budzinski. 2012. Use of mixed-mode ion exchange sorbent for the passive sampling of organic acids by polar organic chemical integrative sampler (POCIS). *Environ. Sci. Technol.* 46: 13344–13353.

Fontanals, N., F. Borrull, and R.M. Marcé. 2013. On-line weak cationic mixed-mode solid-phase extraction coupled to liquid chromatography–mass spectrometry to determine illicit drugs at low concentration levels from environmental waters. *J. Chromatogr. A* 1286: 16–21.

Fontanals, N., P.A.G. Cormack, R.M. Marcé, and F. Borrull. 2010a. Mixed-mode ion-exchange polymeric sorbents: Dual-phase materials that improve selectivity and capacity. *Trends Anal. Chem.* 29: 765–779.

Fontanals, N., P.A.G. Cormack, and D.C. Sherrington. 2008. Hypercrosslinked polymer microspheres with weak anion-exchange character. Preparation of the microspheres and their applications in pH-tuneable, selective extractions of analytes from complex environmental samples. *J. Chromatogr. A* 1215: 21–29.

Fontanals, N., P.A.G. Cormack, D.C. Sherrington, R.M. Marcé, and F. Borrull. 2010b. Weak-anion exchange hypercrosslinked sorbent in on-line solid-phase extraction-liquid chromatography coupling to achieve automated determination with an effective clean-up. *J. Chromatogr. A* 1217: 2855–2861.

Fontanals, N., R.M. Marcé, and F. Borrull. 2007. New materials in sorptive extraction techniques for polar compounds. *J. Chromatogr. A* 1152: 14–31.

Fontanals, N., S. Ronka, F. Borrull, A.T. Trochimczuk, and R.M. Marcé. 2009. Supported imidazolium ionic liquid phases: A new material for solid-phase extraction. *Talanta* 80: 250–256.

Fontanals, N., B.C. Trammell, M. Galià, R.M. Marcé, P.C. Iraneta, F. Borrull, and U.D. Neue. 2006. Comparison of mixed-mode anion-exchange performance of N-vinylimidazole-divinylbenzene sorbent. *J. Sep. Sci.* 29: 1622–1629.

Fountain, K.J., Z. Yin, and D.M. Diehl. 2009. Simultaneous analysis of morphine-related compounds in plasma using mixed-mode solid phase extraction and UltraPerformance liquid
chromatography-mass spectrometry. *J. Sep. Sci.* 32: 2319–2326.

García-López, M., I. Rodríguez, and R. Cela. 2010. Mixed-mode solid-phase extraction followed by liquid chromatography-tandem mass spectrometry for the determination of
tri- and di-substituted organophosphorus species in water samples. *J. Chromatogr. A*
1217: 1476–1484.

Ge, L., C.Y.C. Peh, J.W.H. Yong, S.N. Tan, L. Hua, and E.S. Ong. 2007. Analyses of
gibberellins by capillary electrophoresis-mass spectrometry combined with solid-phase
extraction. *J. Chromatogr. A* 1159: 242–249.

Ge, L., J.W.H. Yong, S.N. Tan, X.H. Yang, and E.S. Ong. 2006. Analysis of cytokinin nucleotides in coconut (*Cocos nucifera* L.) water using capillary zone electrophoresis-tandem
mass spectrometry after solid-phase extraction. *J. Chromatogr. A* 1133: 322–331.

Gheorghe, A., A. van Nuijs, B. Pecceu, L. Bervoets, P. Jorens, R. Blust, H. Neels, and
A. Covaci. 2008. Analysis of cocaine and its principal metabolites in waste and surface
water using solid-phase extraction and liquid chromatography-ion trap tandem mass
spectrometry. *Anal. Bioanal. Chem.* 391: 1309–1319.

Gil-Garcia, M.D., M.J. Culzoni, M.M. De Zan, R. Santiago-Valverde, M. Martínez-Galera, and
H.C. Goicoechea. 2008. Solving matrix effects exploiting the second-order advantage in
the resolution and determination of eight tetracycline antibiotics in effluent wastewater
by modelling liquid chromatography data with multivariate curve resolution-alternating
least squares and unfolded-partial least squares followed by residual bilinearization
algorithms: II. Prediction and figures of merit. *J. Chromatogr. A* 1179: 115–124.

Gilart, N., R.M. Marcé, F. Borrull, and N. Fontanals. 2012. Determination of pharmaceuticals
in wastewaters using solid-phase extraction-liquid chromatography-tandem mass spectrometry. *J. Sep. Sci.* 35: 875–882.

González-Mariño, I., J.B. Quintana, I. Rodríguez, M. González-Díez, and R. Cela. 2011.
Screening and selective quantification of illicit drugs in wastewater by mixed-mode
solid-phase extraction and quadrupole-time-of-flight liquid chromatography–mass
spectrometry. *Anal. Chem.* 84: 1708–1717.

Gros, M., M. Petrovic, and D. Barceló. 2006. Development of a multi-residue analytical methodology based on liquid chromatography–tandem mass spectrometry (LC-MS/MS) for
screening and trace level determination of pharmaceuticals in surface and wastewaters.
Talanta 70: 678–690.

Gros, M., M. Petrovic, and D. Barceló. 2009. Tracing pharmaceutical residues of different
therapeutic classes in environmental waters by using liquid chromatography/quadrupole-
linear ion trap mass spectrometry and automated library searching. *Anal. Chem.* 81:
898–912.

Harris, S.R., J.I. Gedge, A.N.R. Nedderman, S.J. Roffey, and M. Savage. 2004. A sensitive
HPLC-MS-MS assay for quantitative determination of midazolam in dog plasma.
J. Pharm. Biomed. Anal. 35: 127–134.

Heinig, K. and T. Wirz. 2009. Determination of taspoglutide in human and animal plasma
using liquid chromatography-tandem mass spectrometry with orthogonal column-
switching. *Anal. Chem.* 81: 3705–3713.

Hewitt, D., M. Alvarez, K. Robinson, J. Ji, Y.J. Wang, Y.H. Kao, and T. Zhang. 2011. Mixed-
mode and reversed-phase liquid chromatography-tandem mass spectrometry methodologies to study composition and base hydrolysis of polysorbate 20 and 80. *J. Chromatogr. A*
1218: 2138–2145.

Huq, S., M. Garriques, and K.M.R. Kallury. 2006. Role of zwitterionic structures in the solid-
phase extraction based method development for clean up of tetracycline and oxytetracy-
cline from honey. *J. Chromatogr. A* 1135: 12–18.

Ivanov Dobrev, P. and M. Kaminek. 2002. Fast and efficient separation of cytokinins from auxin and abscisic acid and their purification using mixed-mode solid-phase extraction. *J. Chromatogr. A* 950: 21–29.

Izumi, Y., A. Okazawa, T. Bamba, A. Kobayashi, and E. Fukusaki. 2009. Development of a method for comprehensive and quantitative analysis of plant hormones by highly sensitive nanoflow liquid chromatography-electrospray ionization-ion trap mass spectrometry. *Anal. Chim. Acta* 648: 215–225.

Johansen, K.T., S.J. Ebild, S.B. Christensen, M. Godejohann, and J.W. Jaroszewski. 2012. Alkaloid analysis by high-performance liquid chromatography-solid phase extraction-nuclear magnetic resonance: New strategies going beyond the standard. *J. Chromatogr. A* 1270: 171–177.

Josefsson, M. and A. Sabanovic. 2006. Sample preparation on polymeric solid phase extraction sorbents for liquid chromatographic-tandem mass spectrometric analysis of human whole blood—A study on a number of beta-agonists and beta-antagonists. *J. Chromatogr. A* 1120: 1–12.

Kakimoto, K., A. Toriba, T. Ohno, M. Ueno, T. Kameda, N. Tang, and K. Hayakawa. 2008. Direct measurement of the glucuronide conjugate of 1-hydroxypyrene in human urine by using liquid chromatography with tandem mass spectrometry. *J. Chromatogr. B* 867: 259–263.

Kanaujia, P.K., D. Pardasani, A.K. Purohit, V. Tak, and D.K. Dubey. 2011. Polyelectrolyte functionalized multi-walled carbon nanotubes as strong anion-exchange material for the extraction of acidic degradation products of nerve agents. *J. Chromatogr. A* 1218: 9307–9313.

Karlonas, N., A. Padarauskas, A. Ramanavicius, and A. Ramanaviciene. 2013. Mixed-mode SPE for a multi-residue analysis of benzodiazepines in whole blood using rapid GC with negative-ion chemical ionization MS. *J. Sep. Sci.* 36: 1437–1445.

Kaserzon, S.L., K. Kennedy, D.W. Hawker, J. Thompson, S. Carter, A.C. Roach, K. Booij, and J.F. Mueller. 2012. Development and calibration of a passive sampler for perfluorinated alkyl carboxylates and sulfonates in water. *Environ. Sci. Technol.* 46: 4985–4993.

Kasprzyk-Hordern, B., R.M. Dinsdale, and A.J. Guwy. 2007. Multi-residue method for the determination of basic/neutral pharmaceuticals and illicit drugs in surface water by solid-phase extraction and ultra performance liquid chromatography-positive electrospray ionisation tandem mass spectrometry. *J. Chromatogr. A* 1161: 132–145.

Kharbouche, H., F. Sporkert, S. Troxler, M. Augsburger, P. Mangin, and C. Staub. 2009. Development and validation of a gas chromatography-negative chemical ionization tandem mass spectrometry method for the determination of ethyl glucuronide in hair and its application to forensic toxicology. *J. Chromatogr. B* 877: 2337–2343.

Kirchner, B. 2009. *Ionic Liquids*. Springer, Berlin, Germany.

Klinke, H.B. and K. Linnet. 2007. Performance of four mixed-mode solid-phase extraction columns applied to basic drugs in urine. *Scand. J. Clin. Lab. Invest.* 67: 778–782.

Kojima, M., S. Tsunoi, and M. Tanaka. 2004. High performance solid-phase analytical derivatization of phenols for gas chromatography-mass spectrometry. *J. Chromatogr. A* 1042: 1–7.

Kollroser, M. and C. Schober. 2002. Determination of amiodarone and desethylamiodarone in human plasma by high-performance liquid chromatography-electrospray ionization tandem mass spectrometry with an ion trap detector. *J. Chromatogr. B* 766: 219–226.

Kusch, P., G. Knupp, M. Hergarten, M. Kozupa, and M. Majchrzak. 2006. Solid-phase extraction-gas chromatography and solid-phase extraction-gas chromatography-mass spectrometry determination of corrosion inhibiting long-chain primary alkyl amines in chemical treatment of boiler water in water-steam systems of power plants. *J. Chromatogr. A* 1113: 198–205.

Lai, S.S.L., H.S. Yeung, W.O. Lee, C. Ho, and Y.T. Wong. 2011. Determination of closantel and rafoxanide in animal tissues by online anionic mixed-mode solid-phase extraction followed by isotope dilution liquid chromatography tandem mass spectrometry. *J. Sep. Sci.* 34: 1366–1374.

Landberg, R., P. Aman, and A. Kamal-Eldin. 2009. A rapid gas chromatography-mass spectrometry method for quantification of alkylresorcinols in human plasma. *Anal. Biochem.* 385: 7–12.

Lara, F.J., A.M. García-Campaña, F. Ales-Barrero, J.M. Bosque-Sendra, and L.E. García-Ayuso. 2006. Multiresidue method for the determination of quinolone antibiotics in bovine raw milk by capillary electrophoresis-tandem mass spectrometry. *Anal. Chem.* 78: 7665–7673.

Lavén, M., T. Alsberg, Y. Yu, M. Adolfsson-Erici, and H. Sun. 2009. Serial mixed-mode cation- and anion-exchange solid-phase extraction for separation of basic, neutral and acidic pharmaceuticals in wastewater and analysis by high-performance liquid chromatography-quadrupole time-of-flight mass spectrometry. *J. Chromatogr. A* 1216: 49–62.

Lee, H.-B., T.E. Peart, and M.L. Svoboda. 2007. Determination of ofloxacin, norfloxacin, and ciprofloxacin in sewage by selective solid-phase extraction, liquid chromatography with fluorescence detection, and liquid chromatography-tandem mass spectrometry. *J. Chromatogr. A* 1139: 45–52.

Lehtonen, P., H. Siren, I. Ojanpera, and R. Kostiainen. 2004. Migration behaviour and separation of tramadol metabolites and diastereomeric separation of tramadol glucuronides by capillary electrophoresis. *J. Chromatogr. A* 1041: 227–234.

Li, K., H. Wang, C.O. Brant, S. Ahn, and W. Li. 2011. Multiplex quantification of lamprey specific bile acid derivatives in environmental water using UHPLC-MS/MS. *J. Chromatogr. B* 879: 3879–3886.

Li, Y., A.C. Li, H. Shi, H. Junga, X. Jiang, W. Naidong, and J.H. Lauterbach. 2006. Determination of *S*-phenylmercapturic acid in human urine using an automated sample extraction and fast liquid chromatography-tandem mass spectrometric method. *Biomed. Chromatogr.* 20: 597–604.

Lin, Z.J., S.-X. Qiu, A. Wufuer, and L. Shum. 2005. Simultaneous determination of glycyrrhizin, a marker component in radix Glycyrrhizae, and its major metabolite glycyrrhetic acid in human plasma by LC-MS/MS. *J. Chromatogr. B* 814: 201–207.

López, R., E. Gracia-Moreno, J. Cacho, and V. Ferrreira. 2011. Development of a mixed-mode solid phase extraction method and further gas chromatography mass spectrometry for the analysis of 3-alkyl-2-methoxypyrazines in wine. *J. Chromatogr. A* 1218: 842–848.

Luque, N. and S. Rubio. 2012. Extraction and stability of pesticide multiresidues from natural water on a mixed-mode admicellar sorbent. *J. Chromatogr. A* 1248: 74–83.

Malakova, J., M. Nobilis, Z. Svoboda, M. Lisa, M. Holcapek, J. Kvetina, J. Klimes, and V. Palicka. 2007. High-performance liquid chromatographic method with UV photodiode-array, fluorescence and mass spectrometric detection for simultaneous determination of galantamine and its phase I metabolites in biological samples. *J. Chromatogr. B* 853: 265–274.

Marchi, I., S. Rudaz, and J.-L. Veuthey. 2009. Sample preparation development and matrix effects evaluation for multianalyte determination in urine. *J. Pharm. Biomed. Anal.* 49: 459–467.

Martín-Esteban, A. 2013. Molecularly-imprinted polymers as a versatile, highly selective tool in sample preparation. *TrAC Trends Anal. Chem.* 45: 169–181.

Matejicek, D., P. Houserova, and V. Kuban. 2007. Combined isolation and purification procedures prior to the high-performance liquid chromatographic-ion-trap tandem mass spectrometric determination of estrogens and their conjugates in river sediments. *J. Chromatogr. A* 1171: 80–89.

Montes, R., M. García-López, I. Rodríguez, and R. Cela. 2010. Mixed-mode solid-phase extraction followed by acetylation and gas chromatography mass spectrometry for the reliable determination of trans-resveratrol in wine samples. *Anal. Chim. Acta* 673: 47–53.

Msagati, T.A.M. and M.M. Nindi. 2006. Comparative study of sample preparation methods; supported liquid membrane and solid phase extraction in the determination of benzimidazole anthelmintics in biological matrices by liquid chromatography-electrospray-mass spectrometry. *Talanta* 69: 243–250.

Musenga, A. and D.A. Cowan. 2013. Use of ultra-high pressure liquid chromatography coupled to high resolution mass spectrometry for fast screening in high throughput doping control. *J. Chromatogr. A* 1288: 82–95.

Nanita, S.C., A.M. Pentz, J. Grant, E. Vogl, T.J. Devine, and R.M. Henze. 2008. Mass spectrometric assessment and analytical methods for quantitation of the new herbicide aminocyclopyrachlor and its methyl analogue in soil and water. *Anal. Chem.* 81: 797–808.

Nochetto, C.B., R. Reimschuessel, C. Gieseker, C.S. Cheely, and M.C. Carson. 2009. Determination of tricaine residues in fish by liquid chromatography. *J. AOAC Int.* 92: 1241–1248.

Park, J.W., J.S.J. Hong, N. Parajuli, H.S. Koh, S.R. Park, M.-O. Lee, S.-K. Lim, and Y.J. Yoon. 2007. Analytical profiling of biosynthetic intermediates involved in the gentamicin pathway of *Micromonospora echinospora* by high-performance liquid chromatography using electrospray ionization mass spectrometric detection. *Anal. Chem.* 79: 4860–4869.

Pedrouzo, M., F. Borrull, E. Pocurull, and R.M. Marcé. 2011. Drugs of abuse and their metabolites in waste and surface waters by liquid chromatography-tandem mass spectrometry. *J. Sep. Sci.* 34: 1091–1101.

Peru, K.M., S.L. Kuchta, J.V. Headley, and A.J. Cessna. 2006. Development of hydrophilic interaction chromatography-mass spectrometry assay for spectinomycin and lincomycin in liquid hog manure supernatant and run off from cropland. *J. Chromatogr. A* 1107: 152–158.

Ramos, L. 2012. Critical overview of selected contemporary sample preparation techniques. *J. Chromatogr. A* 1221: 84–98.

Ratpukdi, T., J.A. Rice, G. Chilom, A. Bezbaruah, and E. Khan. 2009. Rapid fractionation of natural organic matter in water using a novel solid-phase extraction technique. *Water Environ. Res.* 81: 2299–2308.

Regueiro, J., E. Martín-Morales, G. Álvarez, and J. Blanco. 2011. Sensitive determination of domoic acid in shellfish by on-line coupling of weak anion exchange solid-phase extraction and liquid chromatography–diode array detection–tandem mass spectrometry. *Food Chem.* 129: 672–678.

Roberts, P.H. and P. Bersuder. 2006. Analysis of OSPAR priority pharmaceuticals using high-performance liquid chromatography-electrospray ionisation tandem mass spectrometry. *J. Chromatogr. A* 1134: 143–150.

Rodríguez-Gonzalo, E., R. Carabias-Martínez, E.M. Cruz, J. Domínguez-Álvarez, and J. Hernández-Méndez. 2009. Ultrasonic solvent extraction and nonaqueous CE for the determination of herbicide residues in potatoes. *J. Sep. Sci.* 32: 575–584.

Schonberg, L., T. Grobosch, D. Lampe, and C. Kloft. 2006. New screening method for basic compounds in urine by on-line extraction-high-performance liquid chromatography with photodiode-array detection. *J. Chromatogr. A* 1134: 177–185.

Seitz, W., W. Schulz, and W.H. Weber. 2006. Novel applications of highly sensitive liquid chromatography/mass spectrometry/mass spectrometry for the direct detection of ultra-trace levels of contaminants in water. *Rapid Commun. Mass Spectrom.* 20: 2281–2285.

Sentellas, S., E. Moyano, L. Puignou, and M.T. Galceran. 2004. Optimization of a clean-up procedure for the determination of heterocyclic aromatic amines in urine by field-amplified sample injection-capillary electrophoresis-mass spectrometry. *J. Chromatogr. A* 1032: 193–201.

Shi, Z.-G., F. Chen, J. Xing, and Y.-Q. Feng. 2009. Carbon monolith: Preparation, characterization and application as microextraction fiber. *J. Chromatogr. A* 1216: 5333–5339.

Siwek, M., A. Noubar, R. Erdmann, B. Niemeyer, and B. Galunsky. 2008. Application of
mixed-mode oasis MCX adsorbent for chromatographic separation of selenomethionine
from antarctic krill after enzymatic digestion. *Chromatographia* 67: 305–308.

Sousa, M., C. Gonçalves, E. Cunha, J. Hajslová, and M. Alpendurada. 2011. Cleanup strategies and advantages in the determination of several therapeutic classes of pharmaceuticals in wastewater samples by SPE-LC-MS/MS. *Anal. Bioanal. Chem.* 399: 807–822.

Strahm, E., I. Kohler, S. Rudaz, S. Martel, P.-A. Carrupt, J.-L. Veuthey, M. Saugy, and C. Saudan.
2008a. Isolation and quantification by high-performance liquid chromatography-ion-
trap mass spectrometry of androgen sulfoconjugates in human urine. *J. Chromatogr. A*
1196–1197: 153–160.

Strahm, E., S. Rudaz, J.-L. Veuthey, M. Saugy, and C. Saudan. 2008b. Profiling of 19-norsteroid
sulfoconjugates in human urine by liquid chromatography mass spectrometry. *Anal.
Chim. Acta* 613: 228–237.

Sun, H., F. Wang, and L. Ai. 2007. Validated method for determination of ultra-trace closantel
residues in bovine tissues and milk by solid-phase extraction and liquid chromatography-
electrospray ionization-tandem mass spectrometry. *J. Chromatogr. A* 1175: 227–233.

Taniyasu, S., K. Kannan, M.K. So, A. Gulkowska, E. Sinclair, T. Okazawa, and N. Yamashita.
2005. Analysis of fluorotelomer alcohols, fluorotelomer acids, and short- and long-chain
perfluorinated acids in water and biota. *J. Chromatogr. A* 1093: 89–97.

Tansupo, P., P. Suwannasom, D.L. Luthria, S. Chanthai, and C. Ruangviriyachai. 2010.
Optimised separation procedures for the simultaneous assay of three plant hormones in
liquid biofertilisers. *Phytochem. Anal.* 21: 157–162.

Tauxe-Wuersch, A., L.F. De Alencastro, D. Grandjean, and J. Tarradellas. 2006. Trace
determination of tamoxifen and 5-fluorouracil in hospital and urban wastewaters.
Int. J. Environ. Anal. Chem. 86: 473–485.

Tsyurupa, M.P. and V.A. Davankov. 2002. Hypercrosslinked polymers: Basic principles of
preparing the new class of polymeric materials. *React. Funct. Polym.* 53: 193–203.

Tylová, T., M. Kolařík, and J. Olšovská. 2011. The UHPLC-DAD fingerprinting method for
analysis of extracellular metabolites of fungi of the genus *Geosmithia* (Ascomycota:
Hypocreales). *Anal. Bioanal. Chem.* 400: 2943–2952.

Van De Steene, J.C., K.A. Mortier, and W.E. Lambert. 2006. Tackling matrix effects during
development of a liquid chromatographic-electrospray ionisation tandem mass spectrometric analysis of nine basic pharmaceuticals in aqueous environmental samples.
J. Chromatogr. A 1123: 71–81.

Vidal, L., M.-L. Riekkola, and A. Canals. 2012. Ionic liquid-modified materials for solid-
phase extraction and separation: A review. *Anal. Chim. Acta* 715: 19–41.

Wang, Z., Z. Wang, J. Wen, and Y. He. 2007. Simultaneous determination of three aconitum
alkaloids in urine by LC-MS-MS. *J. Pharm. Biomed. Anal.* 45: 145–148.

Weigel, S., R. Kallenborn, and H. Hühnerfuss. 2004. Simultaneous solid-phase extraction of
acidic, neutral and basic pharmaceuticals from aqueous samples at ambient (neutral) pH
and their determination by gas chromatography-mass spectrometry. *J. Chromatogr. A*
1023: 183–195.

Williams, L. and J. Caulfield. 2009. High throughput extraction of melanin using Evolute CX
mixed-mode SPE plates. *LC·GC North America* Feb, pp. 44–47.

Wu, X., B. Zhu, L. Lu, W. Huang, and D. Pang. 2012. Optimization of a solid phase extraction
and hydrophilic interaction liquid chromatography–tandem mass spectrometry method
for the determination of metformin in dietary supplements and herbal medicines. *Food
Chem.* 133: 482–488.

Wu, Y.Y., W.X. Shi, and S.Q. Chen. 2009. Determination of beta-estradiol, bisphenol A, diethylstilbestrol and salbutamol in human urine by GC/MS. *Zhejiang Da Xue Xue Bao Yi Xue Ban* 38: 235–241.

Xia, X., X. Li, S. Ding, S. Zhang, H. Jiang, J. Li, and J. Shen. 2009. Ultra-high-pressure liquid chromatography-tandem mass spectrometry for the analysis of six resorcylic acid lactones in bovine milk. *J. Chromatogr. A* 1216: 2587–2591.

Xu, Y., L. Du, E.D. Soli, M.P. Braun, D.C. Dean, and D.G. Musson. 2005. Simultaneous determination of a novel KDR kinase inhibitor and its N-oxide metabolite in human plasma using 96-well solid-phase extraction and liquid chromatography/tandem mass spectrometry. *J. Chromatogr. B* 817: 287–296.

Xue, Y.-J., J.B. Akinsanya, J. Liu, and S.E. Unger. 2006. A simplified protein precipitation/mixed-mode cation-exchange solid-phase extraction, followed by high-speed liquid chromatography/mass spectrometry, for the determination of a basic drug in human plasma. *Rapid Commun. Mass Spectrom.* 20: 2660–2668.

Yang, S., R.E. Synovec, M.G. Kalyuzhnaya, and M.E. Lidstrom. 2011. Development of a solid phase extraction protocol coupled with liquid chromatography mass spectrometry to analyze central carbon metabolites in lake sediment microcosms. *J. Sep. Sci.* 34: 3597–3605.

Yeung, H.-S., W.-O. Lee, and Y.-T. Wong. 2010. Screening of closantel and rafoxanide in animal muscles by HPLC with fluorescence detection and confirmation using MS. *J. Sep. Sci.* 33: 206–211.

Zedda, M., J. Tuerk, S. Peil, and T.C. Schmidt. 2010. Determination of polymer electrolyte membrane (PEM) degradation products in fuel cell water using electrospray ionization tandem mass spectrometry. *Rapid Commun. Mass Spectrom.* 24: 3531–3538.

Zhao, M., M.A. Rudek, P. He, C. Hartke, S. Gore, M.A. Carducci, and S.D. Baker. 2004. Quantification of 5-azacytidine in plasma by electrospray tandem mass spectrometry coupled with high-performance liquid chromatography. *J. Chromatogr. B* 813: 81–88.

Zheng, H., L.-G. Deng, X. Lu, S.-C. Zhao, C.-Y. Guo, J.-S. Mao, Y.-T. Wang, G.-S. Yang, and H. Aboul-Enein. 2010. UPLC-ESI-MS-MS determination of three β2-agonists in pork. *Chromatographia* 72: 79–84.

Zheng, M.M., G.D. Ruan, and Y.Q. Feng. 2009. Hybrid organic-inorganic silica monolith with hydrophobic/strong cation-exchange functional groups as a sorbent for micro-solid phase extraction. *J. Chromatogr. A* 1216: 7739–7746.

Zhou, J.-L., J.-J. An, P. Li, H.-J. Li, Y. Jiang, and J.-F. Cheng. 2009. Two-dimensional turbulent flow chromatography coupled on-line to liquid chromatography-mass spectrometry for solution-based ligand screening against multiple proteins. *J. Chromatogr. A* 1216: 2394–2403.

Zhu, L., Y. Deng, J. Zhang, and J. Chen. 2011. Adsorption of phenol from water by *N*-butylimidazolium functionalized strongly basic anion exchange resin. *J. Colloid Interface Sci.* 364: 462–468.

Zorita, S., L. Larsson, and L. Mathiasson. 2008. Comparison of solid-phase sorbents for determination of fluoroquinolone antibiotics in wastewater. *J. Sep. Sci.* 31: 3117–3121.

11 Interpenetrating Polymer Network Composite Hydrogels and Their Applications in Separation Processes

Ecaterina Stela Dragan and Maria Valentina Dinu

CONTENTS

11.1 INTRODUCTION

Interpenetrating polymer networks (IPNs) are "alloys" of cross-linked polymers, at least one of them being synthesized and/or cross-linked within the immediate presence of the other, without any covalent bonds between them, which cannot be separated unless chemical bonds are broken (Myung et al. 2008, Sperling 1994, 2005). According to the chemistry of preparation, IPN can be classified into (1) simultaneous IPN, when both network precursors are mixed and the two networks are synthesized at the same time by independent, noninterfering routs such as chain and stepwise polymerization (Myung et al. 2008, Sperling 2005, Wang and Liu 2013), and (2) sequential IPN, typically performed by swelling of a single network into a solution containing the mixture of monomer, initiator, and activator, with or without a cross-linker. If a cross-linker is present, full-IPN results, while in the absence of a cross-linker, a network having linear polymers embedded within the first network is formed (semi-IPN) (Hoare and Kohane 2008, Myung et al. 2008, Sperling 1994).

Hydrogels are 3D, hydrophilic, polymeric networks capable to retain large amounts of water or biological fluids and have a soft and rubbery consistence, being thus similar with living tissues (Hoffman 2002, Peppas et al. 2000). Hydrogels may be chemically stable or "reversible" (physical gels), stabilized by molecular entanglements and/or secondary forces including ionic, H-bonding or hydrophobic interactions, these hydrogels being nonhomogeneous (Hoffman 2002, Peppas et al. 2000). Covalently cross-linked networks form permanent or chemical gels (Hoffman 2002). However, single-network hydrogels have weak mechanical properties and slow response at swelling. By the IPN strategy, relatively dense hydrogel matrices can be produced, which feature stiffer and tougher mechanical properties, more widely controllable physical properties, and (frequently) more efficient drug loading compared to conventional hydrogels (Hoare and Kohane 2008, Myung et al. 2008, Peak et al. 2013).

The main classes of polymers, which are employed for the formation of IPN hydrogels, are natural polymers and their derivatives (polysaccharides and proteins) and synthetic polymers containing hydrophilic functional groups such as –COOH, –OH, –CONH$_2$, SO$_3$H, amines and R$_4$N$^+$, and ether. An overview on the main synthesis strategies of IPN hydrogels and their applications in the separation processes is presented in this chapter.

11.2 IPN COMPOSITE HYDROGELS BASED ON POLYSACCHARIDES

11.2.1 ALGINATE-BASED IPN COMPOSITE HYDROGELS

Sodium alginate (SA) is a linear polysaccharide, derived from sea algae composed of 1–4-linked β-D-mannuronic acid (M) and α-L-guluronic acid (G), arranged in a blockwise fashion as homopolymer blocks (MM, GG) or alternating blocks of M and G with different M/G ratios (Draget 2000). It can be easily cross-linked by divalent ions (e.g., Ca^{2+}), which bind the G residues with the transformation in hydrogel. For the preparation of IPN composite hydrogels, SA has been combined with various synthetic polymers (Samanta and Ray 2014). IPN composite hydrogels, composed of SA and synthetic polymers containing carboxylic groups, with novel properties like superporosity (Yin et al. 2007b), electrical sensitivity (Kim et al. 2004b), drug-controlled release (Wang et al. 2009), and multiple-responsiveness (Lin et al. 2010), have been designed.

Superporous IPN composite hydrogels were prepared by Yin et al. (2007b) through sequential strategy, by the fast cross-linking polymerization of acrylamide (AAm) and sodium acrylate in the presence of SA as an entrapped polymer and sodium bicarbonate as a blowing agent. CaCl$_2$ was applied to cross-link the SA chains in semi-IPN gels. The IPN hydrogels thus obtained had a fast swelling and a high equilibrium swelling ratio, depending on the external pH and ionic strength, a high mechanical strength, and a good biocompatibility. IPN composite hydrogels based on poly(methacrylic acid) (PMAA) and SA, in HCl solution, showed a significant and quick bending when subjected to an electric field, and therefore the authors assumed that this hydrogel could be useful for artificial organ components (Kim et al. 2004b).

Synthesis of multiresponsive IPN composite hydrogels, based on SA and poly(N-isopropylacrylamide) (PNIPAAm), constitutes one of the strategies adopted to increase the porosity of the gels and a faster response rate as required for the drug release systems. Both semi-IPN (Dumitriu et al. 2011, Mallikarjuna Reddy et al. 2008, Zhang et al. 2005) and full-IPN (Lee et al. 2006, Zadrazil and Stepanek 2010) have been prepared. The pH-/temperature-sensitive release of indomethacin from semi-IPN hydrogel beads composed of Ca alginate and PNIPAAm has been reported by Shi and coworkers (Shi et al. 2006). The drug release was higher at 37°C than at 25°C and showed that the Ca alginate and PNIPAAm beads had potential as pH/temperature drug delivery system. The pulsatile swelling/deswelling behavior of semi-IPN hydrogels composed of cross-linked PNIPAAm and linear SA revealed that the process was repeatable, by alternating both temperature and pH, their mechanical strength making them suitable for stimuli-responsive drug release systems (Zhang et al. 2005). An interesting strategy for the preparation of IPN SA/PNIPAAm recently reported consists of the preparation first of the ionically cross-linked SA beads, which are then soaked in the solution of NIPAAm, cross-linker, and initiator, followed by cross-linking polymerization of NIPAA at a temperature above the LCST of PNIPAAm (Zadrazil and Stepanek 2010). The obtained composite beads changed their transparency in response to the change of temperature but kept their original shape and size.

Synthesis of IPN composite hydrogels based on SA and other synthetic polymers like poly(acrylamide) (PAAm) (Demirel et al. 2006), poly(ethylene glycol) (PEG) (Mahou and Wandrey 2010), poly(vinyl alcohol) (PVA) (Solak 2011), and poly(vinylpyrrolidone) (PVP) (Wang and Wang 2010) and their swelling properties and applications for controlled release of bioactive agents was also recently reported. In situ formed IPN hydrogels based on a physical network of calcium alginate interpenetrated with a chemical cross-linked network based on dextran (Dx) derivatized with 2-hydroxyethyl methacrylate (HEMA) have been evaluated for the protein release, mechanical characteristics, and biocompatibility (Pescosolido et al. 2011).

11.2.2 Chitosan-Based IPN Composite Hydrogels

Chitosan (CS), the single linear cationic polysaccharide, has remarkable biological properties like biodegradability, biocompatibility, and antibacterial activity being thus ideal for biomedical applications. The high content of amino and hydroxyl functional groups endows CS with high potential as a sorbent for proteins, dyes, and metal ions (Crini and Badot 2008, Wan Ngah et al. 2011). Moreover, CS and its derivatives have been used as components in the formation of IPN composite hydrogels with various ionic polymers containing carboxylic groups like poly(acrylic acid) (PAA) and copolymers of acrylic acid (Bocourt et al. 2011, Yin et al. 2007a, 2008), PMAA (Chen et al. 2005, Milosavljevic et al. 2011), or cationic centers like quaternary ammonium groups and amine groups (Guo et al. 2007, Kim et al. 2005). The synthesis of semi-IPN has been carried out either by selective cross-linking of CS in the presence of a preformed polyelectrolyte (Chen et al. 2005, Kim et al. 2005) or by the synthesis of the cross-linked polyelectrolyte in the presence of CS (Bocourt et al. 2011, Guo et al. 2007, Milosavljevic et al. 2011).

IPNs have been also prepared by the post-cross-linking of CS entrapped in a polyelectrolyte matrix (Yin et al. 2007a). The ionic interactions between $-NH_3^+$ groups of CS and $-COO^-$ from the anionic polyelectrolyte contributed to the increase of the mechanical properties of the gels and to the decrease of the swelling degree because they contribute to the relative increase of the cross-linking density of the gel (Chen et al. 2005, Milosavljevic et al. 2011). However, the ionic cross-links make the gels to be reversible responsive to variation of the solution pH and ionic strength (Chen et al. 2005, Yin et al. 2008). Guo et al. have obtained thermo- and pH-responsive semi-IPN polyampholyte hydrogels based on carboxymethyl CS and poly(dimethylaminoethyl methacrylate) (PDMAEM) (Guo et al. 2007). The advantage of this hydrogel was that the release rate of coenzyme A could be modulated as a function of temperature, making the semi-IPN hydrogel pH/temperature responsive and thus suitable for drug delivery systems.

Numerous IPN composite hydrogels were prepared by cross-linking polymerization of nonionic monomers in the presence of CS, the most employed monomers being AAm (Bonina et al. 2004, Zhou and Wu 2011), N-isopropylacrylamide (NIPAAm) (Alvarez-Lorenzo et al. 2005), and HEMA (Han et al. 2008, Ramesh Babu et al. 2010). Modulation of the mechanical properties and the water content of the hydrogels was expected, one main purpose being their use in controlled release systems and as scaffolds in tissue engineering.

In our own research, we have prepared first semi-IPN composed of CS as entrapped polymer in a matrix of PAAm, as conventional composite hydrogels (Drăgan et al. 2012c), by free-radical cross-linking copolymerization of AAm with N,N'-methylenebisacrylamide (BAAm) in the presence of CS. Formation of full-IPN hydrogels was achieved by a selective cross-linking of CS with epichlorohydrin (ECH), in alkaline medium, when a simultaneous generation of anionic sites on the PAAm matrix, by the partial hydrolysis of amide groups (Ilavsky et al. 1984, Zhao et al. 2010), occurred. The formation of PAAm/CS IPN composite hydrogels is schematically presented in Figure 11.1.

Semi-IPN and IPN hydrogels were characterized by FTIR, differential scanning calorimetry (DSC), scanning electron microscopy (SEM), and equilibrium swelling. The freeze-dried hydrogels displayed a porous morphology generated by the sublimation of ice crystals under freeze-drying conditions. To evaluate the sorption mechanism of water by both semi-IPN and IPN hydrogels, the transport of water has been analyzed by means of the semiempirical equation proposed by Franson and Peppas (1983) and was found that the transport of water was the Fickian diffusion for all the gels (Drăgan et al. 2012c).

IPN composite hydrogels have been also prepared by blending CS with preformed synthetic polymers like PAAm (Bonina et al. 2004), polyacrylonitrile (PAN) (Kim et al. 2003), PEG (Martinez et al. 2004), PVA (Liang et al. 2009, Liu et al. 2011), PVP (Marsano et al. 2005), followed by the selective cross-linking of CS. Yang et al. (2013) have recently reported the preparation of novel hydrogels composed of PEG grafted on carboxymethyl CS and SA and found

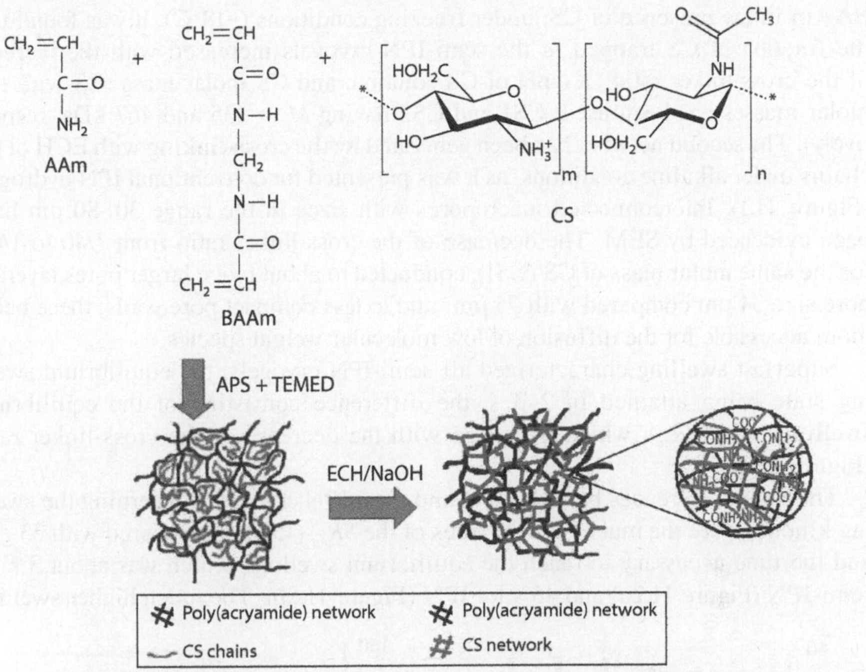

FIGURE 11.1 Schematic representation of the formation of PAAm/CS IPN hydrogels in two steps. (Reprinted from *Carbohydr. Polym.*, 88, Drăgan, E.S., Perju, M.M., and Dinu, M.V., Preparation and characterization of IPN composite hydrogels based on polyacrylamide and chitosan and their interaction with cationic dyes. 2270–2281, Copyright 2012, with permission from Elsevier.)

an improvement in the protein release at pH 7.4, suggesting this composite hydrogel to be promising for protein drug delivery in the intestine.

Mechanical properties and the diffusion of solutes in IPN hydrogels can be further modulated by their preparation below the freezing point of the reaction solutions, when the most part of water forms crystals, the bound water and the soluble substances (monomers, initiators, and polymers) being concentrated in a nonfrozen liquid microphase, where the gel is formed (Baydemir et al. 2009, Dinu et al. 2007, Jain and Kumar 2009, Lozinsky et al. 2003, Plieva et al. 2005, Zhao et al. 2010). By their interconnected pore structure, cryogels allow the unhindered diffusion of solutes or even colloidal particles, making them very attractive in biomedicine and biotechnology including chromatographic materials, carriers for the immobilization of molecules and cells, matrices for cell separations, and cell culture. Cryogels are endowed with a capillary network through which the solvent can flow by convective mass transport and a high mechanical strength and osmotic stability, which make them adequate for various biomedical applications and bioseparations (Jain and Kumar 2009, Lozinsky et al. 2003).

Drăgan et al. have prepared ionic composite cryogels consisting of two independently cross-linked and oppositely charged networks (Drăgan et al. 2012b). Semi-IPN cryogels have been prepared by cross-linking polymerization of AAm with

BAAm in the presence of CS, under freezing conditions (–18°C). It was found that the fraction of CS trapped in the semi-IPN cryogels increased with the increase of the cross-linker ratio (X), pH of CS solution, and CS molar mass (CS with two molar masses has been used: CS1 and CS2 having M_v = 235 and 467 kDa, respectively). The second network has been generated by the cross-linking with ECH of CS chains under alkaline conditions, as it was presented for conventional IPN hydrogels (Figure 11.1). Interconnected macropores with sizes in the range 30–80 μm have been evidenced by SEM. The decrease of the cross-linker ratio from 1/40 to 1/60, for the same molar mass of CS (CS1), conducted to about twice larger pores (average pore size 34 μm compared with 75 μm) and to less compact pore walls, these being more accessible for the diffusion of low molecular weight species.

Superfast swelling characterized all semi-IPN cryogels, the equilibrium swelling state being attained in 2–3 s, the difference consisting of the equilibrium swelling ratio (SR_{eq}), which increased with the decrease of the cross-linker ratio (Figure 11.2a).

The main differences between IPN and semi-IPN cryogels concerning the swelling kinetics were the much higher values of the SR_{eq} (155 g/g compared with 33 g/g) and the time necessary to reach the equilibrium swelling, which was about 3 s for semi-IPN (Figure 11.2a) and 45 s for IPN (Figure 11.2b). The much higher swelling

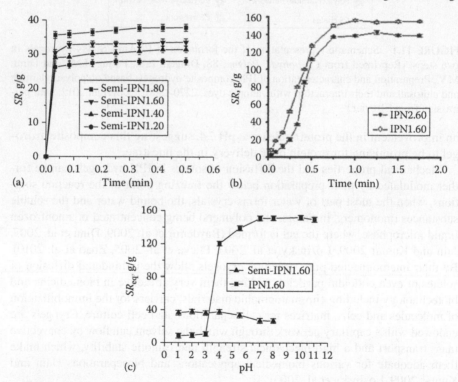

FIGURE 11.2 Swelling ratio of PAAm/CS semi-IPN (a) and IPN (b) cryogels as a function of time and equilibrium swelling ratio of PAAm/CS cryogels as a function of pH (c). (Adapted from Drăgan, E.S. et al., *Chem. Eng. J.*, 204–206, 198, 2012b.)

ratios of IPN cryogels have been attributed to the presence of the anionic matrix, which is bearing $-COO^-$ groups. As can be seen in Figure 11.2c, both semi-IPN and IPN were pH responsive; the SR values of semi-IPN decreased when pH increased from 4 to 7, while the IPN cryogels, having two independent networks responsive at pH, behaved completely different. Thus, the SR values abruptly increased when pH increased from 3 to 4, monotonously increased with the increase of pH up to 7, and remained almost constant when the pH varied in the range 7–10.

11.2.3 IPN HYDROGELS BASED ON STARCH AND DERIVATIVES

Native starch granules are water insoluble, containing two major components: (1) amylose, 20%–30% of the starch granules, which consists of linear chains of α-(1–4-linked-D-glucose) units, and (2) amylopectin, which consists of branched chains of α-(1–4-linked-D-glucose) units interlinked by α-(1–6-linked-D-glucose) linkages, in proportion of 70%–80%. Various modifications of starch were developed to improve its hydrophilicity (Bhuniya et al. 2003, Reis et al. 2008). Multicomponent hydrogels as semi-IPN or IPN showed improved mechanical properties, faster response rate, and diffusion of solutes (Jin et al. 2013, Keshavara Murthy et al. 2006, Li et al. 2008b, 2009). The swelling/diffusion properties of the semi-IPN hydrogels prepared by the cross-linking copolymerization of AAm and sodium methacrylate in the presence of starch have been investigated by Keshavara Murthy et al. (2006). The high equilibrium water content (EWC) of these composite hydrogels recommends them as novel biomaterials in biomedical/pharmaceutical technology or as moisture maintenance materials in agriculture fields. Amphoteric semi-IPN composite hydrogels have been prepared by the graft copolymerization of acrylic acid onto cationic starch in the presence of either poly(methacryloyloxyethylammonium chloride) (PDMC) (Li et al. 2009) or poly(diallyldimethylammonium chloride) (PDADMAC) (Li et al. 2008b). The swelling studies showed a high swelling capacity in distilled water and outstanding pH sensitivity of the semi-IPN hydrogels.

In our investigations, native potato starch (PS) or anionically modified PS (PA) have been entrapped in a PAAm matrix, both conventional semi-IPN hydrogels (Drăgan and Apopei 2011, Drăgan et al. 2012a) and semi-IPN cryogels being prepared (Apopei and Drăgan 2013, Drăgan and Apopei 2013, Drăgan and Apopei Loghin 2013). The synthesis strategy is schematically presented in Figure 11.3.

The swelling data have been analyzed by the empirical equation of Franson and Peppas (1983), the swelling being controlled by a Fickian diffusion when PS has been entrapped ($n < 0.45$), and by an anomalous transport ($0.45 < n < 0.89$) when the PA has been entrapped in PAAm matrix (Drăgan and Apopei 2011).

Multiresponsive semi-IPN composite cryogels as rods were prepared by the radical cross-linking copolymerization of AAm with BAAm in the presence of PS or PA, under the freezing point of the solvent ($-18°C$) (Drăgan and Apopei 2013). Figure 11.4 shows that the gel morphology has been influenced by the composition of the gels and also by the posttreatment consisting of the controlled hydrolysis of the composite gels under alkaline conditions (4 h in 0.5 M NaOH, at 25°C).

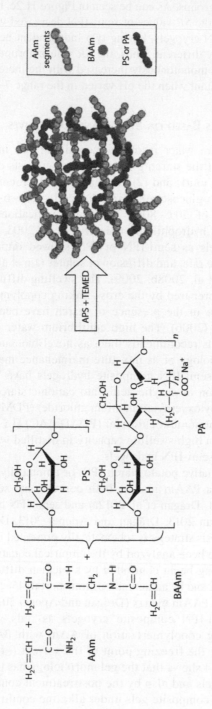

FIGURE 11.3 Schematic representation of the PAAm/PS and PAAm/PA semi-IPN hydrogels.

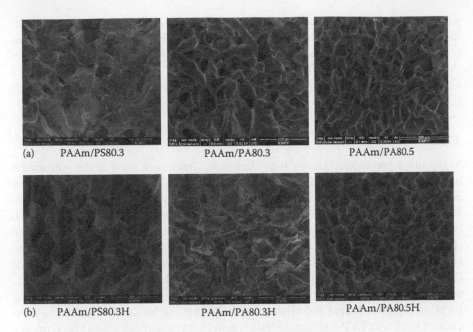

(a)   PAAm/PS80.3   PAAm/PA80.3   PAAm/PA80.5

(b)   PAAm/PS80.3H   PAAm/PA80.3H   PAAm/PA80.5H

FIGURE 11.4 SEM images of PAAm/PS80.3, PAAm/PA80.3, and PAAm/PA80.5 semi-IPN composite cryogels, before (a) and after (b) the controlled hydrolysis (Mag 500×). (Adapted from Drăgan, E.S. and Apopei, D.F., *Carbohydr. Polym.*, 92, 23, 2013.)

As Figure 11.4 shows, the internal morphology (the size and geometry of pores) of the samples prepared with a monomer concentration of 3 wt.% has been strongly influenced by the entrapped polymer structure, the average pore size being around 67 μm for PAAm/PS80.3 and 62 μm for PAAm/PA80.3. The morphology of selected gels changed in a different way after hydrolysis, macropores with sizes in the range 80–82 μm being visible in the SEM image of PAAm/PS80.3H, while the influence of hydrolysis was smaller in the case of the sample PAAm/PA80.3H, pore sizes decreased to around 53 μm. The larger pores found in the case of PAAm/PS80.3H than in the case of PAAm/PA80.3H are attributed to the higher swelling degree of PAAm/PS80.3H, caused by a higher density of the anionic groups (–COO⁻) generated by a higher level of hydrolysis of PAAm matrix of the first gel. The increase of the monomer concentration had a definite influence on the morphology of the gels based on PA (PAAm/PA80.5 compared with PAAm/PA80.3) (Figure 11.4). Polyhedral interconnected pores can be seen in the SEM images of PAAm/PA80.3 and PAAm/PA80.5 before hydrolysis, the pore size decreasing with the increase of the monomer concentration (58 μm for PAAm/PA80.5 compared with 62 μm for PAAm/PA80.3). The influence of the controlled hydrolysis on the pore size and geometry has been more critical at a lower concentration of monomers (PAAm/PA80.3H compared with PAAm/PA80.5H).

The rapid response rate to the external stimuli of the smart hydrogels is the most essential function of their applications, and therefore various methods have been used to increase the response kinetics. The study of the deswelling/reswelling kinetics of the composite gels, having a low cross-linker ratio (1/80), in ethanol–water as poor solvent/good solvent, and in 1 M NaCl/water, showed a higher responsivity of the anionic gels having the monomer concentration, C_o, of 3% than of those having $C_o = 5\%$, this behavior supporting the potential of these gels as "smart" hydrogels (Drăgan and Apopei 2013).

11.2.4 IPN HYDROGELS BASED ON OTHER POLYSACCHARIDES

Many other polysaccharides or their derivatives have been used in the preparation of semi-IPN or IPN composite hydrogels, the most employed being cellulose (Wang et al. 2013a), carboxymethylcellulose (CMC) (Bajpai and Mishra 2004, Bajpai and Mishra 2005, Wang et al. 2011b, Xiao et al. 2009), hyaluronic acid (HA) (Agostino et al. 2012), kappa-carrageenan (Chen et al. 2009, Mahdavinia et al. 2010), xanthan (Hamcerencu et al. 2011), guar gum (Li et al. 2008a), chondroitin sulfate (Agostino et al. 2012), etc. Semi-IPN hydrogels composed of cross-linked PAA and entrapped CMC were prepared by Bajpai and Mishra, the network parameters (average molecular weight between cross-links, cross-link density) being evaluated from the water sorption capacity (Bajpai and Mishra 2004). By grafting NaAA on CMC in the presence of the linear PVP and a cross-linker, novel superabsorbent semi-IPN composite hydrogels have been prepared by Wang et al. (2011b).

HA is a linear polysaccharide of high molecular weight consisting of two alternating disaccharide units of β-1,4-linked D-glucuronic acid and β-1,3-N-acetyl-D-glucosamine (Kim et al. 2004a). HA is behaving like a lubricant by protecting the articular cartilage surface, acting as a scavenger molecule for free radicals (Agostino et al. 2012). One drawback of unmodified HA is the low stability of the resulting construct because of its high water solubility, and therefore some strategies have been developed to get stable constructs like semi-IPN hydrogels (Agostino et al. 2012). Semi-IPN with rapid response rate at temperature composed of kappa-carrageenan and a matrix of poly(N,N-diethylacrylamide) were recently reported (Chen et al. 2009).

In order to prepare macroporous semi-IPN composite hydrogels with superfast responsiveness, Dinu et al. have synthesized semi-IPN hydrogels composed of PAAm as a matrix and either Dx (Dinu et al. 2011a,b, 2013) or dextran sulfate (DxS) (Dinu et al. 2011c, 2012) as entrapped polysaccharide. The characteristics of semi-IPN composite hydrogels were compared with those of the cross-linked PAAm without polysaccharide. The gel preparation temperature and the presence of Dx or DxS were found to be the key factors determining the porous structure of the networks. It was found that the stability of DxS into the composite hydrogels increased with the decrease of the synthesis temperature and with the increase of the cross-linker ratio, the lowest percentage of DxS being released from the composite hydrogels obtained at −18°C and having a cross-linker ratio of 1/40. The swelling ratios of the composite hydrogels were higher than those found for the PAAm gels, irrespective of the gel preparation temperature. Moreover, by conducting the cross-linking polymerization reaction at −18°C, semi-IPNs with superfast responsive rate (some seconds) were obtained (Dinu et al. 2011a,b).

The interior network structures of the semi-IPNs prepared at −18°C (cryogels) exhibited a heterogeneous morphology consisting of pores of sizes around 80 µm, while those formed at +5°C or +25°C showed pores with sizes around 3 µm. The SEM images of the PAAm/DxS semi-IPN hydrogels included in Figure 11.5a and b illustrate the difference between the interior morphology of the composite gels prepared at +5°C (Figure 11.5a) and that of the cryogel (Figure 11.5b) where interconnected pores with sizes of about 120 µm are visible.

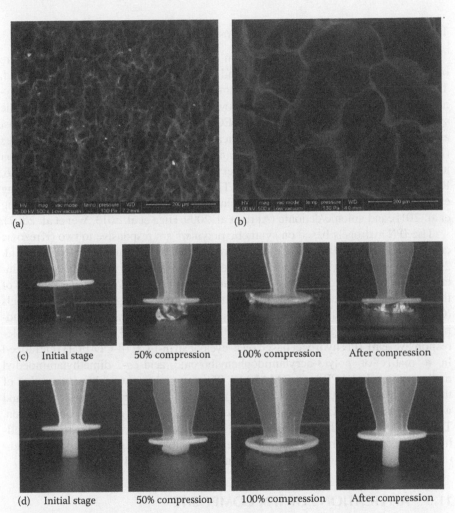

FIGURE 11.5 SEM images of the PAAm/DxS semi-IPN composite hydrogels prepared at +5°C (a) and at −18°C (b), at X = 1/80, at Mag 500×; comparative behavior of PAAm/DxS semi-IPN composite hydrogels at X = 1/40 under uniaxial compression; (c) conventional hydrogels prepared at +20°C, and (d) cryogels prepared at −18°C. (Reprinted from *React. Funct. Polym.,* 71, Dinu, M.V., Perju, M.M., and Drăgan, E.S., Composite IPN ionic hydrogels based on polyacrylamide and dextran sulfate, 881–890, Copyright 2011, with permission from Elsevier.)

The uniaxial compression measurements performed on equilibrium swollen PAAm/DxS semi-IPN composite hydrogels clearly show the influence of the synthesis temperature on their mechanical properties. Thus, the behavior of the gels against the uniaxial compression showed that the composite gels prepared at +5°C (Figure 11.5c) have been broken under compression, while highly stable hydrogels have been obtained by conducting the cross-linking copolymerization at subzero temperature, which recovered their initial shape after the compression (Figure 11.5d).

11.2.5 IPN Hydrogels Based on Synthetic Polymers

The structures of biopolymers depend on many natural factors, while the structure of synthetic polymers can be designed and reproduced whenever they are requested. This could explain the numerous IPN composite hydrogels based on synthetic polymers, which have been grouped as follows: (1) IPN hydrogels based on nonionic synthetic polymers, where PHEMA, PEG, PAAm, and PVA represent the most employed polymers mainly for biomedical applications and separation processes (Lee et al. 2008, Wang and Liu 2012, Zhang et al. 2009); (2) IPN hydrogels based on ionic polymers, such as anionic (Baskan et al. 2013, Jin et al. 2009, Liu et al. 2012a,b, Rodriguez et al. 2006, Tang et al. 2008, Thimma Reddy and Takahara 2009, Wang et al. 2008), cationic (Chen et al. 2010, Huang et al. 2011b, Silan et al. 2012, Zhang et al. 2011), and anionic/cationic (Ajiro et al. 2009, Higa et al. 2013, Wei et al. 2007).

The IPN hydrogels based on synthetic polymers are responsive to two (Krezovic et al. 2013, Liu et al. 2012a,b) or three external stimuli (Huang et al. 2011b, Zhao et al. 2006). The second network or the entrapped linear polymer in semi-IPN could influence the responsivity of the gels. For example, the deswelling/reswelling kinetics of the semi-IPN hydrogels was much faster than that of the single-network hydrogels (Chen et al. 2010, Liu et al. 2012a, Zhang et al. 2009). To increase the drug loading capacity of hydrogels for hydrophobic drugs, Huang et al. prepared a multiple-responsive semi-IPN hydrogel based on β-cyclodextrin-ECH (β-CD-ECH) entrapped in a matrix of poly(3-acrylamidophenylboronic acid-co-2-dimethylaminoethyl methacrylate) [P(AAPBA-co-DMAEM)]. The drug release was slower than that of the conventional P(AAPBA-co-DMAEM) hydrogel due to the presence of β-CD and was influenced by pH, temperature, ionic strength, and the glucose concentration. The method used for the synthesis of IPN could have also an influence on the swelling kinetics and drug release, a higher thermosensitivity being observed for sequential than for simultaneous semi-IPN (Thimma Reddy and Takahara 2009).

11.3 APPLICATIONS OF IPN COMPOSITE HYDROGELS IN SEPARATION PROCESSES

11.3.1 Removal of Dyes

Water pollution with dyes is becoming a serious threat to the environment and human health due to the large variety of dyes used in numerous industries, which discharge a large amount of effluents including dyes. Majority of the dyes are toxic, carcinogenic, and nonbiodegradable. Physical, chemical, and biological methods are usually employed for the dye removal from industrial waste waters, but they are not always effective. Adsorption is considered an effective and economical method to remove (recover) dyes even at high concentrations, having some advantages such as flexibility in the selection of the adequate sorbent and operation and the production of effluents suitable to be reused (Kumar and Sivanesan 2006, Wawrzkiewicz 2013). Therefore, the interest focused on finding novel sorbents with high adsorption capacities, fast adsorption/desorption rate, and easy separation and regeneration strongly increased last decade (Dalaran et al. 2011, Huang et al. 2011a). Lately, multicomponent hydrogels as semi-IPN or IPN incorporating synthetic polymers (Kundakci et al. 2011, Mandal et al. 2012, Tang et al. 2008, Üzüm and Karadağ 2012, 2013) or polymers coming from bioresources (Drăgan and Apopei 2011, Drăgan et al. 2012a,c, Jeon et al. 2008, Kusuktham 2006, Mandal and Ray 2013, Şolpan et al. 2008, Zhao et al. 2012) are gaining more and more interest for application as novel sorbents.

Semi-IPN and IPN hydrogels based on PVA and poly(AA-*co*-HEMA) have been prepared by Mandal et al. and their sorption capacity for rhodamine B (RB) and methyl violet (MV) from dilute aqueous solution have been investigated (Mandal et al. 2012). The dye adsorption of both dyes was lower on IPN than on semi-IPN, situation attributed to the tighter network structure of the IPN. Şolpan et al. prepared SA/AAm semi-IPN conventional hydrogels containing 3 wt.% SA entrapped in PAAm matrix and followed their usability in the removal of some textile dyes (Şolpan et al. 2008). S-type adsorption isotherms were found for all cationic dyes, the adsorption capacities at pH 7 and 25°C being in the order MV > methylene blue (MB) > Safranin O > magenta.

Macroporous hydrogels, characterized by a faster response rate at small changes of the external stimuli and a fast diffusion of solutes compared to the conventional hydrogels gained more and more interest last decade. As it was already shown (Section 11.2.2), cryogelation is one of the most accessible and versatile techniques to generate permanent macropores in hydrogels. Ionic multicomponent cryogels, having enhanced mechanical and chemical resistance, have been tested as novel sorbents in the separation processes of small ionic species and in bioseparations (Drăgan et al. 2012b, Hajizadeh et al. 2010, Perju et al. 2012). In our research, the sorption of two ionic dyes, the anionic dye Direct Blue 1 (DB1) and the cationic dye MB, on the semi-IPN and IPN hydrogels based on PAAm and CS was investigated on both conventional hydrogels (Drăgan et al. 2012c) and cryogels (Drăgan et al. 2012b). It was found that semi-IPN hydrogels preferentially sorbed DB1, owing to the positively charged groups from CS, while the IPN sorbed a high amount of MB owing to the presence of anionic COO^- groups generated during the formation of the second network. The controlling mechanism of adsorption was investigated by fitting three kinetic models on the experimental data: the pseudo-first-order (PFO) kinetic model (Lagergren 1898), the pseudo-second-order (PSO) kinetic model (Ho 2006, Ho and McKay 1998), and the intra-particle diffusion model by Weber and Morris (1963). It was found that the theoretical $q_{e,calc}$ values estimated by PFO model were very close to the experimental values, for both semi-IPN and full-IPN, and this supports physisorption as the main controlling mechanism of sorption.

Analysis of equilibrium data by isotherm models is very important to compare different sorbents under different operational conditions and to design and optimize an operating procedure. Therefore, the relationship between the amount of MB sorbed onto IPN2.60 composite cryogel and the dye concentrations at equilibrium has been described by Langmuir, Freundlich, and Sips sorption isotherm models (Figure 11.6a).

It was found that the Sips isotherm fitted the best the experimental data, and the maximum sorption capacity estimated by fitting Sips model has been 755.5 mg MB/g cryogel (Drăgan et al. 2012b). The PAAm/CS IPN cryogels were endowed with excellent properties in separation of MB from its mixture with methyl orange (MO) (Figure 11.6b) and a high level of reusability as shown in Figure 11.6c.

Semi-IPN composite cryogels having anionically modified PS as entrapped polymer in a matrix of PAAm with enhanced sorption of MB from aqueous solutions were recently reported (Drăgan and Apopei Loghin 2013, Drăgan et al. 2012a). The sorption capacity was further increased by a controlled hydrolysis of PAAm matrix, owing to the increase of the density of active COO^- groups. The shape of the sorption isotherm changed from an isotherm of "L" type, before hydrolysis (Figure 11.6d), to an "H" isotherm, after hydrolysis (Figure 11.6e), which supports a very high affinity of the hydrolyzed composite gel for the cationic dye. The sorption kinetics was better described by the PFO kinetic model and this showed the overall adsorption process appearing to be controlled by physisorption. The cryogels have also showed that their sorption capacity did not decrease after six consecutive sorption/desorption cycles (Figure 11.6f).

Table 11.1 summarizes some of the recently reported results on the sorption of dyes, either on conventional IPN hydrogels or on IPN cryogels.

In general, IPN hydrogels are endowed with a high level of reusability, which is of great practical significance (Drăgan and Apopei 2011, Mandal and Ray 2013), this being further enhanced by the synthesis of hydrogels at subzero temperatures (Drăgan and Apopei Loghin 2013, Drăgan et al. 2012b).

11.3.2 Removal of Heavy Metals

Pollution of water and soil by heavy metal ions is considered extremely dangerous because of their nonbiodegradability and high toxicity. Among heavy metals, lead (Pb), mercury (Hg), cadmium (Cd), chromium (Cr), and arsenic (As) are considered the most dangerous because of their very high toxicity, carcinogenic effects, and persistence in the environment. There are three main categories of the treatment technologies for heavy metal removal: (1) chemical treatment (chemical precipitation, chemical reduction, and electrolysis), (2) separation/enrichment by sorption without alteration of metal state (active carbon adsorption, ion exchange, reverse osmosis, evaporation, etc.), and (3) biological processes. For the recovery/enrichment of heavy metal ions, adsorption is preferred owing to its wide range of applicability, feasibility, and flexibility (Zhao et al. 2011). Biopolymers such as gelatin (Chauhan et al. 2003), CS (Drăgan and Dinu 2013, Gerente et al. 2007, Huang et al. 2012), SA (Şolpan and Torun 2005, Wang et al. 2013b), starch (Apopei et al. 2012), cellulose derivatives (Chauhan and Mahajan 2002), and other polysaccharides (Chauhan et al. 2009) constitute partners in the preparation of semi- or full-IPN hydrogels with promising sorption properties for heavy metal ions. Wang et al. have recently reported preparation of semi-IPN

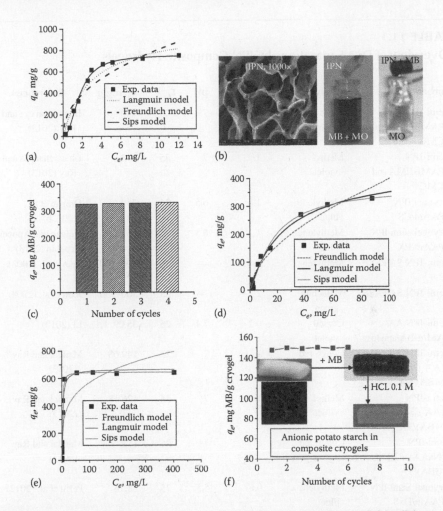

FIGURE 11.6 Sorption of MB on PAAm/CS IPN cryogels (a–c) and on PAAm/PA semi-IPN cryogels (d–f): (a) experimental isotherm and the isotherms obtained by the nonlinear fit of Langmuir, Freundlich, and Sips model isotherms for the sorption of MB onto PAAm/CS IPN cryogel; (b) separation of MB from its mixture with MO by selective sorption on the PAAm/CS IPN cryogel; (c) equilibrium sorption capacity of MB onto PAAm/CS IPN cryogel as a function of the number of cycles; (d) experimental isotherm and the isotherms obtained by the nonlinear fit of Langmuir, Freundlich, and Sips model isotherms for the sorption of MB onto PAAm/PA semi-IPN cryogel, before hydrolysis; (e) experimental isotherm and the isotherms obtained by the nonlinear fit of Langmuir, Freundlich, and Sips model isotherms for the sorption of MB onto PAAm/PA semi-IPN cryogel, after hydrolysis; (f) equilibrium sorption capacity of MB onto PAAm/PA semi-IPN cryogel, after hydrolysis, as a function of the number of cycles. (a–c: Reprinted from *Chem. Eng. J.*, 204–206, Drăgan, E.S., Lazăr, M.M., Dinu, M.V., and Doroftei, F., Macroporous composite IPN hydrogels based on poly(acrylamide) and chitosan with tuned swelling and sorption of cationic dyes, 198–209, Copyright 2012; d–e: *Carbohydr. Polym.*, 92, Drăgan, E.S. and Apopei, D.F., Multiresponsive macroporous semi-IPN composite hydrogels based on native or anionically modified potato starch, 23–32, Copyright 2013, with permission from Elsevier.)

TABLE 11.1
Overview of Dye Separations by IPN Composite Hydrogels

Sorbent	Dye	Sorbent Dosage, g/L	pH	T, °C	q_e, mg/g	Reference
Semi-IPN PAMHEMA and CMC	Basic fuchsin	1	7	25	920	Bhattacharyya and Ray (2013)
Semi-IPN PAMHEMA and CMC	Methyl violet	1	7	25	613.8	Bhattacharyya and Ray (2013)
Cryogel IPN PAAm/CS	Methylene blue	1	6.5	25	750	Drăgan et al. (2012b)
Cryogel semi-IPN PAAm/PA	Methylene blue	1	6.5	25	667.7	Drăgan and Apopei Loghin (2013)
Semi-IPN SA/PASP	Methylene blue	2	—	25	600–700	Jeon et al. (2008)
Semi-IPN SA/PASP	Malachite green	2	—	25	300–350	Jeon et al. (2008)
Semi-IPN AA/ AM/n-BA/amylose	Crystal violet	0.2	7.4	25	35.09	Li (2010)
Semi-IPN (AA-co-HEMA/ MBA)/SA	Congo red	1	7	25	149.68	Mandal and Ray (2013)
Semi-IPN (AA-co-HEMA/ MBA)/SA	Methyl violet	1	7	25	126.18	Mandal and Ray (2013)
Semi-IPN (NaAA-co-HEMA/ MBA)/SA	Congo red	1	7	25	172	Mandal and Ray (2013)
Cryogel Semi-IPN PAAm/DxS	Methylene blue	0.67	5.5	25	18.76	Perju et al. (2012)
Semi-IPN CS/ (AAm-PEG macromer)	Methyl orange	0.6	—	25	185.24	Zhao et al. (2012)
Semi-IPN CS/ (AAm-PEG macromer)	Acid red 18	0.6	—	25	342.54	Zhao et al. (2012)

hydrogels composed of CS-g-PAA as a matrix and gelatin as entrapped protein with very high sorption capacities for Pb^{2+} (Huang et al. 2012) and Cu^{2+} (Wang et al. 2013b), the equilibrium of sorption being attained in about 15 min. The equilibrium adsorption isotherm was fitted by the nonlinear form of the Langmuir isotherm, and the sorption kinetic was well described by the PSO kinetic model indicating the chemisorption by chelation of metal ions as the mechanism of sorption. The presence of gelatin chains enhanced the mechanical strength of semi-IPN composite hydrogels and contributed to the increase of the maximum sorption capacity of metal ions (261.08 mg Cu^{2+}/g gel and up to 736.95 mg Pb^{2+}/g gel) and to the increase of the sorbent reusability. Semi-IPN composite hydrogels composed of SA-g-PAA as a matrix and PVP and gelatin as entrapped chains and their sorption capacity for Ni^{2+}, Cu^{2+}, Zn^{2+}, and Cd^{2+} have been reported by Wang et al., the maximum equilibrium adsorption capacity, in noncompetitive conditions, being 3.16 mmol Ni^{2+}/g, 3.22 mmol Cu^{2+}/g, 3.03 mmol Zn^{2+}/g, and 2.91 mmol Cd^{2+}/g (Wang et al. 2013c).

Our own research has been focused on the investigation of the equilibrium sorption capacity for Cu^{2+}, Cd^{2+}, Ni^{2+}, and Zn^{2+} of the semi-IPN cryogels based on PAAm as a matrix and anionically modified PS as entrapped polymer (Apopei et al. 2012). The experimental data obtained in batch mode have been analyzed by four isotherm models: Langmuir (Langmuir 1918), Freundlich (Freundlich 1906), Sips (Srivastava et al. 2006), and Temkin (Gerente et al. 2007). A comparison of the linear and non-linear regression fitting of these isotherms has been performed because, sometimes, the linearization of a model function has a negative effect on the ability of the model to fit the experimental data (Gerente et al. 2007, Limousin et al. 2007). Based on the nonlinear regression method, it was found that the best fitted isotherm on the experimental data was the Sips isotherm model with a theoretical sorption capacity of 40.72 mg Cu^{2+}/g, 19.72 mg Cd^{2+}/g, 9.31 mg Ni^{2+}/g, and 7.48 mg Zn^{2+}/g. Sorption of Cu^{2+} on the PAAm/CS IPN cryogels is illustrated in Figure 11.7b, for the composite gels prepared with CS1 and CS2.

As can be observed from the images of IPN hydrogel, a strong dehydration occurred after the sorption of Cu^{2+}, the size of gel sample decreasing at about 1/4 in the case of conventional IPN hydrogel (Figure 11.7a) and at about 1/2 in the case of cryogel (Figure 11.7b), from the initial size of the sample. This behavior demonstrates that a strong interaction between metal ions and the IPN hydrogels occurred, attributed mainly to the presence of –COO⁻ sites. Concerning the sorption kinetics

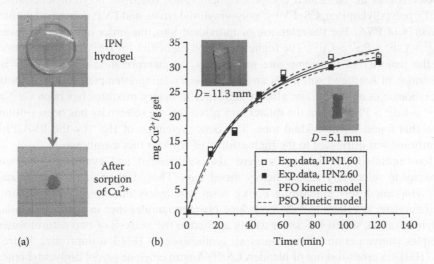

FIGURE 11.7 Sorption of Cu^{2+} on PAAm/CS IPN conventional hydrogels (a) and on PAAm/CS IPN cryogels (b).

of Cu^{2+} on the PAAm/CS IPN cryogels, the PSO kinetic model fitted better the experimental data (Figure 11.7b), and this showed the chemisorption was the rate-determining step controlling the adsorption process.

Stimuli-responsive semi-IPN hydrogels composed of PNIPAAm as matrix and poly(sodium acrylate) as entrapped homopolymer and their sorption properties for Cu^{2+} have been reported by Yamashita et al. (2003). The IPN hydrogels exhibited volume-phase transition behavior in the adsorption conditions, that is, the IPN hydrogel adsorbed sufficiently Cu^{2+} ions below but not above the volume phase transition temperature (VPTT). Wang et al. prepared IPN hydrogels with enhanced adsorption properties for heavy metal ions either by simultaneous, free-radical/cationic photo-polymerization of 2-acrylamido-2-methyl-1-propansulfonic acid (AMPS) and triethylene glycol divinyl ether (DVE-3) (Wang and Liu 2013) or by sequential strategy with poly(ethyleneglycol diacrylate) [poly(PEGDA)] and PMAA as the two independent networks (Wang et al. 2011a). Adsorption properties of the IPN hydrogels for the removal of Cu^{2+}, Cd^{2+}, and Pb^{2+}, compared with single networks, have been examined in batch mode, under noncompetitive conditions. The adsorption capacity of the sequential IPN hydrogels increased with the increase of PMAA content in the IPN hydrogel (Wang et al. 2011a). The adsorption capacity of the IPN gels increased almost linearly with the increase of the initial concentration of metal ion, the experimental isotherms being better described by the Freundlich isotherm, suggesting the adsorption was heterogeneous, in multilayer pattern. Examination of the sorption mechanism revealed a diffusion-controlled transport mechanism of metal ion sorption on that IPN hydrogel. In general, the sorption kinetics was very fast compared with the single-network hydrogels (Wang and Liu 2013, Wang et al. 2011a).

Metal complexing membranes have been prepared by semi-IPN technique, and their sorption properties for Pb^{2+} (Bessbousse et al. 2008, 2012, M'Bareck et al. 2006), Hg^{2+}, Cd^{2+}, and Cu^{2+} (Bessbousse et al. 2008, 2012) have been investigated. Thus, Bessbousse et al. fabricated composite membranes based on poly(ethyleneimine) (PEI), poly(allylamine), CS, PVPy, poly(vinylimidazole), and PVP, entrapped within a matrix of PVA. For the retention of individual ions, the order of selectivity was $Hg^{2+} > Cu^{2+} > Pb^{2+} > Cd^{2+}$. The highest sorption capacities for Hg^{2+} have been found on the majority of the composite membranes, an interpretation of the results in the frame of hard and soft acids and bases theory being attempted by the authors (Bessbousse et al. 2012). The affinity sequence in binary mixtures has been $Cu^{2+} > Hg^{2+} > Cd^{2+} > Pb^{2+}$, and in the quaternary mixtures, the selectivity has been similar with that found for individual ions. The better retention of Hg^{2+} by the PVA/PEI membrane was attributed to the high affinity of Hg^{2+} for this membrane.

Ion-imprinted IPN hydrogels were also synthesized and evaluated for their capacity to selectively adsorb heavy metal ions. Thus, Junyan et al. synthesized Cd^{2+}-imprinted IPN containing epoxy resin, triethylenetetramine, and cadmium methacrylate acrylamide-N,N-methylene-bis-(acrylamide) by in situ sequential polymerization, which was successfully applied to the analysis of two natural water samples (Junyan et al. 2006). Liu et al. synthesized an IPN ion-imprinting hydrogel (IIH) via cross-linking of blended CS/PVA with ethylene glycol diglycidyl ether (EGDE) using uranyl ion as template (Liu et al. 2010). The optimum pH has been in the range 5.0–6.0, and the adsorption process was well described by both Langmuir and Freundlich isotherms. Equilibrium of sorption was attained within 2 h, and the maximum adsorption capacity was 156 mg/g. The most significant results, which support the advantage of the IIH compared to the nonimprinted hydrogel, consist of the selective adsorption of uranyl ion in a mixture with other heavy metals, the distribution ratio of IIH for uranyl ion being sixfold greater than that of the non-imprinted hydrogel but was almost the same for the other heavy metals. A novel thermoresponsive Cu^{2+} ion-imprinted IPN [Cu^{2+}-IIH] has been recently reported by Wang and Liu (2013). The Cu^{2+}-IIH has been prepared by free-radical/cationic polymerization (simultaneous strategy) of NIPAAm and triethylene glycol DVE-3 using Cu^{2+} ion as template. The memory was fixed by shrinking above the VPTT and was deleted by swelling below the VPTT. The Cu^{2+}-IIH showed a stronger affinity for Cu^{2+} ions than for other competitor metal ions compared with the non-imprinted IPN hydrogel.

11.3.3 SEPARATION OF LIQUIDS

Pervaporation is a membrane process extensively used as an energy-saving process for separation of close boiling and azeotropic liquid mixtures, such as dehydration process of various organic solvents. Separation of water from ethanol–water (Amnuaypanich and Kongchana 2009, Amnuaypanich et al. 2009, Buyanov et al. 2004), isopropanol–water (Singha et al. 2009a), and tetrahydrofuran–water (Kuila and Ray 2012) mixtures by IPN membranes has been investigated last decades as alternative to the classical pervaporation membranes. Another strategy has been applied by Amnuaypanich et al. who prepared semi-IPN membranes based on natural rubber (NR) entrapped either in the PAA matrix cross-linked with ethylene glycol or in the cross-linked PVA incorporating also zeolite 4A. Examination of water state in the composite membranes by DSC showed that by increasing the PAA content, more water molecules were bound to the PAA functional groups as nonfreezing bound water. It was found that at low water concentrations of feed mixtures (<30% w/w), increasing the PAA content of the membrane can enhance both water permeation flux and separation factor (Amnuaypanich and Kongchana 2009). For low feed water concentration, the water flux was further increased by rising the temperature. However, by increasing the content of PAA in the membrane and water content in the feed, the water separation factor decreased because the network structure has been expanded and gave the possibility as ethanol molecules to also diffuse into the membrane (Amnuaypanich and Kongchana 2009, Buyanov et al. 2004). Therefore, another approach has been adopted in the preparation of more effective composite membranes by embedding a hydrophilic zeolite in the polymer matrix formed by cross-linked PVA, the NR being entrapped in the PVA matrix, to get a novel mixed matrix (MM) membrane (Amnuaypanich et al. 2009). The incorporated zeolite caused a less degree of swelling in water of the MM membrane compared with the NR/PVA membranes and lower content of the nonfreezing bound water. The MM membranes also showed that for low water concentration of the feed, the increase of zeolite content improved both the total permeation flux and the water separation factor. The estimated activation energies for water flux, lower than those of ethanol flux, suggested that the composite MM IPN membranes were highly water selective.

Separation of organic–organic mixtures by pervaporation is very attractive because the PV is performed at low temperature and the membrane can be reused (Park et al. 1994, Singha et al. 2009b). However, the use of IPN membranes for this task is only seldom mentioned in literature (Singha et al. 2009b). Separation of toluene–methanol mixture has been successfully performed by a full-IPN produced by cross-link copolymerization of AA and HEMA in aqueous solution of PVA, followed by cross-linking of PVA. A conventional PVA membrane cross-linked with GA has also been used for comparison. The flux and selectivity of these IPN membranes were found to be much higher than those of the conventional GA cross-linked PVA membrane. It was found that the optimum content of the cross-linked copolymer AA/HEMA was 50%, this membrane being characterized by optimum performance in terms of flux and methanol selectivity.

11.4 CONCLUSIONS AND PERSPECTIVES

Because single-network hydrogels have poor mechanical properties and slow response at swelling, various strategies have been used to remediate these weak points. By the IPN strategy, relatively dense hydrogel matrices can be produced, which, due to the synergistic effect among the independent networks, are endowed with more widely controllable physical properties like swelling ratio and swelling kinetics and tougher mechanical properties and (frequently) more efficient drug loading and separation properties compared to single-network hydrogels. Such composite hydrogels are promising sorbents for the removal of pollutants in water. Sorption kinetics and reusability of IPN hydrogels have been further enhanced by the synthesis of the IPN hydrogels under the freezing temperature of the solvent.

ABBREVIATIONS

AAm	Acrylamide
AMPS	2-acrylamido-2-methyl-1-propansulfonic acid
BAAm	N,N'-methylenebisacrylamide
CMC	Carboxymethyl cellulose
CS	Chitosan
DSC	Differential scanning calorimetry
Dx	Dextran
DxS	Dextran sulfate
ECH	Epichlorohydrin
EWC	Equilibrium water content
GA	Glutaraldehyde
HA	Hyaluronic acid
HEMA	2-hydroxyethyl methacrylate
IPN	Interpenetrating polymer network
LCST	Lower critical solution temperature
MB	Methylene blue
MO	Methyl orange
MV	Methyl violet

NIPAAm   *N*-isopropylacrylamide
PAA   Poly(acrylic acid)
PAAm   Poly(acrylamide)
PAN   Poly(acrylonitrile)
PASP   Poly(aspartic acid)
PDADMAC   Poly(diallyldimethylammonium chloride)
PDMAEM   Poly(*N*,*N*-dimethylaminoethyl methacrylate)
PDMC   Poly(methacryloyloxyethylammonium chloride)
PEG   Poly(ethylene glycol)
PEI   Poly(ethyleneimine)
PMAA   Poly(methacrylic acid)
PNIPAAm   Poly(*N*-isopropylacrylamide)
PS   Potato starch
PVA   Poly(vinyl alcohol)
PVP   Poly(vinylpyrrolidone)
RB   Rhodamine B
SA   Sodium alginate
SEM   Scanning electron microscopy
SF   Silk fibroin
SR_{eq} Equilibrium swelling ratio
VPTT   Volume phase transition temperature

ACKNOWLEDGMENT

This work was supported by a grant of the Ministry of National Education, CNCS-UEFISCDI, project number PN-II-ID-PCE-2011-3-0300.

REFERENCES

Agostino, A.D., A. La Gatta, T. Busico, M. De Rosa, and C. Schiraldi. 2012. Semi-interpenetrated hydrogels composed of PVA and hyaluronan or chondroitin sulphate: Chemico-physical and biological characterization. *J. Biotechnol. Biomater.* 2: Art. 140.

Ajiro, H., Y. Takemoto, T. Asoh, and M. Akashi. 2009. Novel polyion complex with interpenetrating polymer network of poly(acrylic acid) and partially protected poly(vinylamine) using *N*-vinylacetamide and *N*-vinylformamide. *Polymer* 50: 3503–3507.

Alvarez-Lorenzo, C., A. Concheiro, A.S. Dubovik, N.V. Grinberg, T.V. Burova, and V.Y. Grinberg. 2005. Temperature-sensitive chitosan-poly(*N*-isopropylacrylamide) interpenetrated networks with enhanced loading capacity and controlled release properties. *J. Control. Release* 102: 629–641.

Amnuaypanich, S. and N. Kongchana. 2009. Natural rubber/poly(acrylic acid) semi-interpenetrating polymer network membranes for the pervaporation of water-ethanol mixtures. *J. Appl. Polym. Sci.* 114: 3501–3509.

Amnuaypanich, S., J. Patthana, and P. Phinyocheep. 2009. Mixed matrix membranes prepared from natural rubber/poly(vinyl alcohol) semi-interpenetrating polymer network (NR/PVA semi-IPN) incorporating with zeolite 4A for the pervaporation dehydration of water-ethanol mixtures. *Chem. Eng. Sci.* 64: 4908–4918.

Apopei, D.F., M.V. Dinu, A. Trochimczuk, and E.S. Dragan. 2012. Sorption isotherms of heavy metal ions onto semi-IPN cryogels based on polyacrylamide and anionically modified potato starch. *Ind. Eng. Chem. Res.* 51: 10462–10471.

Apopei, D.F. and E.S. Dragan. 2013. Semi-interpenetrating polymer networks based on polyacrylamide and starch or modified starch. *J. Nanostruct. Polym. Nanocomposites* 9: 16–20.

Bajpai, A.K. and A. Mishra. 2004. Ionizable interpenetrating polymer networks of carboxymethyl cellulose and polyacrylic acid: Evaluation of water uptake. *J. Appl. Polym. Sci.* 93: 2054–2065.

Bajpai, A.K. and A. Mishra. 2005. Preparation and characterization of tetracycline-loaded interpenetrating polymer networks of carboxymethyl cellulose and poly(acrylic acid): Water sorption and drug release study. *Polym. Int.* 54: 1347–1356.

Baskan, T., D.C. Tuncaboylu, and O. Okay. 2013. Tough interpenetrating Pluronic F127/polyacrylic acid hydrolysis. *Polymer* 54: 2979–2987.

Baydemir, G., N. Bereli, M. Andac, R. Say, I.Y. Galaev, and A. Denizli. 2009. Bilirubin recognition via molecularly imprinted supermacroporous cryogels. *Colloids Surf. B* 68: 33–38.

Bessbousse, H., T. Rhlalou, J.-F. Verchere, and L. Lebrun. 2008. Removal of heavy metal ions from aqueous solutions by filtration with a novel complexing membrane containing poly(ethyleneimine) in a poly(vinyl alcohol) matrix. *J. Membr. Sci.* 307: 249–259.

Bessbousse, H., J.-F. Verchere, and L. Lebrun. 2012. Characterisation of metal-complexing membranes prepared by the semi-interpenetrating polymer networks technique. Application to the removal of heavy metal ions from aqueous solutions. *Chem. Eng. J.* 187: 16–28.

Bhattacharyya, R. and S.K. Ray. 2013. Kinetic and equilibrium modeling for adsorption of textile dyes in aqueous solutions by carboxymethyl cellulose/poly(acrylamide-*co*-hydroxyethyl methacrylate) semi-interpenetrating network hydrogel. *Polym. Eng. Sci.* 53: 2439–2453.

Bhuniya, S.P., S. Rahman, A.J. Satyanand, M.M. Gharia, and A.M. Dave. 2003. Novel route to synthesis of allyl starch and biodegradable hydrogel by copolymerizing allyl modified starch with methacrylic acid and acrylamide. *J. Polym. Sci. Part A: Polym. Chem.* 41: 1650–1658.

Bocourt, M., W. Arguelles-Monal, J.V. Cauich-Rodríguez, A. May, N. Bada, and C. Peniche. 2011. Interpenetrated chitosan-poly(acrylic acid-*co*-acrylamide) hydrogels. Synthesis, characterization and sustained protein release studies. *Mater. Sci. Appl.* 2: 509–520.

Bonina, P., Ts. Petrova, and N. Manolova. 2004. pH-Sensitive hydrogels composed of chitosan and polyacrylamide—Preparation and properties. *J. Bioact. Compat. Polym.* 19: 101–116.

Buyanov, A.L., L.G. Revel'skaya, E.Y. Rosova, and G.K. Elyashevich. 2004. Swelling behavior and pervaporation properties of new composite membrane systems: Porous polyethylene film-poly(acrylic acid) hydrogel. *J. Appl. Polym. Sci.* 94: 1461–1465.

Chauhan, G.S., S. Kumar, A. Kumari, and R. Sharma. 2003. Study on the synthesis, characterization, and sorption of some metal ions on gelatin- and acrylamide-based hydrogels. *J. Appl. Polym. Sci.* 90: 3856–3871.

Chauhan, G.S. and S. Mahajan. 2002. Use of novel hydrogels based on modified cellulosics and methacrylamide for separation of metal ions from water systems. *J. Appl. Polym. Sci.* 86: 667–671.

Chauhan, K., G.S. Chauhan, and J.-H. Ahn. 2009. Synthesis and characterization of novel guar gum hydrogels and their use as Cu^{2+} sorbents. *Bioresour. Technol.* 100: 3599–3603.

Chen, J., M. Liu, and S. Chen. 2009. Synthesis and characterization of thermo- and pH-sensitive kappa-carrageenan-*g*-poly(methacrylic acid)/poly(*N,N*-diethylacrylamide) semi-IPN hydrogel. *Mater. Chem. Phys.* 115: 339–346.

Chen, J., M. Liu, H. Liu, L. Ma, C. Gao, S. Zhu, and S. Zhang. 2010. Synthesis and properties of thermo- and pH-sensitive poly(diallyldimethylammonium chloride)/poly(*N,N*-diethylacrylamide) semi-IPN hydrogel. *Chem. Eng. J.* 159:247–256.

Chen, S., M. Liu, S. Jin, and Y. Chen. 2005. Synthesis and swelling properties of pH-sensitive hydrogels based on chitosan and poly(methacrylic acid) semi-interpenetrating polymer network. *J. Appl. Polym. Sci.* 98: 1720–1726.

Crini, G. and P.M. Badot. 2008. Application of chitosan, a natural aminopolysaccharide for dye removal from aqueous solution by adsorption process using batch studies: A review of recent literature. *Prog. Polym. Sci.* 33: 399–447.

Dalaran, M., S. Emik, G. Guclu, T.B. Iyim, and S. Ozgumus. 2011. Study on a novel poly-ampholyte nanocomposite superabsorbent hydrogels: Synthesis, characterization and investigation of removal of indigo carmine from aqueous solution. *Desalination* 279: 170–182.

Demirel, G., G. Özcetin, F. Şahin, H. Tümtürk, S. Aksoy, and N. Hasirci. 2006. Semi-interpenetrating polymer networks (IPNs) for entrapment of glucose isomerase. *React. Funct. Polym.* 66: 389–394.

Dinu, M.V., M. Cazacu, and E.S. Drăgan. 2013. Mechanical, thermal, and surface properties of poly(acrylamide)/dextran semi-interpenetrating network hydrogels tuned by the synthesis temperature. *Cent. Eur. J. Chem.* 11: 248–258.

Dinu, M.V., M.M. Ozmen, E.S. Dragan, and O. Okay. 2007. Freezing as a path to build macroporous structures: Superfast responsive polyacrylamide hydrogels. *Polymer* 48: 195–204.

Dinu, M.V., M.M. Perju, M. Cazacu, and E.S. Drăgan. 2011a. Polyacrylamide-dextran polymeric networks: Effect of gel preparation temperature on their morphology and swelling properties. *Cellulose Chem. Technol.* 45: 197–203.

Dinu, M.V., M.M. Perju, and E.S. Drăgan. 2011b. Porous semi-interpenetrating hydrogel networks based on dextran and polyacrylamide with superfast responsiveness. *Macromol. Chem. Phys.* 212: 240–251.

Dinu, M.V., M.M. Perju, and E.S. Drăgan. 2011c. Composite IPN ionic hydrogels based on polyacrylamide and dextran sulfate. *React. Funct. Polym.* 71: 881–890.

Dinu, M.V., S. Schwarz, I.A. Dinu, and E.S. Drăgan. 2012. Comparative rheological study of ionics-IPN composite hydrogels based on polyacrylamide and dextran sulfate and of polyacrylamide hydrogels. *Colloid Polym. Sci.* 290: 1647–1657.

Drăgan, E.S. and D.F. Apopei. 2011. Synthesis and swelling behavior of pH-sensitive semi-interpenetrating polymer network composite hydrogels based on native and modified potatoes starch as potential sorbent for cationic dyes. *Chem. Eng. J.* 178: 252–263.

Drăgan, E.S. and D.F. Apopei. 2013. Multiresponsive macroporous semi-IPN composite hydrogels based on native or anionically modified potato starch. *Carbohydr. Polym.* 92: 23–32.

Drăgan, E.S. and D.F. Apopei Loghin. 2013. Enhanced sorption of Methylene Blue from aqueous solutions by semi-IPN composite cryogels with anionically modified potato starch entrapped in PAAm matrix. *Chem. Eng. J.* 234: 211–222.

Drăgan, E.S. and M.V. Dinu. 2013. Design, synthesis and interaction with Cu^{2+} of ice templated composite hydrogels. *Res. J. Chem. Environ.* 17: 4–10.

Drăgan, E.S., M.V. Dinu, and D.F. Apopei. 2012a. Macroporous anionic interpenetrating polymer networks composite hydrogels and their interaction with Methylene Blue. *Int. J. Chem.* 1: 548–569.

Drăgan, E.S., M.M. Lazăr, M.V. Dinu, and F. Doroftei. 2012b. Macroporous composite IPN hydrogels based on poly(acrylamide) and chitosan with tuned swelling and sorption of cationic dyes. *Chem. Eng. J.* 204–206: 198–209.

Drăgan, E.S., M.M. Perju, and M.V. Dinu. 2012c. Preparation and characterization of IPN composite hydrogels based on polyacrylamide and chitosan and their interaction with cationic dyes. *Carbohydr. Polym.* 88: 2270–2281.

Draget, K.I. 2000. Alginates. In *Handbook of Hydrocolloids*, G.O. Philips and P.A. Williams (eds.). Woodhead Publishing, Cambridge, U.K., pp. 379–395.

Dumitriu, R.P., G.R. Mitchell, and C. Vasile. 2011. Multi-responsive hydrogels based on N-isopropylacrylamide and sodium alginate. *Polym. Int.* 60: 222–233.

Franson, N.M. and N.A. Peppas. 1983. Influence of copolymer composition on non-Fickian water transport through glassy copolymers. *J. Appl. Polym. Sci.* 28: 1299–1310.

Freundlich, H.M.F. 1906. Über die adsorption in lösungen. *Z. Phys. Chem.* 57(A): 385–470.

Gerente, C., V.K.C. Lee, P. Le Cloirec, and G. McKay. 2007. Application of chitosan for the removal of metals from wastewaters by adsorption—Mechanisms and models review. *Crit. Rev. Environ. Sci. Technol.* 37: 41–127.

Guo, B., J. Yuan, L. Yao, and Q. Gao. 2007. Preparation and release profiles of pH/temperature responsive carboxymethyl chitosan/P(2-(dimethylamino)ethyl methacrylate) semi-IPN amphoteric hydrogel. *Colloid Polym. Sci.* 285: 665–671.

Hajizadeh, S., H. Kirsebom, I.Y. Galaev, and B. Mattiasson. 2010. Evaluation of selective composite cryogel for bromate removal from drinking water. *J. Sep. Sci.* 33: 1752–1759.

Hamcerencu, M., J. Desbrieres, A. Khoukh, M. Popa, and G. Riess. 2011. Thermodynamic investigation of thermoresponsive xanthan-poly(N-isopropylacrylamide) hydrogels. *Polym. Int.* 60: 1527–1534.

Han, Y.A., E.M. Lee, and B.C. Ji. 2008. Mechanical properties of semi-interpenetrating polymer network hydrogels based on poly(2-hydroxyethyl methacrylate) copolymer and chitosan. *Fibers Polym.* 9: 393–399.

Higa, M., M. Kobayashi, Y. Kakihana, A. Jikihara, and N. Fujiwara. 2013. Charge mosaic membranes with semi-interpenetrating polymer network structures prepared from a polymer blend of poly(vinyl alcohol) and polyelectrolytes. *J. Membr. Sci.* 428: 267–274.

Ho, Y.S. 2006. Review of second-order models for adsorption systems. *J. Hazard. Mater.* 136: 681–689.

Ho, Y.S. and G. McKay. 1998. Sorption of dye from aqueous solution by peat. *Chem. Eng. J.* 70: 115–124.

Hoare, T.R. and D.S. Kohane. 2008. Hydrogels in drug delivery: Progress and challenges. *Polymer* 49: 1993–2007.

Hoffman, A.S. 2002. Hydrogels for biomedical applications. *Adv. Drug Deliv. Rev.* 43: 3–12.

Huang, D., W. Wang, Y. Kang, and A. Wang. 2012. Efficient adsorption and recovery of Pb(II) from aqueous solution by a granular pH-sensitive chitosan-based semi-IPN hydrogel. *J. Macromol. Sci. Part A: Pure Appl. Chem.* 49: 971–979.

Huang, X.Y., X.Y. Mao, H.T. Bu, X.Y. Yu, G.B. Jiang, and M.H. Zeng. 2011a. Chemical modification of chitosan by tetraethylenepentamine and adsorption study for anionic dye removal. *Carbohydr. Res.* 346: 1232–1240.

Huang, Y., M. Liu, L. Wang, C. Gao, and S. Xi. 2011b. A novel triple-responsive poly(3-acrylamidephenylboronic acid-co-2-(dimethylamino) ethyl methacrylate)/(b-cyclodextrin-epichlorohydrin) hydrogels: Synthesis and controlled drug delivery. *React. Funct. Polym.* 71: 666–673.

Ilavsky, M., J. Hrouz, J. Stejskal, and K. Bouchal. 1984. Phase transition in swollen gels. 6. Effect of aging on the extent of hydrolysis of aqueous polyacrylamide solutions and on the collapse of gels. *Macromolecules* 17: 2868–2874.

Jain, E. and A. Kumar. 2009. Designing supermacroporous cryogels based on polyacrylonitrile and a polyacrylamide-chitosan semi-interpenetrating network. *J. Biomater. Sci.* 20: 877–902.

Jeon, Y.S., J. Lei, and J.-H. Kim. 2008. Dye adsorption characteristics of alginate/polyaspartate hydrogels. *J. Ind. Eng. Chem.* 14: 726–731.

Jin, S., F. Bian, M. Liu, S. Chen, and H. Liu. 2009. Swelling mechanism of porous P(VP-co-MAA)/PNIPAM semi-IPN hydrogels with various pore sizes prepared by a freeze treatment. *Polym. Int.* 58: 142–148.

Jin, S., Y. Wang, J. He, Y. Yang, X. Yu, and G. Yue. 2013. Preparation and properties of a degradable interpenetrating polymer network based on starch with water retention, amelioration of soil, and slow release of nitrogen and phosphorus fertilizer. *J. Appl. Polym. Sci.* 128: 407–415.

Junyan, P., W. Sui, and Z. Ruifeng. 2006. Ion-imprinted IPNs for preconcentration and determination of Cd(II) by flame atomic absorption spectrometry. *Chem. Anal. (Warsaw)* 51: 701–713.

Keshavara Murthy, P.S., Y. Murali Mohan, J. Sreeramulu, and K. Mohana Raju. 2006. Semi-IPNs of starch and poly(acrylamide-co-sodium methacrylate): Preparation, swelling and diffusion characteristics evaluation. *React. Funct. Polym.* 66: 1482–1493.

Kim, S.J., C.K. Lee, and S.I. Kim. 2004a. Characterization of the water state of hyaluronic acid and poly(vinyl alcohol) interpenetrating polymer networks. *J. Appl. Polym. Sci.* 92: 1467–1472.

Kim, S.J., S.R. Shin, Y.M. Lee, and S.I. Kim. 2003. Swelling characterizations of chitosan and polyacrylonitrile semi-interpenetrating polymer network hydrogels. *J. Appl. Polym. Sci.* 87: 2011–2015.

Kim, S.J., S.R. Shin, G.M. Spinks, I.Y. Kim, and S.I. Kim. 2005. Synthesis and characteristics of a semi-interpenetrating polymer network based on chitosan/polyaniline under different pH conditions. *J. Appl. Polym. Sci.* 96: 867–873.

Kim, S.J., S.G. Yoon, Y.H. Lee, and S.I. Kim. 2004b. Bending behavior of hydrogels composed of poly(methacrylic acid) and alginate by electrical stimulus. *Polym. Int.* 53: 1456–1460.

Krezovic, B.D., S.I. Dimitrijevic, J.M. Filipovic, R.R. Nikolic, and S.Lj. Tomic. 2013. Antimicrobial P(HEMA/IA)/PVP semi-interpenetrating network hydrogels. *Polym. Bull.* 70: 809–819.

Kuila, S.B. and S.K. Ray. 2012. Sorption and permeation studies of tetrahydrofuran-water mixtures using full interpenetrating network membranes. *Sep. Purif. Technol.* 89: 39–50.

Kumar, K.V. and S. Sivanesan. 2006. Equilibrium data, isotherm parameters and process design for partial and complete isotherm of methylene blue onto activated carbon. *J. Hazard. Mater.* B134: 237–244.

Kundakci, S., E. Karadag, and O.B. Üzüm. 2011. Investigation of swelling/sorption characteristics of highly swollen AAm/AMPS hydrogels and semi IPNs with PEG as biopotential sorbent. *J. Encapsul. Adsorpt. Sci.* 1: 7–22.

Kusuktham, B. 2006. Preparation of interpenetrating polymer network gel beads for dye adsorption. *J. Appl. Polym. Sci.* 102: 1585–1591.

Lagergren, S. 1898. Kungliga svenska vetenskapsakademiens. *Handlingar* 24: 1–39.

Langmuir, I. 1918. The adsorption of gases on plane surfaces of glass, mica and platinum. *J. Am. Chem. Soc.* 40: 1361–1403.

Lee, S.B., E.K. Park, Y.M. Lim, S.K. Cho, S.Y. Kim, Y.M. Lee, and Y.C. Nho. 2006. Preparation of alginate/poly(n-isopropylacrylamide) semi-interpenetrating and fully interpenetrating polymer network hydrogels with γ-ray irradiation and their swelling behaviors. *J. Appl. Polym. Sci.* 100: 4439–4446.

Lee, Y., D.N. Kim, D. Choi, W. Lee, J. Park, and W.-G. Koh. 2008. Preparation of interpenetrating polymer network composed of poly(ethylene glycol) and poly(acrylamide) hydrogels as a support of enzyme immobilization. *Polym. Adv. Technol.* 18: 852–858.

Li, S. 2010. Removal of crystal violet from aqueous solution by sorption into semi-interpenetrated networks hydrogels constituted of poly(acrylic acid-acryamide-methacrylate) and amylose. *Bioresour. Technol.* 101: 2197–2202.

Li, X., W. Wu, and W. Liu. 2008a. Synthesis and properties of thermo-responsive guar gum/poly(N-isopropylacrylamide) interpenetrating polymer network hydrogels. *Carbohydr. Polym.* 71: 394–402.

Li, X., S. Xu, Y. Pen, and J. Wang. 2008b. The swelling behaviors and network param-
    eters of cationic starch-g-acrylic acid/poly(dimethyldiallylammonium chloride) semi-
    interpenetrating networks hydrogels. *J. Appl. Polym. Sci.* 110: 1828–1836.

Li, X., S. Xu, J. Wang, X. Chen, and S. Feng. 2009. Structure and characterization of
    amphoteric semi-IPN hydrogel based on cationic starch. *Carbohydr. Polym.* 75:
    688–693.

Liang, S., L. Liu, Q. Huang, and K.L. Yam. 2009. Preparation of single or double-network
    chitosan/poly(vinyl alcohol) gel films through selectively cross-linking method.
    Carbohydr. Polym. 77: 718–724.

Limousin, G., J.-P. Gaudet, L. Charlet, S. Szenknect, V. Barthes, and M. Krimissa. 2007.
    Sorption isotherms: A review on physical bases, modeling and measurement. *Appl.
    Geochem.* 22: 249–275.

Lin, H., J. Zhou, C. Yingde, and S. Gunasekaran. 2010. Synthesis and characterization of ph-
    and salt-responsive hydrogels based on etherificated sodium alginate. *J. Appl. Polym.
    Sci.* 115: 3161–3167.

Liu, J., W. Wang, and A. Wang. 2011. Synthesis, characterization, and swelling behaviors of
    chitosan-g-poly(acrylic acid)/poly(vinyl alcohol) semi-IPN superabsorbent hydrogels.
    Polym. Adv. Technol. 22: 627–634.

Liu, M., H. Su, and T. Tan. 2012a. Synthesis and properties of thermo- and pH-sensitive
    poly(N-isopropylacrylamide)/polyaspartic acid IPN hydrogels. *Carbohydr. Polym.* 87:
    2425–2431.

Liu, X., H. Guo, and L. Zha. 2012b. Study of pH/temperature dual stimuli-responsive nanogels
    with interpenetrating polymer network structure. *Polym. Int.* 61: 1144–1159.

Liu, Y., X. Cao, R. Hua, Y. Wang, Y. Liu, C. Pang, and Y. Wang. 2010. Selective adsorption of
    uranyl ion on ion-imprinted chitosan/PVA cross-linked hydrogel. *Hydrometallurgy* 104:
    150–155.

Lozinsky, V.I., I.Y. Galaev, F.M. Plieva, I.N. Savina, H. Jungvid, and B. Mattiasson. 2003.
    Polymeric cryogels as promising materials of biotechnological interest. *Trends
    Biotechnol.* 21: 445–451.

Mahdavinia, G.R., G.B. Marandi, A. Pourjavadi, and G. Kiani. 2010. Semi-IPN carrageenan-
    based nanocomposite hydrogels: Synthesis and swelling behavior. *J. Appl. Polym. Sci.*
    118: 2989–2997.

Mahou, R. and C. Wandrey. 2010. Alginate-poly(ethylene glycol) hybrid microspheres with
    adjustable physical properties. *Macromolecules* 43: 1371–1378.

Mallikarjuna Reddy, K., V. Ramesh Babu, K.S.V. Krishna Rao, M.C.S. Subha, K. Chowdoji
    Rao, M. Sairam, and T.M. Aminabhavi. 2008. Temperature sensitive semi-IPN micro-
    spheres from sodium alginate and n-isopropylacrylamide for controlled release of
    5-fluorouracil. *J. Appl. Polym. Sci.* 107: 2820–2829.

Mandal, B. and S.K. Ray. 2013. Synthesis of interpenetrating network hydrogel from
    poly(acrylic acid-co-hydroxyethyl methacrylate) and sodium alginate: Modeling and
    kinetics study for removal of synthetic dyes from water. *Carbohydr. Polym.* 98: 257–269.

Mandal, B., S.K. Ray, and R. Bhattacharyya. 2012. Synthesis of full and semi interpenetrating
    hydrogel from polyvinyl alcohol and poly (acrylic acid-co-hydroxyethylmethacrylate)
    copolymer: Study of swelling behavior, network parameters, and dye uptake properties.
    J. Appl. Polym. Sci. 124: 2250–2268.

Marsano, E., E. Bianchi, S. Vicini, L. Compagnino, A. Sionkowska, J. Skopinska, and
    M. Wisniewski. 2005. Stimuli responsive gels based on interpenetrating network of chi-
    tosan and poly(vinylpyrrolidone). *Polymer* 46: 1595–1600.

Martinez, L., F. Agnely, R. Bettini, M. Besnard, P. Colombo, and G. Couarraze. 2004.
    Preparation and characterization of chitosan based micro networks: Transposition to a
    prilling process. *J. Appl. Polym. Sci.* 93: 2550–2558.

M'Bareck, C.O., Q.T. Nguyen, S. Alexandre, and I. Zimmerlin. 2006. Fabrication of ion-exchange ultrafiltration membranes for water treatment I. Semi-interpenetrating polymer networks of polysulfone and poly(acrylic acid). *J. Membr. Sci.* 278: 10–18.

Milosavljevic, N.B., N.Z. Milasinovic, I.G. Popovic, J.M. Filipovic, and M.T.K. Krusic. 2011. Preparation and characterization of pH-sensitive hydrogels based on chitosan, itaconic acid and methacrylic acid. *Polym. Int.* 60: 443–452.

Myung, D., D. Waters, M. Wiseman, P.-E. Duhamel, J. Noolandi, C.N. Ta, and C.W. Frank. 2008. Progress in the development of interpenetrating polymer network hydrogels. *Polym. Adv. Technol.* 19: 647–657.

Park, H.C., R.M. Meertens, M.H.V. Mulder, and C.A. Smolders. 1994. Pervaporation of alcohol–toluene mixtures through polymer blend membranes of poly(acrylic acid) and poly(vinyl alcohol). *J. Membr. Sci.* 90: 265–274.

Peak, C.W., J.J. Wilker, and G. Schmidt. 2013. A review on tough and sticky hydrogels. *Colloid Polym. Sci.* 291: 2031–2047.

Peppas, N.A., P. Bures, W. Leobandung, and H. Ichikawa. 2000. Hydrogels in pharmaceutical formulations. *Eur. J. Pharm. Biopharm.* 50: 27–46.

Perju, M.M., M.V. Dinu, and E.S. Drăgan. 2012. Sorption of Methylene Blue onto ionic composite hydrogels based on polyacrylamide and dextran sulfate: Kinetics, isotherms and thermodynamics. *Sep. Sci. Technol.* 47: 1322–1333.

Pescosolido, L., T. Vermonden, J. Malda, R. Censi, W.J.A. Dhert, F. Alhaique, W.E. Hennink, and P. Matricardi. 2011. In situ forming IPN hydrogels of calcium alginate and dextran-HEMA for biomedical applications. *Acta Biomater.* 7: 1627–1633.

Plieva, F.M., M. Karlsson, M.R. Aguilar, D. Gomez, S. Mikhalovsky, and I.Y. Galaev. 2005. Pore structure in supermacroporous polyacrylamide based cryogels. *Soft Matter* 1: 303–309.

Ramesh Babu, V., C. Kim, S. Kim, C. Ahn, and Y.-I. Lee. 2010. Development of semi-interpenetrating carbohydrate polymeric hydrogels embedded silver nanoparticles and its facile studies on *E. coli. Carbohydr. Polym.* 81: 196–202.

Reis, A.V., M.R. Guilherme, T.A. Moia, L.H.C. Mattoso, E.C. Muniz, and E.B. Tambourgi. 2008. Synthesis and characterization of a starch-modified hydrogel as potential carrier for drug delivery system. *J. Polym. Sci. Part A: Polym. Chem.* 46: 2567–2574.

Rodriguez, D.E., J. Romero-Garcia, E. Ramirez-Vargas, A.S. Ledezma-Perez, and E. Arias-Marin. 2006. Synthesis and swelling characteristics of semi-interpenetrating polymer network hydrogels composed of poly(acrylamide) and poly(γ-glutamic acid). *Mater. Lett.* 60: 1390–1393.

Samanta, H.S. and S.K. Ray. 2014. Synthesis, characterization, swelling and drug release behavior of semi-interpenetrating network hydrogels of sodium alginate and polyacrylamide. *Carbohydr. Polym.* 67: 666–678.

Shi, J., N.M. Alves, and J.F. Mano. 2006. Drug release of pH/temperature-responsive calcium alginate/poly(N-isopropylacrylamide) semi-IPN beads. *Macromol. Biosci.* 6: 358–363.

Silan, C., A. Akcali, M.T. Otkun, N. Ozbey, S. Butun, O. Ozay, and N. Sahiner. 2012. Novel hydrogel particles and their IPN films as drug delivery systems with antibacterial properties. *Colloids Surf. B: Biointerfaces* 89: 245–253.

Singha, N.R., S. Kar, S. Ray, and S.K. Ray. 2009a. Separation of isopropyl alcohol-water mixtures by pervaporation using crosslink IPN membranes. *Chem. Eng. Process.: Process Intensification* 48: 1020–1029.

Singha, N.R., S.B. Kuila, P. Das, and S.K. Ray. 2009b. Separation of toluene-methanol mixtures by pervaporation using crosslink IPN membranes. *Chem. Eng. Process.: Process Intensification* 48: 1560–1565.

Solak, E.K. 2011. Preparation and characterization of IPN microspheres for controlled delivery of naproxen. *J. Biomater. Nanobiotechnol.* 2: 445–453.

Şolpan, D. and M. Torun. 2005. Investigation of complex formation between (sodium alginate/ acrylamide) semi-interpenetrating polymer networks and lead, cadmium, nickel ions. *Colloids Surf. A* 268: 12–18.

Şolpan, D., M. Torun, and O. Güven. 2008. The usability of (sodium alginate/acrylamide) semi-interpenetrating polymer networks on removal of some textile dyes. *J. Appl. Polym. Sci.* 108: 3787–3795.

Sperling, L.H. 1994. Interpenetrating polymer networks: An overview. In *Interpenetrating Polymer Networks*, D. Klempner, L.H. Sperling, and L.A. Utracki (eds.). American Chemical Society, Washington, DC, pp. 3–38.

Sperling, L.H. 2005. Interpenetrating polymer networks. In *Encyclopedia of Polymer Science and Technology*, H.F. Mark (ed.). John Wiley & Sons, New York, Vol. 10, pp. 272–311.

Srivastava, Y.C., M.M. Swamy, I.D. Mall, B. Prasad, and I.M. Mishra. 2006. Adsorptive removal of phenol by bagasse fly ash and activated carbon: Equilibrium, kinetics and thermodynamics. *Colloids Surf. A* 272: 89–104.

Tang, Q., J. Wu, and J. Lin. 2008. A multifunctional hydrogel with high conductivity, pH-responsive, thermo-responsive and release properties. *Carbohydr. Polym.* 73: 315–321.

Thimma Reddy, T. and A. Takahara. 2009. Simultaneous and sequential micro-porous semi-interpenetrating polymer network hydrogel films for drug delivery and wound dressing applications. *Polymer* 50: 3537–3546.

Üzüm, U.B. and E. Karadağ. 2012. Equilibrium swelling studies and dye sorption characterization of AAm/SA hydrogels cross-linked by PEGDMA and semi-IPNs with PEG. *Adv. Polym. Technol.* 31: 141–153.

Üzüm, U.B. and E. Karadağ. 2013. Water and dye sorption studies of novel semi-IPNs: Acrylamide/4-styrenesulfonic acid sodium salt/PEG hydrogels. *Polym. Eng. Sci.* 53: 1262–1271.

Wan Ngah, W.S., L.C. Teong, and M.A.K.M. Hanafiah. 2011. Adsorption of dyes and heavy metals by chitosan composites: A review. *Carbohydr. Polym.* 83: 1446–1456.

Wang, B., M.-Z. Liu, R. Liang, S.-L. Ding, Z.-B. Chen, S.-L. Chen, and S.-P. Jin. 2008. MMTCA recognition by molecular imprinting in interpenetrating polymer network hydrogels based on poly(acrylic acid) and poly(vinyl alcohol). *Macromol. Biosci.* 8: 417–425.

Wang, J., F. Liu, and J. Wei. 2011a. Enhanced adsorption properties of interpenetrating polymer network hydrogels for heavy metal ion removal. *Polym. Bull.* 67: 1709–1720.

Wang, J., X. Zhou, and H. Xiao. 2013a. Structure and properties of cellulose/poly(*N*-isopropylacrylamide) hydrogels prepared by SIPN strategy. *Carbohydr. Polym.* 94: 749–754.

Wang, J.J. and F. Liu. 2012. UV-curing of simultaneous interpenetrating network silicone hydrogels with hydrophilic surface. *Polym. Bull.* 69: 685–697.

Wang, J.J. and F. Liu. 2013. Enhanced adsorption of heavy metal ions onto simultaneous interpenetrating polymer network hydrogels synthesized by UV irradiation. *Polym. Bull.* 70: 1415–1430.

Wang, Q., J. Zhang, and A. Wang. 2009. Preparation and characterization of a novel pH-sensitive chitosan-*g*-poly(acrylic acid)attapulgite/sodium alginate composite hydrogel bead for controlled release of diclofenac sodium. *Carbohydr. Polym.* 78: 731–737.

Wang, W., D. Huang, Y. Kang, and A. Wang. 2013b. One-step in situ fabrication of granular semi-IPN hydrogel based on chitosan and gelatin for fast and efficient adsorption of Cu^{2+} ion. *Colloids Surf. B: Biointerfaces* 106: 51–59.

Wang, W., Y. Kang, and A. Wang. 2013c. One-step fabrication in aqueous solution of a granular alginate-based hydrogel for fast and efficient removal of heavy metal ions. *J. Polym. Res.* 20: 110.

Wang, W. and A. Wang. 2010. Synthesis and swelling properties of pH-sensitive semi-IPN superabsorbent hydrogels based on sodium alginate-*g*-poly(sodium acrylate) and polyvinylpyrrolidone. *Carbohydr. Polym.* 80: 1028–1036.

Wang, W., Q. Wang, and A. Wang. 2011b. pH-responsive carboxymethylcellulose-g-poly(sodium acrylate)/poly(vinylpyrrolydone) semi-IPN hydrogels with enhanced responsive and swelling properties. *Macromol. Res.* 19: 57–65.

Wawrzkiewicz, M. 2013. Removal of C.I. Basic Blue 3 dye by sorption onto cation exchange resin, functionalized and non-functionalized polymeric sorbents from aqueous solutions and wastewaters. *Chem. Eng. J.* 217: 414–425.

Weber Jr., W.J. and J.C. Morris. 1963. Kinetics of adsorption on carbon from solution. *J. Sanit. Eng. Div. Am. Soc. Chem. Eng.* 89: 31–59.

Wei, J., S. Xu, R. Wu, J. Wang, and Y. Gao. 2007. Synthesis and characterization of an amphoteric semi-IPN hydrogel composed of acrylic acid and poly(diallyldimethylammonium chloride). *J. Appl. Polym. Sci.* 103: 345–350.

Xiao, C., H. Li, and Y. Gao. 2009. Preparation of fast pH-responsive ferric carboxymethylcellulose/poly(vinyl alcohol) double-network microparticles. *Polym. Int.* 58: 112–115.

Yamashita, K., T. Nishimura, and M. Nango. 2003. Preparation of IPN-type stimuli-responsive heavy-metal-ion adsorbent gel. *Polym. Adv. Technol.* 14: 189–194.

Yang, J., J. Chen, D. Pan, Y. Wan, and Z. Wang. 2013. pH-sensitive interpenetrating network hydrogels based on chitosan derivatives and alginate for oral drug delivery. *Carbohydr. Polym.* 92: 719–725.

Yin, L., L. Fei, F. Cui, C. Tang, and C. Yin. 2007a. Superporous hydrogels containing poly(acrylic-*co*-acrylamide)/*O*-carboxymethyl chitosan interpenetrating polymer networks. *Biomaterials* 28: 1258–1266.

Yin, L., L. Fei, F. Cui, C. Tang, and C. Yin. 2007b. Synthesis, characterization, mechanical properties and biocompatibility of interpenetrating polymer network-superporous hydrogel containing sodium alginate. *Polym. Int.* 56: 1563–1571.

Yin, L., Z. Zhao, Y. Hu, J. Ding, F. Cui, C. Tang, and C. Yin. 2008. Polymer-protein interaction, water retention, and biocompatibility of a stimuli-sensitive superporous hydrogel containing interpenetrating polymer network. *J. Appl. Polym. Sci.* 108: 1238–1248.

Zadrazil, A. and F. Stepanek. 2010. Investigation of thermo-responsive optical properties of a composite hydrogel. *Colloids Surf. A* 372: 115–119.

Zhang, G.Q., L.S. Zha, M.H. Zhou, J.H. Ma, and B.R. Liang. 2005. Preparation and characterization of pH- and temperature responsive semi-interpenetrating polymer network hydrogels based on linear sodium alginate and crosslinked poly(*N*-isopropylacrylamide). *J. Appl. Polym. Sci.* 97: 1931–1940.

Zhang, J.-T., R. Bhat, and K.D. Jandt. 2009. Temperature-sensitive PVA/PNIPAAm semi-IPN hydrogels with enhanced responsive properties. *Acta Biomater.* 5: 488–497.

Zhang, N., M. Liu, Y. Shen, J. Chen, L. Dai, and C. Gao. 2011. Preparation, properties, and drug release of thermo- and pH-sensitive poly(2-dimethylamino)ethyl methacrylate)/poly(*N*,*N*-diethylacrylamide) semi-IPN hydrogels. *J. Mater. Sci.* 46: 1523–1534.

Zhao, G., X. Wu, X. Tan, and X. Wang. 2011. Sorption of heavy metal ions from aqueous solutions: A review. *Open Colloid Interface J.* 4: 19–31.

Zhao, Q., J. Sun, Y. Lin, and Q. Zhou. 2010. Study of the properties of hydrolyzed polyacrylamide hydrogels with various pore structures and rapid pH-sensitivities. *React. Funct. Polym.* 70: 602–609.

Zhao, S., F. Zhou, L. Li, M. Cao, D. Zuo, and H. Liu. 2012. Removal of anionic dyes from aqueous solutions by adsorption of chitosan-based semi-IPN hydrogel composites. *Composites Part B: Eng.* 43: 1570–1578.

Zhao, Y., J. Kang, and T. Tan. 2006. Salt-, pH-, and temperature-responsive semi-interpenetrating polymer network hydrogel based on poly(aspartic acid) and poly(acrylic acid). *Polymer* 47: 7702–7710.

Zhou, C. and Q. Wu. 2011. A novel polyacrylamide nanocomposite hydrogel reinforced with natural chitosan nanofibers. *Colloids Surf. B: Biointerfaces* 84: 155–162.

12 Toward Adaptive Self-Informed Membranes

Ioana Moleavin and Mihail Barboiu

CONTENTS

12.1 INTRODUCTION

In nature, biological pore proteins represent excellent examples of structure-directed transport function. Their active pores rely on various combinations of the well-defined spatial distribution of molecular functional groups, to generate highly selective pore-type devices, controlling ionic diffusion along polar transduction pathways (Agre 2004, MacKinnon 2004).

Artificial molecular/supramolecular systems, functioning as carriers or as channel-forming superstructures in liquid, bilayer, or solid membranes, have been extensively developed during the last decades (Gokel and Abel 1996). Molecular self-organization and self-assembly to supramolecular structures is the basis for the construction of such functional systems for selective solute translocation. The transition from biomimetic to artificial membranes with recognition-driven transport functions could be ensured by a well-defined inclusion within the membranes of multivalent recognition receptors or active complex pores with high affinity for specific solutes (Barboiu et al. 1997, 1999, 2000a,b).

Within this context, hybrid organic–inorganic materials produced by sol-gel process are the subject of various investigations (Sanchez et al. 2005), offering the opportunity to achieve such nanostructured functional membrane systems (Barboiu 2010). Many groups have found new methods for the elaboration of hybrid materials based on silsesquioxanes in which the functional molecular/supramolecular and inorganic/siloxane networks are covalently connected (Sanchez et al. 2005, van Bommel et al. 2003).

Controlled formation of highly self-organized materials makes hybrid materials of interest for the development of directional transporting membrane systems of increasing addressability (Cazacu et al. 2008).

Of particular interest is the potential ability of the hybrid membrane films to present polyfunctional properties. Important targets are related to achieve *structure-directed function strategy* and to control their build-up by self-organization. The *fixed-site complexant hybrid membranes* highlight the potential ability to present properties such as multivalent molecular recognition and the self-organized directional conduction pathways (Figure 12.1).

In the *fixed-site complexant membranes*, the grafted receptor is not a carrier; it just selectively assists the solute diffusion (D) in the membrane by selective complexation–decomplexation reactions (CDRs) at the receptor-site level. It results in a "fixed-site jumping" transport mechanism, in which the diffusion of solutes into the dense hybrid material is controlled by the specific interactions with the fixed receptor and by the ion–matrix interactions and phenomena. One next premise is to minimize the distance between the grafted receptors *via* self-organization, in order to reduce the nonselective solute-diffusion (D) between receptor sites and to create favorable and continual diffusion patterns for the solute diffusion through the functional hybrid dense material (Figure 12.2b) (Barboiu et al. 2004).

The hierarchical generation of functional hybrid materials may be achieved in two steps: (a) the supramolecular oligomers were generated in solution or gels *via* dynamic self-assembly of monomers followed by (b) the sol-gel transcription (polymerization) step into solid hybrid matrixes self-organized at the nanoscopic level (Figure 12.2).

This chapter will be divided into four sections. The first part will focus on implementation of *self-organization* emphasizing the more recent developments toward ion channels and their transcription in hybrid materials. It will provide examples of molecular networks leading to selection of nanostructured hybrid systems of increased dimensionality/complexity. The second and the third parts will introduce *hybrid nanostructured materials* with particular emphasis on self-assembly approaches used for synthetic construction of self-organized supramolecular and constitutional nanomaterials for ion conduction. We will describe the development of the *dynamic intrapore resolution* toward *dynamic hybrid materials and membranes*. Such systems evolve to form the fittest insidepore architecture, demonstrating flexible functionality and adaptation in confined conditions. The last part will be devoted to *sol-gel resolution of dynamic molecular/supramolecular libraries*, emphasizing recent developments, as pursued especially in our laboratory in the field of G-quadruplex-based materials.

12.2 SELF-ORGANIZATION OF SUPRAMOLECULAR ION CHANNELS AND THEIR TRANSCRIPTION IN SOL-GEL HYBRID MATERIALS

The self-organized hybrid materials reveal great potentialities as well on the level of their chemical composition or organization as to that of the concerned applications. Organogels resulting by multiple self-assembly processes and acting as robust

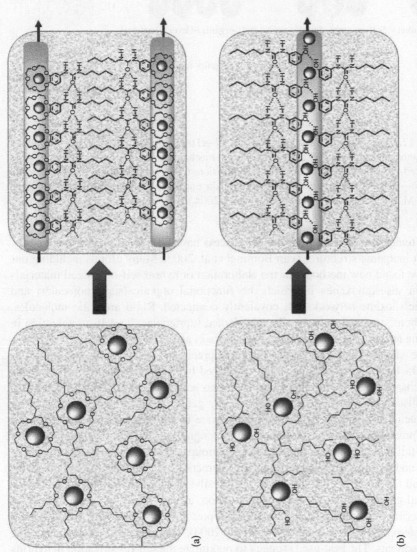

FIGURE 12.1 The generations of randomly ordered molecular recognition-based and of self-organized fixed-site hybrid membranes using (a) molecular receptors and (b) molecular moieties that form specific transport devices by collective self-assembly. (Adapted from Michau, M. et al., *Chem. Eur. J.*, 14, 1776, 2008.)

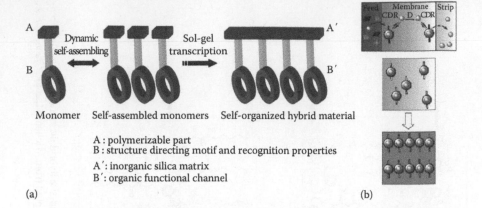

A : polymerizable part
B : structure directing motif and recognition properties
A': inorganic silica matrix
B': organic functional channel

(a)   (b)

FIGURE 12.2 (a) Hierarchical generation and sol-gel transcription of tubular self-assembled hybrid nanomembranes and (b) assisted-diffusion mechanism of the solute by CDRs in solid dense fixed-site complexant membranes (D, diffusion; CDR, complexation–decomplexation reactions)—from random to self-organized fixed-site complexant membranes. (Adapted from Barboiu, M. et al., *J. Am. Chem. Soc.*, 126, 3545, 2004.)

organic templates for the TEOS sol-gel process have been used for the synthesis of different inorganic structures (van Bommel et al. 2003). Many groups including our own have found new methods for the elaboration of hybrid self-organized materials based on silsesquioxanes in which the functional organic (supramolecular) and inorganic siloxane networks are covalently connected. Rigid aromatic molecules, urea ribbons, are used to transcribe an oriented supramolecular self-organization in a siloxane matrix by a sol-gel process (Barboiu et al. 2003, 2004).

Three classes of heteroditopic receptors/precursors used as molecular building blocks for the construction of self-organized hybrid materials will be discussed in this chapter: the crown ethers, **1**; the amino acid conjugates, **2**; and nucleobase ureido-silsesquioxanes, **3** (Figure 12.3). They generate self-organized exchanging superstructures in solution and in the solid state based on three encoded features: (1) they present both cation/anion molecular recognition sites; (2) the supramolecular head-to-tail H-bond interactions of the urea groups, assisted by π–π stacking, were used (**1** and **2**) as assembling directional interactions to generate directional pathways; and (3) the covalently bonded triethoxysilyl moiety can be used as precursor reactional groups for the sol-gel polymerization, allowing *via* the sol-gel processes to transcribe the dynamic self-organization present in solution in "frozen" solid dense heteropolysiloxane materials (Barboiu 2004). These compounds can gelate the organic solvents. The best-used strategy to maximize the self-organization in hybrids is to minimize the cross-linking of the siloxane network, favored by hydrolysis, and to maximize the self-organization of molecular organic moieties. Therefore, the sol-gel experiments are usually conducted under mild hydrolysis–condensation conditions: (a) the first strategy is to use heterogeneous acidic catalysis in aprotic solvents, or by others; (b) the second method uses the dichlorosilane derivatives (R_2SiCl_2) as organic initiator after reacting with the residual water in the reaction medium.

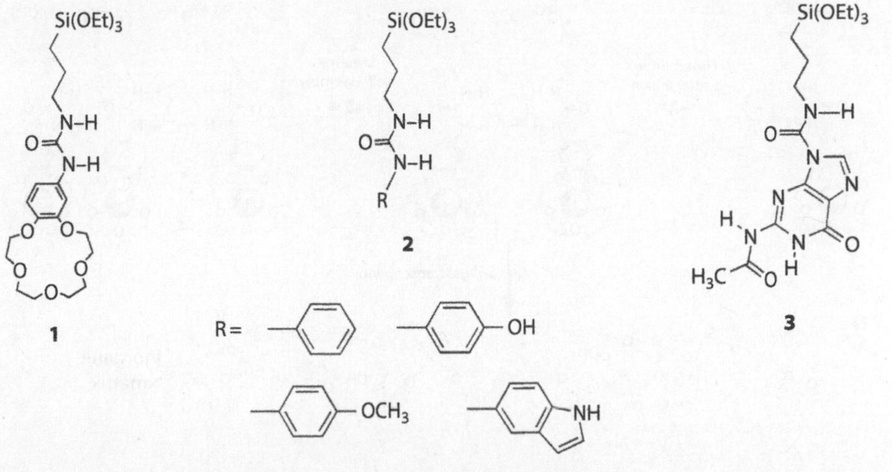

FIGURE 12.3 Molecular structures of molecular receptors 1–3.

The hybrid membranes can be obtained by coating the sol onto planar porous membrane supports using a tape-casting method (Michau et al. 2008).

From the mechanistic point of view, we use carriers that self-assemble in functional aggregates, which would present combined (hybrid) intermediate features between the former carrier monomers and the resulted pseudo-channel-forming superstructures (Barboiu 2004). Within this context, the self-organization properties in the membrane phase have provided the evidence for the possible hybrid transport carrier versus channel mechanisms in correlation with self-assembly properties of the heteroditopic receptors.

In general, the hybrid membrane materials reported by now are composed of nanodomains randomly ordered in the hybrid matrix. These oriented nanodomains resulted from the controlled self-assembly of simple molecular components that encodes the required information for ionic assisted-diffusion within hydrophilic pathways. Although these pathways do not merge to cross the micrometric films, they are well defined along nanometric distances (Michau and Barboiu 2009).

12.3 SUPRAMOLECULAR ION CHANNELS FOR IONIC DIFFUSION

Convergent multidimensional self-assembly strategies have been used on the synthesis of noncovalent self-organized devices, designed to mimic natural ion-channel proteins. Despite their thermodynamic stability, they are in dynamic equilibrium between monomer and supramolecular oligomers and only few of them clearly showed single-channel activity in lipid bilayers (Cazacu et al. 2006).

As an example, a schematic representation of the dynamic self-assembly of crown ethers in solution and the sol-gel transcription in the hybrid membrane is illustrated in Figure 12.4.

The hierarchical generation of functional hybrid materials based on self-organized ribbons of macrocyclic ether stacks can be realized in two steps (Cazacu et al. 2006). First, the self-assembling properties of 1 in the aprotic solvents

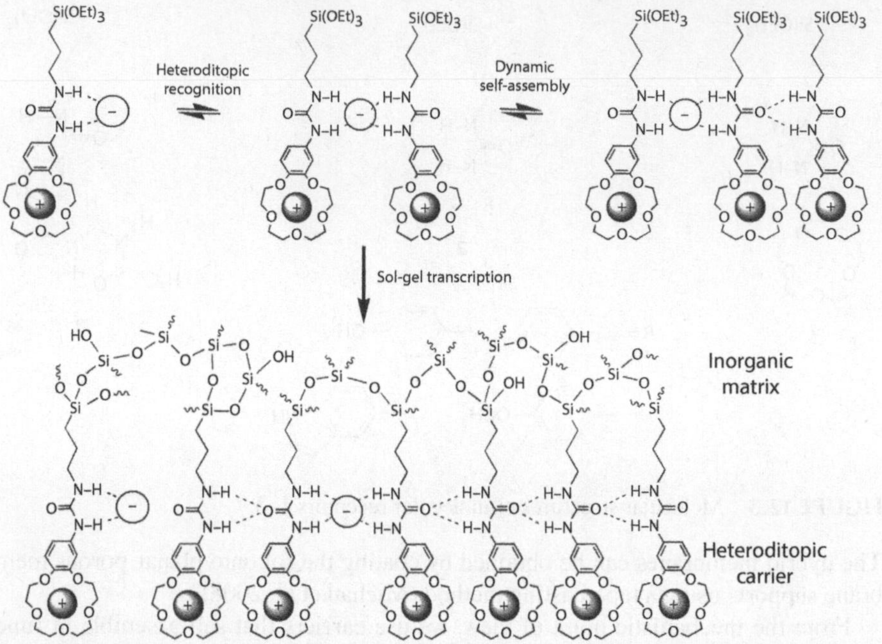

FIGURE 12.4 Schematic representation of the hierarchical organized system **1**: (top) self-organization in solution and (bottom) sol-gel transcription of encoded molecular features into a hybrid heteropolysiloxane matrix.

were determined, revealing the formation of supramolecular oligomers (Barboiu et al. 2003). In homogenous solution, at least two types of hydrogen-bonded aggregates form, which can be modeled as a dimer followed by a discrete higher oligomer, perhaps as large as a hexadecamer. In the solid state, a dominant antiparallel ribbon of urea self-association leads to extended stacks of crown ethers. A minor polymorph consisting of the previously observed parallel urea ribbons is also observed. In a bilayer membrane system, the compounds form channels reluctantly at low concentration, preferring to give disruption over regular behavior. At higher concentration where channels do form, there are apparently two types of structures formed and they generate organogels in chloroform. They show a basic regularity suggesting structural constraints, but they show an immense range of conductance values and lifetimes. Then, in a second *sol-gel transcription* step, these gels can be used to obtain hybrid membrane materials with lamellar organization at the nanoscopic scale. The adenosine triphosphate (ATP^2) transport across these membranes and against its thermodynamic gradient was demonstrated using ion-driven pumping conditions (Barboiu 2004).

Intermolecular interactions involving aromatic rings are key processes in both chemical and biological recognition. Among these interactions, cation–π interactions between positively charged species (alkali, ammonium, and metal ions) and aromatic systems with delocalized π-electrons are now recognized as important noncovalent binding forces of increasing relevance (Gallivan and Dougherty 1999).

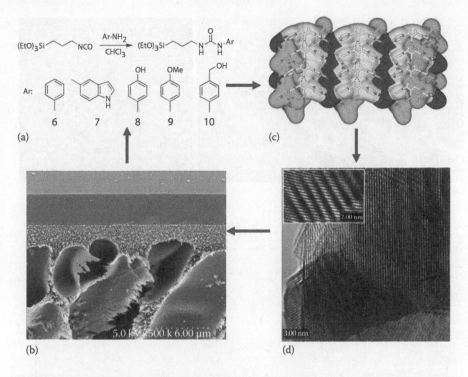

(a)

(b)

(c)

(d)

FIGURE 12.5 (a) Dynamic self-assembly of molecular precursors; (b) SEM cross section of the hybrid membrane; (c) single-crystal structure of the supramolecular channel; (d) crystal-line fields at the surface of the hybrid membrane (TEM image). (Adapted from Michau, M. et al., *Chem. Eur. J.*, 14, 1776, 2008.)

The importance of interactions between alkali cations and aromatic side chains of aromatic amino acids has been known for many years and they are of particular biological significance. Several heterocomplex superstructures, emphasizing particular K^+–π contacts with phenyl, phenol, and indole rings, have been reported in literature (Arnal-Herault et al. 2005).

Convergent functional self-assembly can be obtained also by using ureido-silsesquioxane compounds bearing aromatic moieties of natural aromatic amino acids, which can be used for the preparation of hybrid self-organized materials (Figure 12.5) (Michau et al. 2008). The hybrid membrane materials are composed of nanodomains randomly ordered in the hybrid matrix. These oriented nanodomains result from the controlled self-assembly of simple molecular components that encode the required information for ionic assisted-diffusion within hydrophilic or hydrophobic *aromatic cation–π conduction* pathways (Figure 12.6). The ionic transport across the organized domains illustrates the power of the supramolecular approach for the design of continual hydrophilic transport devices in hybrid materials by self-organization.

These results imply that the control of molecular interactions can define the self-organized supramolecular architectures such as hydrophobic and hydrophilic transporting pathways of different chemical properties. They are essential in the diffusion process and in the selectivity of the transport of hydrated alkali cations.

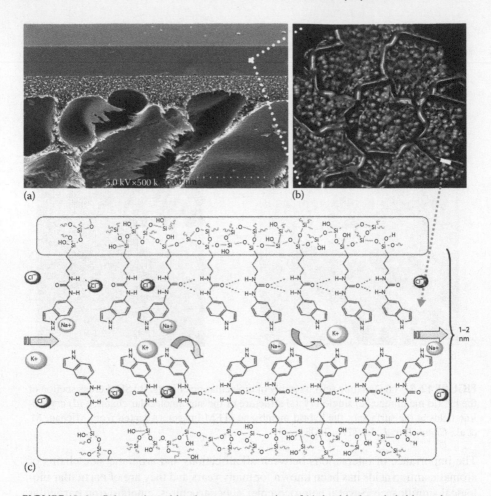

FIGURE 12.6 Schematic multiscale nanostructuration of (a) the thin-layer hybrid membranes in (b) oriented nanodomains of self-organized hybrid materials containing (c) *aromatic cation–π and urea–anion conduction pathways*. The cations and anions are shown without any hydration or solvation for simplicity reasons.

Although these pathways do not merge to cross the micrometric films, they are well defined along nanometric distances. This is reminiscent with the supramolecular organization of binding sites in channel-type proteins collectively contributing to the selective translocation of solutes along the hydrophilic ways.

12.4 CONSTITUTIONAL ION CHANNELS FOR IONIC DIFFUSION

Periodic mesoporous materials have attracted considerable attention during the last decades because of their promising applications as catalyst support or as hosts for nanostructured materials. Many of these applications benefit from arrangements of preferentially aligned, ordered arrays of certain mesostructures. The evaporation-induced self-assembly method has been established as an efficient process for the preparation of thin films with monooriented materials. However, the most frequently obtained films display hexagonally ordered channels that are aligned in a nonfavorable parallel orientation to the surface of the substrate, inside the perpendicular orientation, favorable for the transport through the membrane.

Another point of interest in this field of hybrid solid membranes is related to systems in which the molecular recognition driven transport function could be ensured by a dynamic reversible incorporation of specific organic receptors (Cazacu et al. 2009). These constitutional membranes have the potential abilities to combine functional properties such as solute molecular recognition and stimuli-responsive generation of self-assembled directional conduction pathways within confined conditions. It was then demonstrated that porous anodic alumina membranes, AAO, can serve as support material to form silica–surfactant MCM41 nanocomposite with a desirable orientation of nanochannels, perpendicular to the surface of the support and, consequently, parallel to (along) the alumina pores. Using the channel-forming macrocyclic systems described before, but confined in a solid mesoporous membrane matrix, remarkable results were obtained, in terms of selectivity and most importantly evidencing the adaptability of the material toward its environment and more specifically toward selective ionic diffusion. Constitutional hybrid membranes have been prepared (Figures 12.7 and 12.8) by embedding self-organized ureido crown ethers 15-crown-5, **4** or 18-crown-6, **5** into MCM 41 silica mesoporous hybrid materials, regularly oriented along the pores of the Anodisc 47 (0.02 μm) alumina membranes used as support, which allow preparing very promising dynamic molecular channels (Cazacu et al. 2009).

More precisely, the MCM41-type mesostructured powders were used as hydrophobic host matrix for physical confinement of 15-crown-5- and 18-crown-6-based receptors; based on hydrophobic and hydrogen bonds such as urea–urea or urea–anion interactions, carrier superstructures can be noncovalently trapped in an inorganic porous matrix.

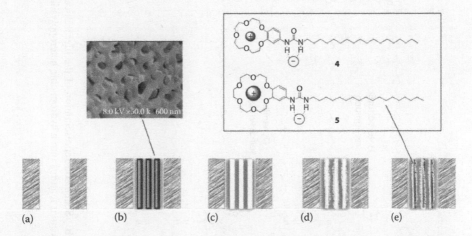

FIGURE 12.7 Schematic representation of the synthetic route to obtain functionalized mesostructured silica–receptor nanocomposite in the AAMs: (a) anodic alumina membrane, mesostructured silica–surfactant before (b) and after (c) calcination, and ODS-hydrophobized silica before (d) and after (e) inclusion of the hydrophobic carrier **4** or **5**.

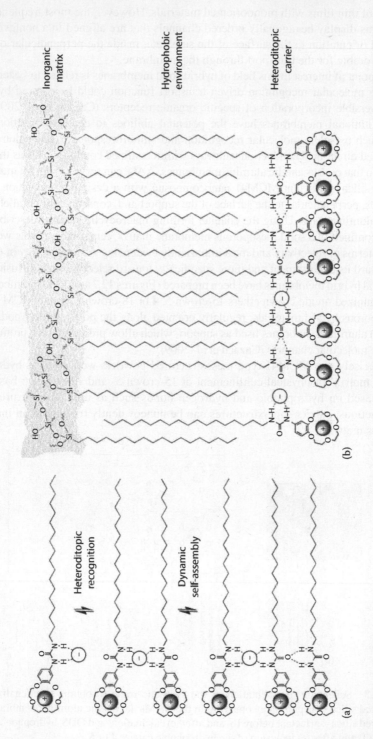

FIGURE 12.8 Schematic representation of the hierarchical organized system 4: (a) self-organization in solution and (b) dynamic transcription of encoded molecular features into a hydrophobic heteropolysiloxane matrix.

TABLE 12.1
Characteristics of Na⁺/K⁺ Transport across Mesoporous Functionalized Membranes

AAM	Membrane Preparation				Permeability (cm²/s × 10⁸)				Diffusion Coefficient (cm²/s × 10⁷)					
					Na^+		K^+		Na^+			K^+		
	A	B	C	D	P_1	P_2	P_1	P_2	D1	D2	D3	D1	D2	D3
S	—	—	—	—	8.1	0.7	25	0.8	300	20	46	552	24	48
M1	x	x	x	—	34	0.5	29	0.4	340	6	26	174	8	25
M2	x	x	x	4	40	1.5	45	1.1	3	0.003	0.09	383	25	22
M3	x	x	x	5	45	1.3	15	1.7	350	20	35	101	5	3
M4	x	—	—	—	8.8	0.2	26	0.3	310	11	18	367	16	21

Notes:

A: CTAB/TEOS sol-gel filling of AAM (except for **M4** where **4** was introduced in the sol-gel solution precursor).

B: Thermal removal of the surfactant.

C: Octadecyltrichlorosilane (ODS) functionalization of silica nanotubes.

D: Functionalization with ureido crown ether derivative **4** or **5**, respectively.

These membranes have been tested in selective Na⁺/K⁺ transport. In the absence of the silica–surfactant–receptor nanocomposite in the alumina membrane, Na⁺ and K⁺ cations are transported through the membrane in a similar proportion. In contrast, the hybrid crown–alumina membranes, including the silica–surfactant composite, show a selective transport of salts depending on the receptor **1** or **2** selectivity toward Na⁺ and K⁺ cations, respectively (Cazacu et al. 2009).

The Fick law diffusion theory allows us to determine the transport parameters such as diffusion coefficients and permeability across the membrane (Table 12.1). Moreover, in every case, we can distinguish two stages for the transport mechanism: (1) a simple and (2) a facilitate diffusion (Figure 12.9). In the first step, the membranes are functioning like a *sponge*, and the simple rapid diffusion through the membrane is accompanied by the selective complexation of the fittest cation (Na⁺ for **1** and K⁺ for **2**, respectively); it is the so-called membrane self-preparing step. The selective transport of the specific cation (Na⁺ for **4**, **M2** and K⁺ for **5**, **M3**, respectively) occurs in the second stage, much faster. Thus, one can conclude that the membrane with the **15C5** receptor, **4** facilitates in the second step, "the membrane self-responding step" the transport of Na⁺, whereas the **18C6** receptor, **5** facilitates the transport of K⁺. These experimental results suggest that the self-assembly of receptors inside the surfactant-templated silica nanochannels can be reorganized during the molecular transport, the limiting transport rate being determined by the initial self-preparing step.

This led to the discovery of the functional supramolecular architecture evolving from a mixture of reversibly insidepore exchanging devices *via* ionic stimuli so as to improve membrane ion transport properties. These phenomena might be

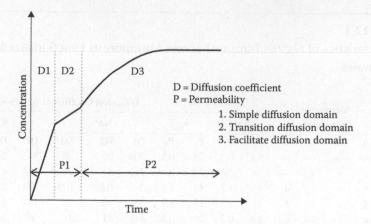

FIGURE 12.9 Generalized concentration versus time profile and diffusion regimes of transport of Na^+ and K through mesoporous constitutional membranes.

considered as an upregulation of the most adapted 3D "insidepore" superstructure, enhancing the membrane efficiency and the selectivity by the binding of the ion effectors. Finally, these results extend the application of constitutional dynamic chemistry (Barboiu 2012) from materials science to functional dynamic interactive membranes—system membranes (Barboiu 2010). This feature offers to membrane science perspectives toward self-designed materials evolving their own functional superstructure so as to improve their transport performances. Prospects for the future include the development of these original methodologies toward dynamic materials, presenting a greater degree of structural complexity. They might provide new insights into the basic features that control the design of new materials mimicking the protein channels with applications in chemical separations, sensors, or as storage-delivery devices.

12.5 G-QUADRUPLEX MULTICOMPONENT ION CHANNELS TOWARD MATERIALS FOR IONIC DIFFUSION

The control of dynamic supramolecular exchanges may be driven by a phase-change phenomenon *via* the sol-gel process, providing simple methods for synthesis of well-defined self-organized hybrid constitutional materials (Arnal-Herault et al. 2007c). Among these systems, the nucleobases and nucleosides are well-known fascinating compounds generating multiple H-bonding of complementary nature, hydrophobic and stacking interactions. In this context, the supramolecular H-bonded macrocycle of four guanine building blocks, the G-quartet, has been proposed as powerful scaffold for the construction of synthetic supramolecular devices and materials. The G-quartet architecture represents a dynamic supramolecular system in which G-quartet is reversibly exchanging with linear ribbons (Figure 12.10). It is stabilized by cations, templating the eight carbonyl oxygens of two sandwiched G-quartets in the G-quadruplex, the columnar device formed by the vertical stacking of four G-quartets (Mihai et al. 2010).

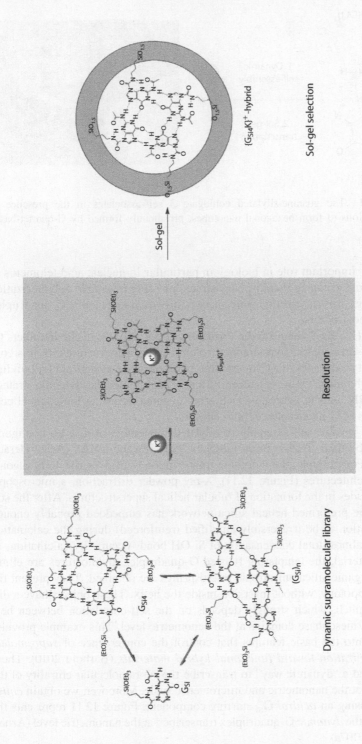

FIGURE 12.10 Cation-template resolution of a dynamic supramolecular guanine system in which G-quartet is reversibly exchanging with linear ribbons followed by a secondary irreversible sol-gel selection of G-quadruplex hybrid materials. (Adapted from Arnal-Herault, C. et al., *Angew. Chem. Int. Ed.*, 46, 8409, 2007a.)

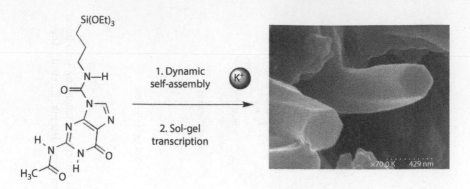

FIGURE 12.11 The guanine-silylated conjugate **3** self-associates in the presence of alkali-metal cations to form hexagonal nanotubes, presumably formed by G-quartet-based interactions.

It plays a very important role in biology in particular in nucleic acid telomeres of potential interest to cancer therapy. New strategies using reversible polymerization or phase-change-driven selection were successfully used to generate G-quadruplex functional nanostructures.

Metal ions (K^+, Ba^{2+}) template the formation of not only the cyclic tetramers G_4 but also the G-quadruplex, four-stranded column-like superstructures. In this context in the last two decades, the G-quartet has been proposed as scaffold for building synthetic ion channels. Only very recently, a new supramolecular dynamic strategy was successfully used to generate a rich array of interconverting ion-channel conductance states of G-quadruplex in lipid bilayers.

Our efforts involve self-assembly of silylated monomers, **3** in a G_4 configuration followed by their fixation in an inorganic polysiloxane matrix (Arnal-Herault et al. 2007a). For example, the inorganic transcription of **3** gives rise to hexagonal helical-like architectures (Figure 12.11). X-ray powder diffraction, a microscopic method, concludes in the formation of tubular helical superstructures. After the sol-gel process, the preformed helical silica network has embedded probably enough chiral information to be irreversibly amplified (reinforced) during the calcination process when almost total condensation of Si-OH bonds occurs. By calcinations of the hybrid material, the templating twisted G-quadruplex architectures are eliminated and inorganic silica anisotropic microsprings are obtained. They present the same helical topology, without inversion inside the helix. These objects have a different helical pitch, which strongly depends on the self-correlation between hexagonal twisted mesophase domains at the nanometric level. This example provides new insights into the basic features that control the convergence of *supramolecular self-organization toward functional hybrid materials* (Barboiu 2010). These findings showed a "dynamic way" to transcribe the supramolecular chirality of the G-quadruplex at the nanometric and micrometric scale. Moreover, we obtain *chiral materials* by using an *achiral* $\mathbf{G_{Si}}$ starting component. Figure 12.11 represents the first picture of the *dynamic* G-quadruplex transcribed at the nanometric level (Arnal-Herault et al. 2007a).

Stacked G-quartet superstructures stabilized by templating K^+ cations induce an important increase of the conductivity by a factor of 1000 of G-quartet membrane films compared to the nontemplated G-ribbon membrane, which presents a very low conductivity (Arnal-Herault et al. 2007b). Base stacking accompanied by π–π interactions may induce coherent charge mobility (Calzolari et al. 2004) or polarizability (Cohen et al. 2007) in stable G-quartet materials.

12.6 CONCLUSION AND PERSPECTIVES

In conclusion, in this chapter, we firstly described one rational approach for building molecular channels in hybrid organic–inorganic materials *via* the inorganic (sol-gel) transcription of dynamic self-assembled superstructures (Barboiu 2010). The basic and specific molecular information encoded in the molecular precursors (crown ether, amino acid, and guanine ureido-silsesquioxanes) results in the generation of tubular and continual superstructures in solution and in the solid state, which *can be "frozen" in a polymeric hybrid matrix by sol-gel process*. These systems have been successfully employed to design solid dense membranes, functioning as ion channels, and illustrate how a self-organized hybrid material performs interesting and potentially useful transporting functions.

Then we showed that the combined features of structural adaptation in a specific hybrid mesoporous nanospace and of dynamic supramolecular selection process make the membranes presented here of general interest for the development of a specific approach toward nanomembranes of increasing structural selectivity.

From the conceptual point of view, these membranes express a synergistic adaptive behavior: the addition of the fittest alkali ion drives a constitutional evolution of the membrane toward the selection and amplification of a specific transport crown-ether superstructure in the presence of the solute that promoted its generation for the first time. It embodies a constitutional self-reorganization (self-adaptation) of the membrane configuration producing an adaptive response in the presence of its solute. This is the first example of dynamic "smart" membranes where a solute induces the upregulation of (prepare itself) its own selective membrane (Cazacu et al. 2009).

The resolution of dynamic supramolecular libraries may be achieved *via* the sol-gel process, resulting in the total amplification of the most compact architectures, as the combined hydrophobic/H-bonding affinities in the final constitutional structure of resulted hybrid materials, compared with the unpolymerized powders (Arnal-Herault et al. 2007c). This strategy has been also used to transcribe the supramolecular chirality of the G-quadruplex, at the nanometric and micrometric scale to obtain *chiral hybrid materials*.

Dynamic self-assembly of supramolecular systems described here, prepared under thermodynamic control, may in principle be connected to kinetically controlled sol-gel or dynamic covalent polymerization processes in order to perform selective transport functions. Such "dynamic marriage" between supramolecular self-assembly and inorganic sol-gel polymerization processes, which synergistically communicate, leads to higher self-organized hybrid channel materials with increased micrometric scales. More generally, applying such consideration to hybrid materials leads to the definition of *constitutional hybrid materials*, in which organic (supramolecular)/inorganic domains are reversibly communicating over the large distances. This might provide new insights into the basic features that control the design of functional constitutional architectures (Barboiu 2012, Barboiu and Lehn 2002). Considering the simplicity of this strategy, possible applications on the synthesis of more complex hybrid architectures might be very effective, reaching close to novel expressions of complex bioinspired matter.

ACKNOWLEDGMENTS

The research leading to these results has received funding from the Romanian National Authority for Scientific Research, CNCS—UEFISCDI grant, project number PN-II-ID-PCCE-2011-2-0028, contract 4/30.05.2012.

REFERENCES

Agre, P. 2004. Aquaporin water channels. *Angew. Chem. Int. Ed.* 43: 4278–4290.

Arnal-Hérault, C., M. Barboiu, A. Pasc, M. Michau, P. Perriat, and A. van der Lee. 2007c. Constitutional self-organization of adenine-uracil-derived hybrid materials. *Chem. Eur. J.* 1: 6792–6800.

Arnal-Herault, C., M. Barboiu, E. Petit, M. Michau, and A. van der Lee. 2005. Cation-π interaction: A case for macrocycle-cation π-interaction by its ureidoarene counteranion. *New J. Chem.* 29: 1535–1539.

Arnal-Herault, C., A. Pasc-Banu, M. Barboiu, M. Michau, and A. van der Lee. 2007a. Amplification and transcription of the dynamic supramolecular chirality of the G-quadruplex. *Angew. Chem. Int. Ed.* 46: 4268–4272.

Arnal-Herault, C., A. Pasc-Banu, M. Michau, D. Cot, E. Petit, and M. Barboiu. 2007b. Functional G-quartet macroscopic membrane films. *Angew. Chem. Int. Ed.* 46: 8409–8413.

Barboiu, M. 2004. Supramolecular polymeric macrocyclic receptors—Hybrid carrier *vs.* channel transporters in bulk liquid membranes. *J. Inclusion Phenom. Macrocyclic Chem.* 49: 133–137.

Barboiu, M. 2010. Dynamic interactive systems—Dynamic selection in hybrid organic-inorganic constitutional networks. *Chem. Commun.* 46: 7466–7476.

Barboiu, M. (ed.). 2012. Constitutional dynamic chemistry. In *Topics in Current Chemistry*. Berlin/Heidelberg: Springer-Verlag.

Barboiu, M., S. Cerneaux, G. Vaughan, and A. van der Lee. 2004. Ion-driven ATP-pump by self-organized hybrid membrane materials. *J. Am. Chem. Soc.* 126: 3545–3550.

Barboiu, M., C. Guizard, N. Hovnanian, J. Palmeri, C. Reibel, C. Luca, and L. Cot. 2000a. Facilitated transport of organics of biological interest I. A new alternative for the amino acids separations by fixed-site crown-ether polysiloxane membranes. *J. Membr. Sci.*, 172: 91–103.

Barboiu, M., C. Guizard, C. Luca, A. Albu, N. Hovnanian, and J. Palmeri. 1999. A new alternative to amino acid transport: Facilitated transport of L-phenylalanine by hybrid siloxane membrane containing a fixed site macrocyclic complexant. *J. Membr. Sci.* 161: 193–206.

Barboiu, M., C. Guizard, C. Luca, N. Hovnanian, J. Palmeri, and L. Cot. 2000. Facilitated transport of organics of biological interest II. Selective transport of organic acids by macrocyclic fixed site complexant membranes. *J. Membr. Sci.* 174: 277–286.

Barboiu, M. and J.-M. Lehn. 2002. Dynamic chemical devices. Modulation of contraction/extension molecular motion by coupled ion binding/pH change induced structural switching. *Proc. Natl. Acad. Sci. USA* 99: 5201–5206.

Barboiu, M., C. Luca, C. Guizard, N. Hovnanaian, L. Cot, and G. Popescu. 1997. Hybrid organic-inorganic fixed site dibenzo-18-crown complexant membranes. *J. Membr. Sci.* 129: 197–207.

Barboiu, M., G. Vaughan, and A. van der Lee. 2003. Self-organised heteroditopic macrocyclic superstructures. *Org. Lett.* 5: 3073–3076.

Calzolari, A., R. Di Felice, and E. Molinari. 2004. Electronic properties of guanine-based nanowires. *Solid State Commun.* 131: 557–567.

Cazacu, A., M. Barboiu, M. Michau, R. Caraballo, C. Arnal-Herault, and A. Pasc-Banu. 2008. Functional organic–inorganic hybrid membranes. *Chem. Eng. Process.* 47: 1044–1052.

Cazacu, A., Y.M. Legrand, A. Pasc, G. Nasr, A. van der Lee, E. Mahon, and M. Barboiu. 2009. Dynamic hybrid materials for constitutional selective membranes. *Proc. Natl. Acad. Sci. USA* 106: 8117–8122.

Cazacu, A., C. Tong, A. van der Lee, T.M. Fyles, and M. Barboiu. 2006. Columnar self-assembled ureidocrown-ethers—An example of ion-channel organization in lipid bilayers. *J. Am. Chem. Soc.* 128: 9541–9548.

Cohen, H., T. Sapir, N. Borovok, T. Molotosky, R. Di Felice, A.B. Kotylar, and D. Porath. 2007. Polarizability of G4-DNA observed by electrostatic force microscopy measurements. *Nano Lett.* 7: 981–986.

Gallivan, J.P. and D.A. Dougherty. 1999. Cation-π interactions in structural biology. *Proc. Natl. Acad. Sci. USA* 96: 9459–9464.

Gokel, G.W. and E. Abel. 1996. Complexation of organic cations. In *Comprehensive Supramolecular Chemistry*, Vol. 1, J.L. Atwood, J.E.D. Davies, D.D. MacNicol, F. Vögtle, and K.S. Suslick (eds.). Oxford, U.K.: Pergamon, pp. 511–534.

MacKinnon, R. 2004. Potassium channels and the atomic basis of selective ion conduction (Nobel Lecture). *Angew. Chem. Int. Ed.* 43: 4265–4289.

Michau, M. and M. Barboiu. 2009. Self-organized proton conductive layers in hybrid proton exchange membranes, exhibiting high ionic conductivity. *J. Mater. Chem.* 19: 6124–6131.

Michau, M., M. Barboiu, R. Caraballo, C. Arnal-Hérault, and A. van der Lee. 2008. Directional ion-conduction pathways in self-organized hybrid membranes. *Chem. Eur. J.* 14: 1776–1783.

Mihai, S., Y. Le Duc, D. Cot, and M. Barboiu. 2010. Sol-gel selection of hybrid G-quadruplex architectures from dynamic supramolecular guanosine libraries. *J. Mater. Chem.* 20: 9443–9448.

Sanchez, C., B. Julian, P. Belleville, and M. Popall. 2005. Applications of hybrid organic-inorganic nanocomposites. *J. Mater. Chem.* 15: 3559–3592.

van Bommel, K.J.C., A. Frigerri, and S. Shinkai. 2003. Organic templates for the generation of inorganic materials. *Angew. Chem. Int. Ed.* 42: 980–999.

Index